I0760161

ELECTRICITY IN GASES

BY

J. S. TOWNSEND

M.A. DUBLIN ; WYKEHAM PROFESSOR OF PHYSICS IN THE UNIVERSITY OF OXFORD ; FELLOW OF NEW COLLEGE ; FELLOW OF THE ROYAL SOCIETY ; FORMERLY FELLOW OF TRINITY COLLEGE, CAMBRIDGE

OXFORD
AT THE CLARENDON PRESS
1915

OXFORD UNIVERSITY PRESS
LONDON EDINBURGH GLASGOW NEW YORK
TORONTO MELBOURNE BOMBAY
HUMPHREY MILFORD M.A.
PUBLISHER TO THE UNIVERSITY

PRINTED IN ENGLAND.

PREFACE

The electrical properties of gases have formed the subjects of many theoretical and experimental researches since Röntgen and Becquerel discovered the rays which render the gases conductors of electricity. These discoveries provided a means of studying the conductivity obtained with small electric forces, and the researches have led to satisfactory theories of many of the phenomena connected with electric discharges. The nature of the rays has also been studied and the remarkable physical properties of the various types of rays have been clearly defined. At the same time new investigations were made of the cathode rays which give rise to Röntgen rays when they collide with solid objects and thus form a connecting link between the currents in gases and the penetrating rays emitted by radio-active substances. Wichert, in 1897, made the important discovery that the charged particles which compose the cathode rays move with a velocity about one-tenth of the velocity of light, and that the mass associated with the atomic charge was about one two-thousandth part of the mass of an atom of hydrogen. The essential difference between the electrons and the positive ions was thus ascertained.

These fundamental discoveries having attracted the attention of many physicists, numerous researches were made which have added to our knowledge of molecular physics, and during the course of the experimental work several books have been written giving accounts of the progress that has been made from time to time. The best-known treatises are those by Sir J. J. Thomson and the collection of original researches, 'Ions, Électrons, and Corpuscules,' published by H. Abraham and P. Langevin.

These branches of electrical science have increased so much in recent years that it has become impossible to give a complete description of the principal researches in one or two ordinary volumes. About three years ago, on the suggestion of Professor Marx, several physicists co-operated in bringing out a treatise in six volumes dealing with the electrical properties of gases and the subject of radiation in general.

This volume contains my contribution to the treatise and is limited to a description of the conductivity obtained in gases at ordinary temperatures and at pressures ranging from atmospheric pressure to pressures of the order of one millimetre in which the discharges may be explained by the theory of ionization by collision. I have also given a brief description of the principal experiments on cathode rays and positive rays obtained in high vacua, as the results are of importance in connection with the general theory of electrical conductivity.

I have been unable to find space to give complete accounts of many interesting investigations, and students who wish for further information are recommended to consult the original memoirs to which I have referred.

While I have been engaged in this work I have received many valuable suggestions from Mr. F. B. Pidduck, and my best thanks are due to him for his assistance and also for having corrected the proofs.

J. S. TOWNSEND.

July, 1914.

CONTENTS

CHAPTER I

INTRODUCTION

CHAPTER II

THE METHODS OF PRODUCING IONS IN GASES

CHAPTER III

THE MOTION OF IONS IN GASES

CHAPTER IV

THE VELOCITY OF IONS IN AN ELECTRIC FIELD

CHAPTER V

DIFFUSION OF IONS

CHAPTER VI

RECOMBINATION

CHAPTER VII

THE FORMATION OF CLOUDS AND THE DETERMINATION OF THE ATOMIC CHARGE

CHAPTER VIII

IONIZATION BY COLLISIONS OF NEGATIVE IONS WITH MOLECULES OF A GAS

CHAPTER IX

IONIZATION BY COLLISIONS OF POSITIVE IONS

CHAPTER X

DISCHARGES BETWEEN CONDUCTORS OF VARIOUS SHAPES

CHAPTER XI

DISCHARGE TUBES

CHAPTER XII

CATHODE RAYS AND POSITIVE RAYS

CHAPTER I

INTRODUCTION

1. Gases as insulators. Under ordinary conditions all gases may be considered to be good insulators: that is, insulated charged bodies surrounded by a gas retain some of their charge for a considerable time. Gases are not, however, perfect insulators, for the rate at which a charged body tends to lose its charge is greater than the small leak along the insulating supports. Also the conductivity gives rise to a very small current between two electrodes in a gas even when the electromotive force between them is only a fraction of a volt. The currents thus obtained can only be measured by very sensitive apparatus, and they are so small that they may be neglected in comparison with the currents that can be obtained by passing Röntgen rays or Becquerel rays through the gas. Currents of the same order may also be sent through a gas by allowing ultra-violet light to fall on a negatively charged metal surface, and it has also been found that a gas in the neighbourhood of a flame or an incandescent metal is a comparatively good conductor. In all these cases the conductivity can be explained on the hypothesis that electricity is atomic in its nature; and that it is carried through the gas by small charged particles which move under electric forces towards the electrodes. These particles are called ions.

2. Currents in gases ionized by rays. The simplest example of an experimental investigation illustrating many of the properties of the ions is the determination of the currents between two electrodes in a gas when different electric forces are used. To fix ideas it will be supposed that the conductivity is produced by a beam of Röntgen rays passing between two parallel-plate electrodes, one electrode being insulated and connected to the insulated quadrants of an electrometer, and the other connected to a battery of cells. The potential difference between the plates may thus be adjusted to any required value, and the corresponding current can be measured either by observing the electrometer deflections or by means of a compensating arrangement such as an electrostatic induction balance.

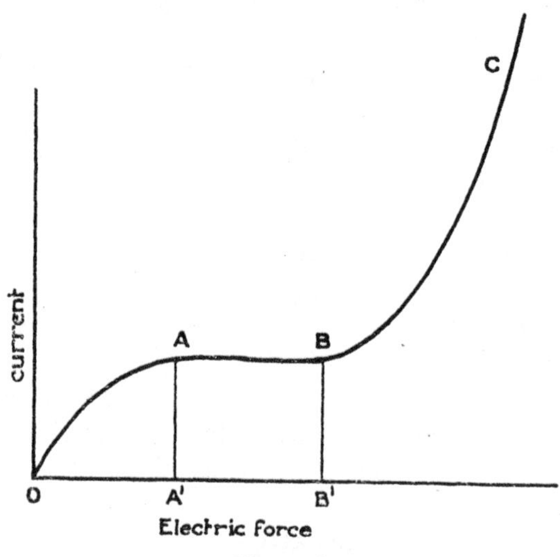

Figure 1.

The results of the experiments may be represented graphically, and the connection between the current and the electric force is found to be given by a curve similar to that shown in figure 1.

The exact form of the curve depends on the conditions under which the experiments are made, such as the pressure of the gas and the initial distribution of the ions, but the curves are all of the same type (except under special conditions when the pressure is very low) and have in common the following distinguishing features. The curve consists of three distinct portions. In the first stage the current increases with the electric force up to a certain point A, in the second stage from A to B the current remains constant for a range of forces between OA' and OB', and in the third stage the current increases rapidly with the electric force. The current AA' or BB' is usually referred to as the saturation current. It may be useful to give here a preliminary outline of the properties of the ions which are called into play under different conditions corresponding to points on the curve, and to show how to account for the nature of the conductivity which is produced in gases.

When the rays pass through the gas a certain number of positive and negative ions are produced from the molecules of the gas. As the ions are small particles they acquire a velocity of agitation by their collisions with the molecules of the gas, so that they are continually moving about even when no electric force is acting. The mean kinetic energy of this motion of the ions is equal to the mean kinetic energy of an equal number of molecules of the gas. Owing to the motion of agitation the ions of one sign come within the field of force of the ions of opposite sign, when they are attracted to each other and recombine to form a neutral body which does not move under an electric force. Also as the ions move about, by the ordinary process of diffusion, they come into contact with the electrodes, or with the boundary of the vessel in which the gas is contained, and become discharged: that is, they do not return to the gas as charged bodies.

When an electric force is acting the ions acquire a mean velocity along the lines of force, which is small compared with the velocity of agitation when the forces correspond to the portion OA of the current-electric-force curve. When the ions are generated in an electric field, therefore, the positive ions move towards the negative electrode and the negative ions towards the positive electrode, and the gas is said to conduct electricity.

3. Effect of recombination and diffusion. There are thus three processes taking place which cause the ions to disappear: recombination, diffusion to the sides, and the motion of the ions towards the electrodes under the electric force which constitutes the current in the gas. It can be seen that the connection between the current and the electric force over the part OAB of the current-electric-force curve may be explained by taking these effects into consideration. The saturation current AA' corresponds to the total number of ions produced per second by the rays. When the electric force is sufficiently great (exceeding OA') the ions move quickly towards the electrodes, and the length of time they are in the gas is so small that only an inappreciable number are lost by recombination or diffusion. Practically all the ions that are being generated by the rays are collected on the electrodes for forces larger than OA', and the current has a maximum value corresponding to the number of ions generated per second by the rays.

When lower forces are used the ions are in the gas for a longer time and the number per cubic centimetre at any point increases. This accelerates the process of recombination, so that although ions are produced at a constant rate by the rays the proportion that reaches the electrodes diminishes as the force is reduced, and currents corresponding to the portion OA of the curve are obtained. In some cases the loss of ions by diffusion is greater than that arising from recombination. Thus with ultra-violet light, where the conductivity is due to the escape of negative ions from the surface on which the light falls, there is no recombination as there are only negative ions passing through the gas. In this case some of the ions set free from the negative electrode diffuse back and are discharged by that electrode. When the force is small a considerable number are lost in this way, and the saturation current is not attained until the electric force is sufficient to make all the ions move with such velocity that only an inappreciable number of them come into contact with the negative electrode after they have been set free. Several investigations have been made to determine the conductivity under various conditions, and it can be shown that the experimental results are in good agreement with the theories that have been suggested.

The principal properties of the ions which account for the nature of the conductivity obtained with small forces have also been made the subject of special researches, and direct measurements have been made of the rate of recombination, the velocity under an electric force, the rate of diffusion, the charge on the ion, and the relation between the charges on the ions in gases and liquids.

4. Ionization by collisions. As regards the third stage of the current-electric-force curve, it can be shown that the increase in the current is due to new ions generated in the gas by the motion of those produced initially by the rays. At first the new ions are produced by the collisions of negative ions with molecules of the gas, and as the force increases and approaches the value required to produce a spark the positive ions also acquire the property of generating others by collisions. These effects are most conveniently obtained in gases at low pressures (of the order of a hundredth of an atmosphere), since sufficiently high velocities of the negative ions can then be obtained with potentials of about 50 volts when the electrodes are a few millimetres apart.

5. Ions generated in discharges. In addition to the ionization produced by collisions, ionization is also produced by the Röntgen rays that are generated when the negative ions impinge on the positive electrode with a high velocity. The penetrating power of these rays depends on the pressure of the gas, and in order to observe their effects outside the vessel containing the gas it is necessary to have very high exhaustion. The penetrating power diminishes as the pressure increases, and for some small pressures rays are produced which ionize the gas in the neighbourhood of the positive electrode, but are not sufficiently penetrating to be observed outside. There are thus several processes of ionization coming into play as the pressure is reduced, and it becomes difficult to explain accurately the currents obtained at very low pressures. When the gas is not too highly exhausted it is possible to arrange comparatively simple experiments to measure the increase of conductivity by collision for potentials less than the sparking potential.

Results have thus been obtained which supply sufficient data to form a theory of some of the phenomena that take place when sparking occurs in a gas and currents are maintained without the aid of external rays. Currents of the latter kind will be referred to as discharges, which may be continuous, as when a constant high potential is maintained between the electrodes by a battery of cells, or the discharge may be in the form of a spark of short duration, as when the potential is obtained from an induction coil. The phenomena connected with the various kinds of discharge are complicated, and they present many interesting problems for which some more or less complete solutions have been suggested.

The passage of an electric current through a gas is frequently accompanied by chemical effects, as for instance the formation of water vapour by a discharge through a mixture of oxygen and hydrogen. The action of the current in promoting the formation or the disintegration of molecules is of particular interest and importance. In many cases the phenomena have not been fully investigated and the actions that take place have not been satisfactorily explained. Possibly when more systematic quantitative investigations are made of some of the typical chemical actions the knowledge which will be gained may throw some light on the formation of molecules.

6. Negative ions and electrons. It may be of advantage to call attention here to the properties of the ions that have been determined with some degree of accuracy. The most striking result perhaps is one in connection with negative ions, and it establishes the theory of the atomic nature of electricity, as far as gases are concerned, when the elementary charges are obtained by various processes.

The experiments show that the negative ions produced in different gases by the action of rays or by collisions, and the negative ions set free by ultra-violet light from metal surfaces, all have the same atomic charge. This charge is the same as the charge on a univalent ion in a liquid electrolyte. Investigations at different pressures of the gas have not revealed any alteration in the charge on negative ions, but the apparent mass of the ion undergoes considerable changes. The mass seems to increase when the motion of the ion is slow: that is, when the ratio of the electric force to the pressure of the gas is small. This is due to the formation of a somewhat unstable group of neutral molecules round the ion which adds to its apparent mass. It can be shown that some molecules, such as those of water vapour, have a greater tendency of collecting round an ion than others, so that the apparent mass of the ion depends on various circumstances.

When the pressure of the gas is low and the electric force increased, the ions move about more quickly in the gas and collide with increased velocity with the molecules. This has the effect of preventing the formation of unstable groups round the negative ions, so that they travel freely as bodies whose masses are all the same, and are small compared with the molecules of hydrogen. The negative ions may then be said to be in the electronic state, and it will be seen that after further increases are made in the ratio of the electric force to the pressure of the gas the electrons begin to produce other ions by collisions with the molecules.

7. Positive ions. With regard to the positive ions, the charge is in most cases exactly the same as that of the negative ions, but there is evidence to show that, under certain circumstances depending on the nature of the rays that produce them, positive ions with double charges are generated. Positive ions also form groups of molecules round them, but the mass definitely associated with each is of the same order as the masses of the atoms. It is difficult to compare the real masses of the positive ions in different gases. All that the experiments show is that one or possibly two electrons, or negative ions with small masses, are separated from a molecule when it is ionized. The remainder of the molecule or some part of it forms the positive ion. Experiments on the chemical effects that take place when a current is passing through a gas show that when ions collide with molecules with certain velocities they dissociate them into their constituent atoms. Thus in the formation of water vapour the simplest explanation of the actions that take place is based on the hypothesis that oxygen molecules are divided into atoms when negative ions collide with them with sufficient velocity, and that the hydrogen molecules combine subsequently with the atoms of oxygen. Thus it is quite possible that when the velocity of the negative ions is sufficiently great to ionize a molecule by detaching from it a negative ion of electronic dimensions, other changes are also produced in the molecule, such as dividing it into atoms. Much of the uncertainty that exists with regard to positive ions will no doubt be removed by the researches which are being made on the properties of positive rays.

CHAPTER II

THE METHODS OF PRODUCING IONS IN GASES

8. Ions generated by rays. Mass associated with elementary charge. In describing the nature of the conductivity of gases it is desirable to deal first with the currents obtained with electromotive forces that are small compared with the sparking potentials, so that the phenomena will not be complicated by the process of ionization by collision. Even when the difference of potential between the electrodes in a gas is of the order of a volt, comparatively large currents may be obtained by various methods, and the gas is temporarily transformed from the state of an insulator to that of a conductor. The ions generated by all these methods have one property in common: that is, the charge on any ion is either equal to or an exact multiple of the atomic charge associated with a univalent ion in a liquid electrolyte. All the negative ions have one atomic charge, and in most cases the positive ions also have a single atomic charge. The other physical properties of ions differ according to the circumstances under which they are produced and depend on the mass associated with the ion. Thus the ions generated by Röntgen rays move much faster under an electric force than those obtained by the oxidation of phosphorus owing to the condensation of acid and water vapour round the ions in the latter case. The ions associated with the smallest mass in gases at ordinary pressures are the simplest, as it is easy to reproduce the circumstances under which constant values of the velocity and other physical quantities depending on the mass may be obtained. These may be generated in a gas by the action of Röntgen or Becquerel rays or by the secondary radiations. Also the negative ions set free from a metal surface by the action of ultra-violet light are among the class of small ions that move with a comparatively large velocity.

9. Ions generated during chemical actions. Increase of mass associated with elementary charge. Slowly moving ions such as those produced by the oxidation of phosphorus may also be obtained by other methods, and they frequently appear in gases when chemical actions take place and vapours are formed which easily condense on small particles. An interesting case is that of a flame. In the hot part of the flame the ions are associated with small masses and have very high velocities under an electric force. In the cool gas round the flame several slowly moving ions are to be found, being the original ions of small mass on which water vapour and other products of combustion have condensed. The number of ions generated in a hot gas may be greatly increased by spraying salt solutions into the gas, and very high conductivities are then obtained. The ions in the air surrounding an incandescent metal are also of this type.

Slowly moving ions are also obtained by bubbling gases through salt or acid solutions or by splashing, and newly prepared gases given off from solutions contain in some cases a large number of positive and negative ions associated with comparatively large masses.

It is thus quite possible for ions to be occasionally produced in gases under ordinary conditions, and, as might be suspected, the open air is by no means a perfect insulator. Also gases even in their normal state when contained in closed spaces are never perfect insulators, and the conductivity can be measured accurately when sufficiently sensitive apparatus is used.

10. Small conductivity in normal state. Effect of moisture on insulating supports. The electrical conductivity of a gas in its ordinary state has for a long time been the subject of experimental investigation, and it has generally been concluded that charged bodies lost some of their charge gradually through the surrounding air in addition to the small leakage that takes place along the insulating supports. It was also known that the rate at which the charge disappears depends on various circumstances: for example, when a flame is burning in the neighbourhood a conductor loses its charge very quickly. For some time moist air was supposed to conduct better than dry air, but subsequent investigation showed that the only effect of the moisture was to increase the leakage over the surface of the support. The principal source of error in all the earlier experiments arose from this leakage, as it is very variable. Different methods have been employed to remedy this defect. With glass high insulation may be obtained by having the surface dry or by coating it with paraffin wax, which is a very good insulator as moisture is not deposited over it in a continuous film. High insulation may also be obtained with ebonite supports when the surface is freshly cut. Exposure to light, however, increases the oxidation of the sulphur on the surface of ebonite, and when the air is moist a conducting film is formed. It is necessary therefore to consider carefully the methods by which the conductors are supported in order to measure small currents through a gas.

11. Electroscope designed by Elster and Geitel. Measurement of small conductivities. Two different methods have been devised by Elster and Geitel* and by C. T. R. Wilson,† to measure the small conductivity of air in its natural state.

The principle of the apparatus used by Elster and Geitel is illustrated in the diagram, figure 2. It consists of a modified form of electroscope in which aluminium leaves are fixed to a stiff metal strip *A*, which is supported in a vertical position on a short ebonite cylinder *E* fixed to the lower part of the case. In this position the ebonite

* J. Elster and H. Geitel, Phys. Zeitschr. **1**, p. 11, 1899; H. Geitel, Phys. Zeitschr. **2**, p. 116, 1900.

† C. T. R. Wilson, Proc. Camb. Phil. Soc. **11**, p. 32, 1900; and Proc. Roy. Soc. **68**, p. 151, 1901.

can easily be kept dry, so that it remains a good insulator during the experiments. To measure the conductivity of the air in a closed vessel, the electroscope is placed on an iron base *C*. A cylinder *Z* is supported above the case by a rod which passes through the opening *B* in the upper part of the case and fits into a hole bored in the top of the conductor *A*. This cylinder is surrounded by a larger cylinder *Y* at zero

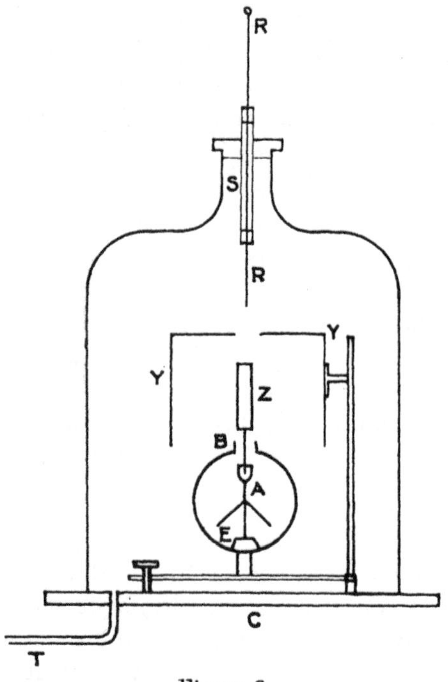

Figure 2.

potential supported from the base and covered on top with a disc having an aperture in the centre. The apparatus is covered with a bell-jar, and the air can be exhausted through the tube *T* and different gases admitted. Contact is made with the insulated system by means of a rod *R* that moves up and down in an air-tight fitting *S*. To investigate currents in the open air the same electrometer may be used and any form of electrode projecting into the air may be substituted for the cylinder *Z*. When a current with the electrode in position has been measured by observing the rate of fall of the leaves of the electroscope, it may be corrected for the current taking place through the gas inside the case of the electroscope and the leakage over the ebonite support by removing the electrode and covering the aperture at *B* with a metal cap after charging the leaves. The divergence of the leaves diminishes slowly and the corresponding current gives the required correction.

When the apparatus is arranged as in figure 2, with air inside the bell-jar, the change in the divergence of the leaves obtained in 15 minutes can easily be measured. The total charge per minute lost by the cylinder *Z* depends on the quantity of air that surrounds it: that is, on the volume of the outer cylinder *Y* and the pressure of the air inside the bell-jar. Also the currents are the same for the positive and negative directions, and are independent of the magnitude of the potential of *Z* over a range of potentials between 80 and 240 volts. This characteristic feature of the conductivity of air in enclosed spaces had previously been observed by Matteucci,* but, as Elster and Geitel point out, the result had not attracted any attention. The constant current that was observed corresponds to the saturation current obtained when rays from an external source ionize the gas.

12. Electroscope designed by C. T. R. Wilson. The same conclusions were arrived at by C. T. R. Wilson using a somewhat different apparatus. It consisted of an electroscope furnished with a special arrangement for preventing any leakage from the leaves along their support. This method is shown in figure 3.

The leaves are very small and are carried on a brass strip fixed to an insulator *A*, the other side of the insulator being fixed to the end of a rod *R* which comes through the air-tight case of the electroscope. The rod is maintained at

* C. Matteucci, Annales de Chimie et de Physique, **28**, p. 390, 1850.

a fixed potential V by means of a battery of cells, or by connecting it to a condenser and an auxiliary electroscope of the ordinary pattern which can be kept at a fixed potential by a replenisher. A movable wire W is used to touch the leaves and charge them to the same potential as the rod R. When the leaves are insulated any loss of charge must take place through the gas inside the case of the electroscope, since a leakage over the surface of the insulating support would tend to keep the leaves at the fixed potential V. A simple means is thus provided of finding the small currents that take place in the space round the leaves. The case of the electroscope itself can easily be made air-tight so that gases at different pressures may be used.

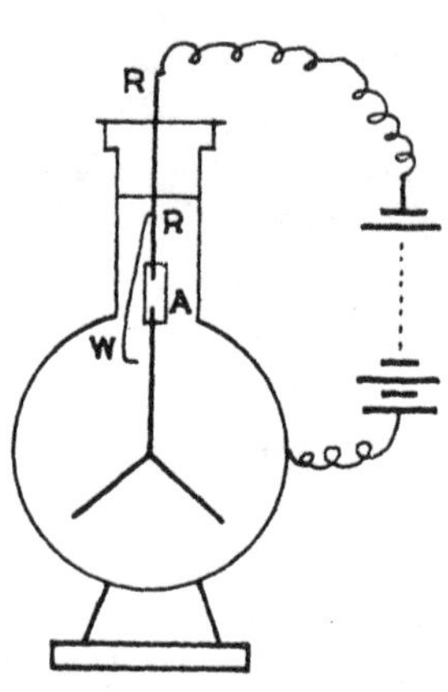

Figure 3.

13. Number of ions per c.c. in normal state. The experiments made with this electroscope have also shown that the current is proportional to the quantity of gas inside the case. When filled with air at atmospheric pressure the current is of the order 10^{-8} electrostatic units per cubic centimetre of the case, and is independent of the potential of the leaves. The latter condition would only hold when the forces are above a certain limit, otherwise some of the ions would be lost by diffusion to the sides of the case, instead of reaching the insulated system under the action of the electric force. It is not to be expected that there would be an appreciable effect due to recombination in these cases, since the number of ions per cubic centimetre is so small.

14. Effect of sides of vessel on conductivity of the gas. Radio-active emanations in ordinary air. Recently it has been found that the current depends on the material of the sides of the vessel in which the gas is enclosed, and that for large vessels the current is not proportional to the volume of the vessels, but increases less rapidly than volume.

An interesting observation was made by Geitel on the effect of leaving air inside a closed vessel. He found that the conductivity increased day by day, and after four days it became about four times as great as it was when first admitted into the vessel. Subsequently the rate of increase diminished and after some time the conductivity attained a constant value which was about five times as great as the initial conductivity. It was afterwards found that the air in cellars and caves conducts very much better than the open air in the neighbourhood,* and air that had been drawn from the earth by a tube sunk in the ground had a conductivity that was more than twenty times the normal. These results lead to the hypothesis that all elements give out traces of radio-active emanation † which diffuses into the air in their neighbourhood, and that some of the ions found in air in its normal state are due to a radio-active gas.

15. Effect of natural penetrating radiations. Experiments have, however, shown that all the ions in the atmosphere and in enclosed gases are not due to an action of this kind, but that some are produced by an action originating from an external source. Rutherford and Cooke ‡ and McLennan § independently observed the effects of a very penetrating radiation inside buildings. Rutherford and Cooke found that the rate of discharge of a sealed brass electroscope was diminished by placing metal screens around it. Cooke ‖

* J. Elster and H. Geitel, Phys. Zeitschr. **2**, p. 560, 1901 ; **3**, p. 574, 1902.

† The emanation from radio-active substances was discovered by Rutherford, and a complete account of its properties is to be found in his treatise on Radio-activity.

‡ E. Rutherford and H. L. Cooke, American Phys. Soc., Dec. 1902.

§ J. C. McLennan, Phys. Rev., No. 4, 1903.

‖ H. L. Cooke, Phil. Mag. (6) **6**, p. 403, 1903.

made a more detailed investigation and found that the conductivity could be reduced to 70 per cent. of its original value by a lead covering 5 cm. thick, but further increases in the thickness of the screens had no effect on the conductivity.

McLennan's experiments were on the same principle. He observed the effect on the conductivity of air in a closed cylinder when it was immersed in a large cylinder filled with water. For a thickness of water between the cylinders of 25 cm. the conductivity was reduced to 63 per cent. of its initial value. There are thus different processes in action which affect the conductivity of the open air, some of which are of such a nature as to ionize also gases contained in closed spaces.

16. Conductivity produced by Röntgen rays. Penetrating power of rays. The method of producing ions by the action of Röntgen or Becquerel rays affords the simplest means of increasing the conductivity of a gas. The Röntgen rays are emitted when a discharge passes through a vacuum tube of almost any shape provided the gas in the tube is reduced to a very low pressure.

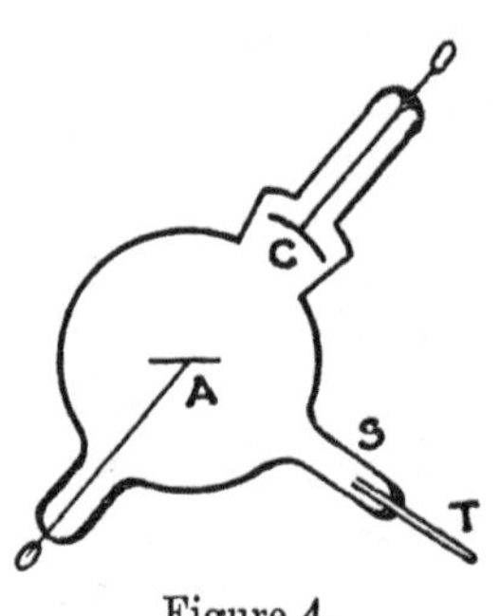

Figure 4.

The tubes that are best suited for producing the rays are nearly all made somewhat after the pattern illustrated in figure 4.

It consists essentially of a concave cathode *C* of aluminium and a flat anode *A* of platinum or platinum-iridium. When a discharge is passed through the tube and the pressure reduced until a green fluorescence appears on the glass, Röntgen rays are given out from the surface of the anode. It is necessary that the cathode rays, which consist of negatively charged particles, should impinge on the anode *A* or some other solid obstacle with a very high velocity, and the points on the surface where the cathode particles are stopped abruptly become the origins from which the Röntgen rays are given out in all directions. There is a great difference in the penetrating power of rays obtained from bulbs that are exhausted to a moderate degree and those which are very highly exhausted. The rays that appear at first when the tube is being exhausted are of the non-penetrating type, or what is often termed 'soft'; the intensity of these rays is reduced to a small fraction of its original value when they pass through thin sheets of metal, such as aluminium one-fifth of a millimetre thick. As the pressure of the gas in the tube is reduced the rays become more penetrating, or 'harder', and it is easy to arrive at a stage where rays are reduced by only five or ten per cent. when passed through the same metal sheets. With very high exhaustion it is possible to obtain appreciable effects through copper or zinc sheets several millimetres thick.

17. Potential required to excite the rays. Adjustment of pressure in vacuum tubes. These variations in the penetrating power are due to the increase in the velocity of the cathode rays as the pressure is reduced, for a reduction in the pressure increases the potential required to produce a discharge in the vacuum tube. The potential required to excite the rays thus gives a measure of their penetrating power, and it can be roughly estimated by finding the distance between electrodes in air at atmospheric pressure, connected in parallel with the Röntgen-ray tube, when a spark begins to pass across the air-gap. It is usually found that the softer rays, given out by a tube which can be excited by a potential difference between the electrodes that would give a spark across an air-gap of three or four centimetres, are the most satisfactory for general experimental purposes. The rays given out under these conditions remain more constant than the harder rays when the tube is run for some minutes at a time.

It is desirable to have some method of adjusting the pressure inside the vacuum tube in order to bring it to a state in which it acts as a constant source of radiation, since owing to continual use the gas becomes absorbed by the glass or by the electrodes and an increasingly high potential is required to send a discharge through the tube. Various methods have been employed to supply a little gas to the tube without connecting it to an air-pump. Thus tubes have been made with some carbon, or powder that absorbs gases, in a side tube, and when this is heated slightly a sufficient amount of gas is given off to make a great reduction in the potential required to excite the rays. Another method which is very simple and effective is to have a fine platinum tube sealed into a side tube, one end being closed and allowed to project four or five centimetres from the tube, so that it can be heated with a Bunsen burner. Some of the hydrogen in the gas supplied to the burner diffuses through the platinum when it is at a red heat and the pressure of the gas inside the vacuum tube is increased. The vacuum tube shown in figure 4 with the platinum tube T sealed into the side tube S is constructed on this principle.

18. Preliminary investigations of conductivity produced by Röntgen rays. Giese's theory. The preliminary investigations of the properties of Röntgen rays and the nature of the conductivity obtained by their passage through gases were undertaken simultaneously by many physicists * shortly after the discovery of the rays and their first investigation by Röntgen himself.

It was generally recognized that the currents could be explained by a theory similar to that which had previously been applied by Giese † to explain the conductivity of gases coming from flames or hot wires. According to this

* L. Benoist and D. Hurmuzescu, Comptes rendus, **122**, p. 235, Feb. 3, 1896; J. J. Thomson, Proc. Roy. Soc. **59**, p. 274, Feb. 13, 1896; J. J. Thomson and E. Rutherford, Phil. Mag. (5) **42**, p. 392, 1896; A. Righi, Rend. della R. Accad. di Bologna, Feb. 9, 1896; id., Comptes rendus, **122**, Feb. 17, 1896; Jean Perrin, L'Éclairage électrique, June 1896; id., Thèse, 1897; id., Annales de Chimie et de Physique, **11**, p. 496, 1897; E. Villari, Rend. della R. Accad. di Napoli, (3 a) **2**, p. 157, April 4, 1896.

† W. Giese, Wied. Ann. **17**, 1882, and **37**, 1889.

hypothesis the molecules of the gas are divided into charged atoms or ions either by the high temperature or by the effect of the rays, and under the action of an electric force the ions move with velocities proportional to the force and give rise to electric currents.

19. Arrangement of electrodes in experiments on conductivity. Parallel-plate electrodes. In order to detect and measure the effects of the rays under various conditions, several types of electrodes have been used to determine the currents. For simplicity it is most convenient to use

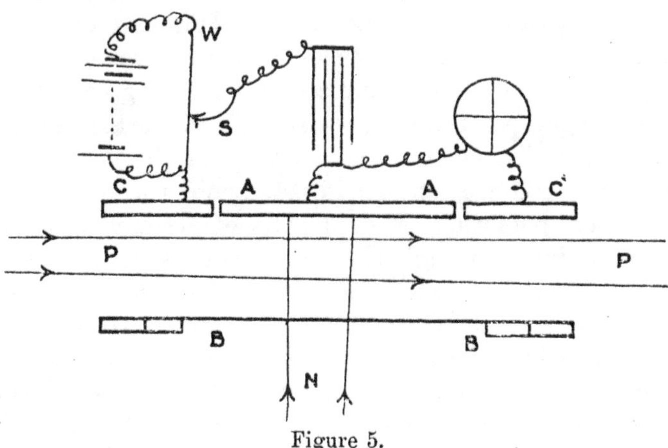

Figure 5.

parallel-plate electrodes, one of which, *A* (figure 5), is insulated and connected to an electrometer, and the other, *B*, connected to a battery of cells so that its potential *V* may be raised to any required number of volts.

In some experiments it is desirable to have a guard ring *C*, connected to earth, round the insulated plate *A*, in order that the current arriving at that plate should contain only the ions generated in the well-defined volume between *A* and *B* which is traversed by the lines of force that terminate on the surface of *A*. To ionize the gas a beam of rays may be sent either in the direction *P* parallel to the plates, or normally in the direction *N*. In order to allow the rays to enter the space between the plates in the latter case it is necessary to have the lower plate very thin at the centre.

The insulated plate A should be of thick metal to prevent the rays from passing through and generating ions in the air above the plate.

20. Measurement of currents. Electrostatic induction balance. The number of ions that come into contact with the plate A may be measured by the electrometer deflection; but a better method is to use some compensating arrangement to measure the current and to keep the plate A as nearly as possible at zero potential. This may be done by connecting the inner plates of a small air-condenser to the plate A, and the outer plates by a sliding contact S to a point on the potentiometer wire W. When the insulated system gets charged by ions arriving at the surface of A the sliding contact can be adjusted and the electrometer kept at zero potential. The quantity of electricity acquired by conductivity through the gas is then given accurately by the product KE, where K is the capacity of the inner plates of the condenser and E the potential to which the outer plates have been raised. This result is independent of the capacity of the electrometer and the plate A. It is, therefore, a very useful method of measuring the quantities of electricity passing through the gas when the plates A and B are set at different distances apart.

21. Ionization as a function of the distance from the origin. When the rays from the tube are limited by lead screens so as to obtain a narrow beam, the number of ions produced per unit length of the beam in a given time may be examined by means of the parallel-plate electrodes. It has been observed that the number of ions produced per unit length of the beam is practically the same for all distances from the tube. This point has recently been investigated by Gill,* and he has found that, when allowance is made for the small absorption by the air as the beam traverses the larger distances, there is no reduction in the ionizing

* E. W. B. Gill, Phil. Mag. (6) **23**, p. 114, 1912.

power per unit length due to the fact that the beam spreads out, and the energy of the radiation is inversely as the square of the distance from the origin. It appears from this that the number of molecules per cubic centimetre in a gas that can be ionized by a beam of intensity I is directly proportional to I. The actual number of the molecules that are ionized is a very small fraction of the whole, even when very intense rays are acting. The conductivity of the gas would be considered very high when 10^9 ions are generated per cubic centimetre, and it will be seen that the number of molecules in a gas at atmospheric pressure is about 3×10^{19}. Thus only one molecule in every 10^{11} or 10^{12} is affected by the rays.

22. Effects of temperature on conductivity. The number of ions generated by Röntgen rays in a gas is independent of the temperature when the density is constant, and is proportional to the number of molecules per cubic centimetre. The proportionality varies with different gases, there being fewer ions generated in hydrogen than in the heavier gases. These results have been established by the experiments of Perrin and McClung. Perrin * showed that when precautions were taken to avoid secondary radiations the conductivity is proportional to the pressure of the gas. He also made some experiments on the effect of temperature which seemed to show that there was an effect when the density was constant, but the more complete investigations subsequently made by McClung † showed that the ionization produced by a beam of rays in a given volume of a gas is proportional to the number of molecules per cubic centimetre and is independent of the temperature.

23. Constitution of Röntgen rays. These results are easily understood if the Röntgen rays are a stream of particles travelling with a high velocity, since both the intensity of the stream and the number of molecules ionized by it would be proportional to the number of particles in the stream. As there is no deflection of the Röntgen rays due

* J. Perrin, Annales de Chimie et de Physique, (7) **11**, p. 496, 1897.

† R. K. McClung, Phil. Mag. (6) **7**, p. 81, 1904.

to a magnetic or electric force, it would be necessary to suppose that the particles are either uncharged or that they have a velocity approximately equal to that of light. This is one theory of the constitution of the rays. The other theory, suggested by Schuster * and Wiechert,† is that the rays are the electromagnetic pulses given out when the cathode rays are stopped abruptly by the anode in the Röntgen ray tube.‡ It is now generally recognized that this latter theory affords the best explanation of the phenomena. The polarization effect observed by Barkla § can only be explained by some form of wave theory, and this conclusion is confirmed by the recent experiments of Laue ‖ on the diffraction by crystals, and the determinations of the velocity, which has been found by Marx ¶ to be the same as the velocity of light.

24. Secondary Röntgen rays. Penetrating rays; Sagnac's experiments. In considering the currents which are generated by Röntgen rays it is necessary to take into consideration the secondary rays which are given out when the primary rays fall upon metal surfaces, or when they pass through any kind of matter. The effects are best observed by allowing the primary rays to fall on a metal surface. The secondary rays that are given out are very complicated and consist of groups of rays differing widely in penetrating power. A small proportion of the secondary rays are as penetrating as the primary rays that excite them, but the greater part is much less penetrating, and one group which produces a large ionizing effect is absorbed by a layer of air about half a centimetre thick near the surface on which the primary rays fall. The penetrating power of the secondary rays depends on that of the primary rays which excite them. The effects produced by the secondary rays were first observed by Sagnac,** who found

* A. Schuster, Nature, **53**, p. 268, Jan. 1896.

† E. Wiechert, Abh. d. physikal.-ökonom. Gesellsch. zu Königsberg, **37**, p. 1, 1896.

‡ See also Stokes, Proc. Manchester Lit. and Phil. Soc., 1897.

§ C. G. Barkla, Phil. Trans. A, **204**, p. 467, 1905.

‖ M. Laue, Ber. der K. Bayr. Akad. d. Wiss., 1912.

¶ E. Marx, Ann. der Phys. **20**, p. 677, 1906; **33**, p. 305, 1910.

** G. Sagnac, Comptes rendus, **125**, p. 230, 1897.

that an electroscope completely screened from the direct rays lost its charge rapidly when a solid body near the side of the electroscope was placed in the path of the primary beam. This effect is produced by the more penetrating rays as they extend to several centimetres from the source in air and pass through a thin sheet of aluminium over an aperture in the case of the electroscope.

25. Secondary Röntgen rays. Perrin's experiments. The less penetrating rays were observed by Perrin * when measuring the conductivity between parallel plates. He found that the number of ions produced between the plates by equal lengths of a beam of rays is much greater when the rays enter normally and fall on the electrodes as the beam N in figure 5, than when the same rays pass through the air in the direction of P parallel to the plates. In order to get the same intensity of the primary in each case a sheet of aluminium should be placed in the path of the beam P of the same thickness as the sheet forming the part of the electrode B through which the beam N enters normally.

When the plates are at a centimetre apart the number of ions generated between them by the normal rays may be two or three times as great as the number produced per centimetre by the primary beam alone.

26. Rays deflected by a magnetic field. The non-penetrating rays which are absorbed by about half a centimetre of air at atmospheric pressure can be made to extend a greater distance from their origin by reducing the pressure. Dorn † has observed that these rays are deflected by a magnet in the same direction as negative electricity moving away from the plate on which the primary beam falls. Thus possibly the greater part of the non-penetrating secondary rays are charged particles moving with a high velocity, and in this respect they differ from the more penetrating secondary rays which are unaffected by a magnet.

* Perrin, loc. cit.
† E. Dorn, Abhandl. d. Naturf. Gesell. zu Halle, **22**, p. 40, 1900.

27. Corpuscular rays negatively charged. This result is confirmed by some experiments made by Curie and Sagnac,* who found that metals such as platinum and lead which give a large secondary ionization near their surfaces, become positively charged when primary Röntgen rays fall on them. In order to observe this effect it is necessary to enclose a plate of the metal in an air-tight vessel having an aluminium window, and to reduce the pressure to a small fraction of a millimetre so that there should not be sufficient gas between the metal and the aluminium window to provide enough ions to discharge the plate when it becomes positively charged due to the escape of negative electricity.

28. Investigation of secondary rays. Owing to the difference in the penetrating power of the two kinds of

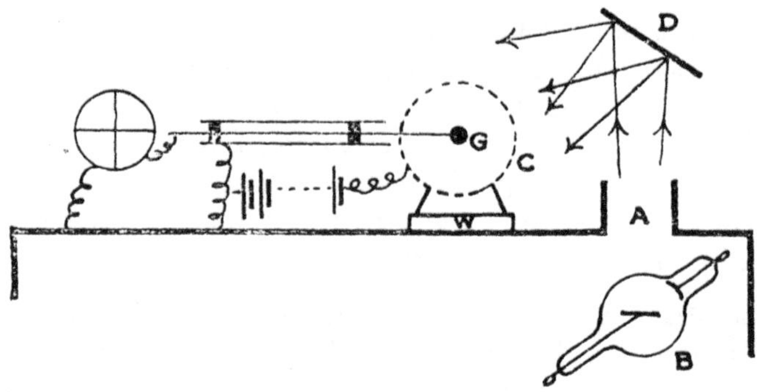

Figure 6.

secondary rays it is necessary to use different methods † in order to measure the ionization they produce in gases at atmospheric pressure.

A form of apparatus which was found convenient for determining the ionization produced by the secondary radiations given out by different bodies is shown in figure 6. It is arranged in a manner suitable for examining

* P. Curie et G. Sagnac, Comptes rendus, **130**, p. 1013, 1900.
† Proc. Camb. Phil. Soc. **10**, p. 217, 1899.

the intensity of the more penetrating secondary rays which extend to a distance of several centimetres from their origin. The Röntgen-ray tube *B* is contained inside a lead box with an aperture at *A*, over which a short lead tube open at both ends was placed so as to confine the primary rays to a small portion of the space above the box. This space was explored with the apparatus consisting of a gauze cylinder *C* and an axial electrode *G* fixed to a movable base *W* by means of insulating supports. The cylinder was raised to a potential of about 80 volts by connecting it to a terminal of a battery of small accumulators. The central electrode was connected to the insulated quadrants of an electrometer, and the connecting wire was surrounded by a metal tube so that ions produced in the air in the neighbourhood should not affect the electrometer. None of the ions produced in the air outside *C* can arrive at the electrode *G*; for if the gauze is positively charged it will repel positive ions and attract negative, but none of the latter, even if they should happen to get through the gauze, would move towards the inner electrode, since the force inside the gauze repels negative ions from the electrode. The only ions which can affect the insulated system are those generated by rays inside *C*, so that the electrometer deflections will be proportional to the intensity of the radiation falling on the cylinder. When a plate of heavy metal is placed at *D* and the primary rays fall on it, a large electrometer deflection is obtained. When there is no radiator at *D* except air the electrometer gives a small deflection which is probably due to the secondary rays given out by the air.

29. Intensity of penetrating secondary rays emitted by different substances. The following table gives the deflections obtained in ten seconds with various substances at *D*. The numbers in the first column were obtained when the secondary rays passed through a column of air about 6 centimetres long (the distance from the centre of the object *D* to the detector). The second column gives the corresponding deflections when the rays are cut down by a sheet of aluminium ¼ millimetre thick placed between *D* and *C*.

Radiator placed at *D*.	Secondary rays passing through 6 cm. of air.	Rays cut down with aluminium ¼ mm. thick.	Radiator placed at *D*.	Secondary rays passing through 6 cm. of air.	Rays cut down with aluminium ¼ mm. thick.
Air	2	1	Paraffin block	30	15·5
Aluminium	6	3·5	Brass	66	2·5
Glass	7·5	3	Zinc	68	3
Lead	24	6	Copper	70	2·5

The table shows that different kinds of radiations are given out by different substances. Thus brass, copper, and zinc give out rays of which only a small proportion pass through a sheet of aluminium ¼ millimetre thick, whereas the rays from aluminium and paraffin are comparatively penetrating. A similar result was obtained by Sagnac.

In order to obtain a rough estimate of the relative intensities of the primary and secondary rays the detector was placed at *D* in the path of the primary beam, and it was found that when the rays were on for half a second an electrometer deflection of over 500 divisions was obtained.

By placing the detector in various positions with respect to the radiator *D* it can be shown that the secondary rays are given out in all directions from the surface.*

30. Effect of corpuscular rays on conductivity between parallel plates. In order to examine the properties of the non-penetrating secondary rays a parallel plate apparatus

* An interesting investigation of the penetrating secondary rays obtained under various conditions has been made by Barkla. His results have reference principally to the nature of the radiation and do not deal directly with the properties of the ions generated in the gas. A description of the experiments would take up so much space that it is impossible to include them in this book. The reader is referred to the original papers, Phil. Trans. A, **204**, p. 467, 1905, and Phil. Mag. (6) **11**, p. 812, 1906.

such as that illustrated in figure 5 may be used. The primary beam N passes through the lower plate of thin aluminium and falls on the upper plate A, which can be adjusted to any required distance from the lower plate by means of a screw on the shaft that supports it. In measuring the currents it is necessary to have the lower plate charged to a sufficiently* high potential in order to collect all the ions that are generated in the space between the plates on the electrodes.

31. Conductivity with different electrodes. The charges acquired by the electrode A, while rays of uniform intensity were acting for 15 seconds, were determined in a set of experiments † in which plates of brass, zinc, copper, and aluminium were used for the electrode A, the lower electrode being the same aluminium sheet in each case. The results of the experiments are given in the accompanying table. The distance between the plates x in millimetres is given in the first column, and the currents in arbitrary units are given in the other columns corresponding to the different metals used at A.

x	Brass.	Copper.	Zinc.	Aluminium.
1	55	54·4	49	15
2	81	84	66	23·7
5	109·5	107·5	87	40·8
10	126	128	103	57
15	142	144	119	73

* To obtain a high degree of accuracy in measuring the relative values of the currents in this case it would be necessary to charge the lower plate to potentials approximately proportional to the product of the square of the distance between the plates and the square root of the maximum current. Usually a constant potential of about 80 volts may be used, as the results agree within one or two per cent. with those obtained by the more accurate method, when the intensity of the rays is not exceptionally great. On this point the reader is referred to Section 71, where a description of the variations of the current with the electromotive force is given. It is also to be borne in mind that when the ordinary method of measuring conductivities is used by observing the electrometer deflections, it is necessary in these experiments to multiply the deflection by the capacity of the insulated system in each case, as the latter varies considerably when the distance between the plates is altered. The best method is to use a compensating induction balance such as is described above and illustrated in figure 5.

† Proc. Camb. Phil. Soc. **10**, p. 222, 1899.

If there were no secondary radiation the numbers in the last four columns would be the same and proportional to the distance x.

These results may be explained on the hypothesis that the primary rays produce an equal number of ions in each millimetre of their path and the secondary rays have a large effect near the surfaces of the plates and are nearly all absorbed by a layer of air 5 millimetres thick. If S is proportional to the total number of ions produced by the secondary rays and P proportional to the number produced by the primary per millimetre of their path, then the currents for different distances x will be $S+Px$ when x exceeds about 5 millimetres. The currents obtained for the larger distances give $P=3{\cdot}2$, and $S=94$, 96, 71, and 25 in the experiments with the different metals. This value of P was also found directly by sending the primary beam through the gas in a direction parallel to the plates so that there should be no secondary rays from the electrodes.

The numbers given above for S include the effects of the rays given out both from the lower plate and the upper plate. The ionization produced in any distance x by the rays from each plate may be estimated in the following manner from the numbers given for the currents in the above table.

32. Intensity of non-penetrating rays emitted by different metals. When the quantity $3{\cdot}2x$ is subtracted from the currents corresponding to the distance x the effect of the secondary radiations from both plates is obtained. The last column then gives the effect of the radiations from the two aluminium plates A and B. The effect due to the rays from one of the plates may be obtained approximately by dividing these numbers by two, and the results can be used to correct for the radiations given out by the lower plate in the other cases.

The ordinates of the curves given in figure 7 are proportional to the number of ions produced by secondary rays in a layer of air of thickness x adjacent to each metal surface. The effect of the primary rays is given by the straight line P.

The curves show that the non-penetrating secondary rays produce a large number of ions near the surface. They extend to about 5 millimetres from the plate in each case,

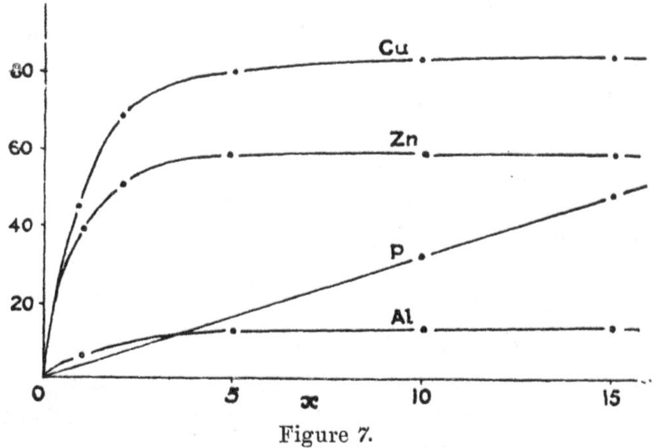

Figure 7.

and about one-half their total effect is produced in the first millimetre.

It must not be concluded from this that a large proportion of the paths of the rays are only one millimetre in length. The high conductivity in the first millimetre is due to the fact that the secondary rays are given out in all directions so that a large number traverse a comparatively long path near the surface. A simple calculation shows that if a number of particles are given out equally in all directions from a point on a surface with a high initial velocity that carries them through a distance d in the gas, and if they produce ions uniformly along their paths, about one-half the total number will be produced within one-fifth of the distance d from the plate. The above experiments, therefore, show that on an average each of the particles of which the secondary rays are composed traverses about 5 millimetres of air at atmospheric pressure.

33. Total ionization between parallel plates. In this examination of the experimental results, the number 3·2 x attributed to the direct effect of the primary rays would more correctly be described as representing the ionization produced by both the primary rays and the penetrating secondary rays. The number of ions, however, produced by the latter is probably not more than a few per cent. of those produced by the primary rays. Also in estimating the effect due to the lower plate it is assumed that the secondary rays emitted by the surface of B from which the primary emerge into the gas are of the same intensity as the secondary from the aluminium plate at A, on which the primary falls. This is not strictly true, as it has been found by Cooksey* that the rays from the lower plate are more intense, so that the numbers given for the different metals for incident rays are probably a few per cent. too large.

It was not considered necessary to correct the curves originally given for these effects, as the intensity of the secondary rays is not a constant quantity, but depends on the nature of the primary rays and on the state of the surface. The recent experiments of Bragg and Porter† are interesting from the point of view of the theory of the nature of the rays. It was observed that the secondary effect produced by a beam emerging from a metal into a gas was larger than the effect obtained by an incident beam: that is, when the beam traverses the gas and then falls on the surface of the metal. Thus the non-penetrating secondary rays given out from a zinc or copper surface are more intense for emerging primary rays than for incident primary rays in the proportion 1·5 : 1, the corresponding ratio for an aluminium surface being 1·8 : 1. Similar effects‡ have been observed in a more marked degree with some metals when the penetrating γ-rays from

* C. D. Cooksey, Nature, April 2, 1908.

† W. H. Bragg and H. L. Porter, Proc. Roy. Soc. A, **85**, p. 349, 1911.

‡ W. H. Bragg and J. P. V. Madson, Phil. Mag. (6) **16**, p. 918, 1908.

radio-active substances are used. Thus with soft γ-rays the secondary rays from aluminium on incidence and emergence are in the ratio of 280 to 1810, and for hard γ-rays in the ratio of 120 to 795; with a zinc surface the incidence and emergence rays are in the ratios 618 to 1160 and 224 to 485 for soft and hard rays respectively. Although these ratios have not been determined with a high degree of accuracy, nevertheless they show what a large difference there is between the effects produced on incidence and emergence.

Similar investigations have recently been made on the photo-electric effect in order to see if there is any difference in the number of ions set free from a metal surface by incident and emerging beams of ultra-violet light. Very thin films of metals can be obtained on quartz plates, by placing them opposite the cathode in an exhausted tube. They are sufficiently transparent to allow rays of light to pass through them so that it is possible to compare the effects of incident and emerging beams. Stuhlmann * has found in this way that the number of ions set free is greater by 17 per cent. on the emergent than on the incident side of a thin platinum film.

34. Velocity of non-penetrating rays. An inspection of the curves shows that the maximum velocity of projection from the different metals must be approximately the same, since the total ionization effect is practically attained within five millimetres of the surface in each case. The velocities of projection of the negative electrons have been deduced by Dorn † from observations of the magnetic deflection of the rays, combined with the known value of the ratio of the charge e to the mass m of the negative electron. For the value ‡ $e/m = 1{\cdot}77 \times 10^7$, the velocities of the electrons are $4{\cdot}7 \times 10^9$ cm. per second for the slowest and $8{\cdot}1 \times 10^9$ for the fastest. The potential required to impart

* O. Stuhlmann, Phil. Mag. (6) **20**, p. 331, 1910.
† E. Dorn, Lorentz Jubilee volume, p. 607, 1900.
‡ Cf. A. Bestelmeyer, Phys. Zeitschr. **12**, p. 972, 1911.

this velocity to the electron starting from rest would be about 20,000 volts.

These results * have been confirmed by Innes,† who made experiments with rays of various penetrating power measured by the equivalent spark gap in air connected in parallel with the Röntgen-ray tube. For 'soft' rays corresponding to a spark length in air of about 3·4 centimetres the velocities ranged from 6×10^9 to $7{\cdot}5 \times 10^9$ centimetres per second, and for the hard rays corresponding to a spark length of 15 to 16 centimetres the velocities ranged from $6{\cdot}3 \times 10^9$ to $8{\cdot}3 \times 10^9$ centimetres per second. The metals lead, silver, zinc, platinum, and gold were used, and it was found that the velocities of the electrons emitted from their surfaces were the same in each case.

If, therefore, the ionization near the surface is due principally to the negatively charged electrons observed by Dorn, it follows that the velocity of $8{\cdot}5 \times 10^9$ centimetres per second is required to make the electron travel a distance of about 5 millimetres in air at atmospheric pressure, and to ionize molecules along that distance.

One result of the investigations which is contrary to what might have been expected is that a definite minimum velocity should have been observed. Even if all the electrons started with the same velocity initially, those coming from points in the metal a little below the surface must have had their velocities reduced by collisions with molecules of the metal before they left the surface, so that it is reasonable to expect that there should be electrons travelling with all velocities up to a certain maximum in the exhausted space outside the surface from which they are given off.

35. Effect of pressure on the range of the non-penetrating rays. Experiments on the ionization due to secondary rays were also made with another apparatus, which showed that the secondary radiation extends to a greater distance as the

* On this subject see also C. D. Cooksey, Amer. Jour. Sci. (4) **24**, p. 285, 1907.

† P. D. Innes, Proc. Roy. Soc. A, **79**, p. 442, 1907.

pressure of the air is reduced, the total ionization being unaltered, until the rays traverse the whole distance between the electrodes. It is unnecessary to quote the exact results as they can easily be explained by considering the numbers given in Section 31 and the curves, figure 7. The total effect of the primary is proportional to the number of molecules in the path of the rays: that is, to the product px of the pressure and the distance between the electrodes. The number of ions produced by the secondary rays in a layer of gas of thickness x at pressure p will be equal to the number produced in a layer of thickness x' at pressure p', if $px = p'x'$, since the secondary radiation largely consists of moving particles. Hence the number of ions produced by the non-penetrating secondary rays is a function of the product px which can be denoted by $f(px)$, and the total number of ions produced between the plates will be $f(px) + Ppx$, Pp being the number produced by the primary rays in unit length of their path. This shows that reducing the distance between the plates has the same effect as reducing the pressure in the same proportion. Hence the currents at different pressures for a fixed distance between the plates may be deduced from the currents at atmospheric pressure corresponding to various distances. For instance, when the pressure is 50·7 millimetres the secondary ionization from copper in a distance of 15 millimetres is 45, and the effect of the primary is 3·2. Thus, with apparatus of ordinary dimensions, where the plates are a few centimetres apart, the greater part of the ionization is produced by secondary rays when the pressure is about a twentieth of an atmosphere or under.

36. Variations in the intensity of secondary rays. Effect of the state of the surface on the non-penetrating rays from a metal. The numbers given above do not accurately represent the relative values of the primary and secondary effects under all conditions, but must only be considered as illustrating the nature of the phenomena. The intensity and penetrating power of the secondary rays that extend to a distance, as well as the properties of the rays which penetrate about half a centimetre of air, depend on the nature of the primary rays. Also, the state of the surface on which the primary beam falls influences the intensity of non-penetrating rays to a great extent. When a very thin layer of water or vaseline is rubbed on the surface, the non-penetrating secondary rays are only a very small fraction of the intensity of the rays coming from a clean bright surface. In this case the non-penetrating rays from the metal surface are completely stopped and another set of smaller ionizing power is given out by the new surface.

37. Becquerel rays. Properties of the α, β, and γ rays. The properties of the rays given out by radio-active substances have been fully described in treatises devoted to that subject by Mme. P. Curie * and E. Rutherford.† The reader is recommended to consult those works for descriptions of the original researches that have been made to solve the many complicated problems that the subject presents. The properties which are of special interest in considering the conductivity produced by the rays are the relative intensity and penetrating power of the different rays, including the secondary rays emitted by various substances. A brief description of some of the characteristic features of the rays will therefore be all that is required in order to explain how radio-active substances can be used to obtain ionization suitable for particular investigations. The rays from radio-active substances are made up of three distinct groups, called α, β, and γ rays, differing widely in their physical properties, but resembling each other in so far as their effects of disintegrating molecules into ions are concerned.

The α rays consist of positively charged particles projected with a velocity about $\frac{1}{20}$th of that of light, and are the least-penetrating rays. They are very slightly deflected in strong magnetic and electric fields. These

* Mme. P. Curie, 'Traité de Radio-activité.'

† E. Rutherford, 'Radio-active Substances and their Radiations.'

rays produce very high conductivity in a gas near the radio-active material, but the most penetrating (those from thorium C_2) only extend to a distance of 8·6 centimetres in air at atmospheric pressure. The charge on the α particle is twice the atomic charge, and the mass of the α particle is of the same order as that of the molecule of helium, since the ratio e/m of the charge to the mass of the α particle is about 6×10^3, and the ratio of the atomic charge to the mass of a molecule of hydrogen is 5×10^3.

The β rays are negatively charged particles travelling with very high velocities ranging from one-third to within a few per cent. of the velocity of light, and are more penetrating than the α rays. They are easily deflected either by an electrostatic force or by a magnetic force. The ratio e/m for a β particle is approximately $1{\cdot}8 \times 10^7$, so that if the charge e is the same as the atomic charge the mass m is about 1/2000 of that of an atom of hydrogen.

The γ rays form a group which produces comparatively small conductivity in a vessel of ordinary dimensions, but they are very penetrating, and resemble Röntgen rays inasmuch as they are not deflected either in a magnetic or electric field.

The relative values of the average penetrating power of the three kinds of rays may be seen from the following numbers given by Rutherford for the thickness of aluminium sheet that will reduce the effect of the rays to one-half their original values: for α rays the thickness is ·0005 cm., for β rays ·05 cm., and for γ rays 8 cm. The ionizing powers of the α, β, and γ rays, in a layer of air between two plates 5 centimetres apart, due to a thin layer of radio-active substance on the lower plate, are roughly proportional to 10,000, 100, and 1.

38. Emanation or radio-active gas. In addition to these rays a radio-active substance gives off a radio-active gas or emanation which diffuses gradually into the surrounding air. When some of the emanation is contained in a gas the latter is a conductor while the emanation lasts. It gradually ceases to be radio-active at different rates, corresponding to the particular substance from which it is derived. The activity of the emanation from radium falls to half its value in about four days, and that from the other substances decays much more rapidly.

The radio-active substances uranium, thorium, radium, and actinium give out penetrating and non-penetrating rays and also emanations, but polonium gives out only the non-penetrating α rays, which extend to a distance of about 4 cm. in air at atmospheric pressure.

39. Ionization due to impact of α particles on a metal surface. The β and γ rays give secondary rays of different types when they fall upon ordinary materials.

With regard to the α rays, it seems to be well established that they do not excite secondary rays that ionize the gas to an appreciable extent, or affect a photographic plate. Negative ions are, however, set free with a small initial velocity when the rays fall on a metal surface,* but they do not escape from the surface when the potential of the surface is 10 or 15 volts above that of the surrounding conductors.

40. Secondary rays excited by β and γ rays. Indirect action of γ rays. The secondary radiations excited by the β and γ rays have been investigated by the electrical method by Eve † and McClelland.‡

Both the β and γ rays give secondary rays consisting of negatively charged particles, of the same type as the β rays, which are deflected by a magnetic field and are sufficiently penetrating to admit of their effects being studied at distances of several centimetres from the surface from which they are given out. The β and γ rays also give rise to secondary rays of the γ type. The deviable secondary rays, however, produce the most notable effects

* These and other effects obtained when α rays emerge from, or are incident on, a metal surface are fully described by Madame Curie in the 'Traité de Radio-activité', vol. ii, p. 170.

† A. S. Eve, Phil. Mag. (6) **8**, p. 669, 1904.

‡ J. A. McClelland, Trans. Roy. Dub. Soc. **8**, pt. 14, and **9**, pt. 1, 1905.

in the gas and can be studied by means of the apparatus illustrated in figure 6, using a specimen of a radio-active substance contained in a thick metal box as the source of the primary radiation. In order to separate the effects of the β and γ rays, the former may be cut out by putting a suitable screen over the aperture in the box which will absorb all the β rays and produce only a small effect on the γ rays. The secondary rays excited by the β and γ rays are very penetrating and produce considerable effects in passing through a gas: in fact, the conductivity produced by γ rays is nearly all due to the secondary rays which are projected from the sides of the vessel on which the rays fall. This result is in accordance with the view suggested by Bragg * that γ rays and Röntgen rays act indirectly by causing a comparatively small number of β particles to be projected with a high velocity through the gas. The experiments recently made by Wilson † on the initial distribution of ions in a gas show the importance of the secondary β rays.

According to Eve, the secondary rays excited by the γ rays form about 15 per cent. of the secondary rays due to the β and γ rays combined. They are each given out in about the same proportion from different reflectors, the total being roughly proportional to the density of the reflector, a result which is also in accordance with McClelland's investigations. The order of the penetrating power of the secondary rays is shown by the reduction of the conductivity produced by passing the rays through a screen of aluminium ·85 mm. thick. The rays from copper or brass are thus reduced to a quarter of their original value, those from lighter materials being less affected. The penetrating power of these secondary rays is, therefore, of the same order as that of the original β rays.

41. Relative conductivity in different gases. When Röntgen rays or Becquerel rays ionize a gas between two

* W. H. Bragg, Phil. Mag. (6) **20**, p. 385, 1910.

† C. T. R. Wilson, Proc. Roy. Soc. A, **85**, p. 285, 1911. An account of these experiments is given in Chapter VII.

electrodes the number of ions produced, as measured by the saturation current, depends on the nature of the rays and on the gas. The relative conductivity obtained in different gases also depends on the pressure at which the conductivity is measured. When the rays are penetrating, the conductivity is proportional to the pressure over a large range of pressures, and the conductivities of different gases are in a constant ratio. With non-penetrating rays the conductivity is proportional to the pressure only when very small pressures are used, and at atmospheric pressures a considerable proportion or even the whole radiation may be absorbed in an apparatus of ordinary dimensions, so that the relative conductivities would depend on the pressure. When the density is so high that all the radiation is absorbed by the gas, further increases of pressure do not affect the number of ions generated between the electrodes. Thus with α rays, or with the non-penetrating secondary Röntgen rays, the maximum current between the electrodes may become constant as far as the pressure is concerned, but the current depends to some extent on the gas. With primary Röntgen rays and with the β and γ rays it would require a very high density of the gas to produce any great diminution of the intensity while the rays passed between electrodes of ordinary dimensions, and it would be impracticable to absorb all these rays by a gas. An interesting comparison of the effects produced by the various kinds of rays is obtained from experiments in which only very slight absorption takes place when the rays traverse the distance between the electrodes. To satisfy this condition, the conductivity produced by the non-penetrating rays should be measured in gases at very low pressures.

The relative conductivities of gases due to the action of Röntgen rays were determined by Perrin,* Rutherford,† and Thomson,‡ and it was found that of the gases that were examined hydrogen had the smallest conductivity, and that

* J. Perrin, Annales de Chimie et de Physique, (7) **11**, p. 496, 1897.
† E. Rutherford, Phil. Mag. (5) **43**, p. 241, 1897.
‡ J. J. Thomson, Proc. Camb. Phil. Soc. **10**, p. 10, 1900.

in general the denser the gas the greater the conductivity. There were, however, large differences in the actual ratios of the conductivities as given by the different observers, and it was subsequently found that the relative conductivity of two gases depends on the nature of the rays.

42. Effect of penetrating power on relative conductivity. The conductivity of different gases compared with that of air was examined by McClung for primary Röntgen rays of various penetrating powers. With the softest of these rays the absorption could not have been very large, so that the conductivity was proportional to the pressure.

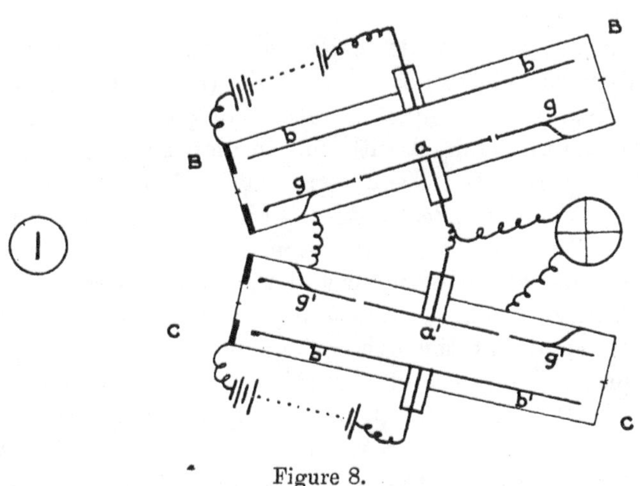

Figure 8.

In these experiments two beams of rays from the same Röntgen-ray tube were passed between the two pairs of parallel plate electrodes without touching the plates, so that no secondary ionization was produced. The principle of the method is illustrated in the diagram, figure 8. The gases to be ionized were contained in two air-tight cylinders *BB* and *CC*. The ends through which the rays entered were made of thick aluminium plates recessed at the centre so that the rays might only have to pass through a small thickness of aluminium. The electrodes were also of aluminium; those in the centre of the field a and a', which are connected to the electrometer, were 7 cm. long by 4 cm. wide, and were surrounded by guard rings g and g'. The opposite electrodes b and b' were 23 cm. long by 6 cm. wide. They were raised to high potentials, one being positive and the other negative. The beams of rays were limited by screens so that the rays did not touch the electrodes, and all the ions produced between the plates were due to primary rays. The non-penetrating secondary radiation from the ends did not reach the space in which the ionization was measured, and the penetrating secondary rays produced an effect which must have been only a small fraction of the primary rays.

When equal numbers of ions are generated in the spaces between the electrodes $a\,b$ and $a'b'$ the electrometer gives no deflexion, since equal quantities of positive and negative ions come into contact with the insulated system connected to the electrometer. The apparatus is first adjusted so that no electrometer deflection is obtained when both cylinders are filled with dry air at the same pressure. The air is then pumped out of one cylinder and a gas G at the same pressure is admitted. The conductivity on the two sides is no longer equal, but the balance can be restored by adjusting the pressure on one of the sides. If the conductivity of the gas G at pressure P_g is exactly the same as the conductivity of air at pressure P_a, then the ratio of the conductivity of G to that of air when both are at the same pressure is P_a/P_g.

Experiments were made with Röntgen-ray tubes at different degrees of exhaustion, and the range of potentials required to excite the rays varied from the potential that produces a spark in air between terminals about 4 centimetres apart, to the potential required to spark between terminals about 20 centimetres apart. The relative conductivities obtained for the different gases are given in the following table: *

* R. K. McClung, Phil. Mag. (6) 8, p. 357, 1904.

Gas.	Ionization with soft rays relative to air.	Ionization with hard rays relative to air.
Air	1·0	1·0
Hydrogen	·105	·177
Oxygen	1·3	1·17
Carbon dioxide	1·46	1·33
Sulphur dioxide	11·05	4·79

The conductivities obtained with soft rays emitted by tubes working with the lower potentials are given in the first column of numbers. The corresponding numbers for the penetrating rays emitted by the highly exhausted tube are given in the last column. Hence the relative conductivity depends on the type of rays which produce the ionization. The experiments also showed that the relative conductivities were independent of the intensity of the rays.

43. Ionization by γ rays and penetrating Röntgen rays. The relative conductivities of gases for very hard Röntgen rays and γ rays from radium were determined by Eve* with a different form of apparatus. The measurements of the conductivity were made by finding the leak of an electroscope connected to a rod which projected into a cylinder containing the gas through which the rays were passed. The ionization in this case included the ionization produced by the secondary rays from the sides of the cylinder, which probably accounts for the high value, ·42, of the relative conductivity of hydrogen and air obtained by this method. The results obtained with γ rays are as follows: air 1·00, hydrogen ·19, sulphuretted hydrogen 1·23, chloroform 4·8, methyl iodide 5·6, and carbon tetrachloride 5·2. These results are practically the same as those obtained by other observers.

44. Relative conductivities produced by rays of different types. Strutt's table of conductivities. A very complete investigation of the relative conductivities of gases traversed by rays of different types has been made by Strutt.†

* A. S. Eve, Phil. Mag. (6) **8**, p. 610, 1904.

† Hon. R. J. Strutt, Phil. Trans. A, **196**, p. 507, 1901, and Proc. Roy. Soc. **72**, p. 208, 1903.

In each case the pressure of the gas was reduced until the current I was directly proportional to the pressure p.

With an unscreened radio-active substance the ionization is almost entirely due to α rays. When the active substance is covered with a sheet of aluminium ·1 millimetre thick the ionization is principally due to β rays, and when covered by a thick sheet of lead the ionization is due to the γ rays. The values of I/p obtained for the different types of Becquerel rays are given in the following table, and also the relative conductivities for the same gases when traversed by soft Röntgen rays.

Gas.	Relative density.	Relative conductivity.			
		α rays.	β rays.	γ rays.	Röntgen rays.
Hydrogen	·0693	·226	·157	·169	·114
Air	1·00	1·00	1·00	1·00	1·00
Oxygen	1·11	1·16	1·21	1·17	1·39
Carbon dioxide	1·53	1·54	1·57	1·53	1·60
Cyanogen	1·86	1·94	1·86	1·71	1·05
Sulphur dioxide	2·19	2·04	2·31	2·13	7·97
Chloroform	4·32	4·44	4·89	4·88	31·9
Methyl iodide	5·05	3·51	5·18	4·80	72·0
Carbon tetrachloride	5·31	5·34	5·83	5·67	45·3

Thus with the exception of hydrogen the ionization produced in gases by α, β, and γ rays is approximately proportional to their densities. It is very remarkable that Röntgen rays give quite different results, and that the γ rays of radium have in this respect a close resemblance to the β rays. This is most probably due to the very penetrating secondary rays which are emitted when γ rays pass through or fall upon the sides of the vessel containing the gas. These secondary rays resemble in all respects the β rays, and the ionization they produce is large compared with the direct effect of the γ rays.

45. Total ionization produced by non-penetrating rays. The total number of ions generated in gases by non-penetrating radiation may be determined by increasing the pressure of the gas between the electrodes until a point is reached when further increases of pressure are not accompanied by any increase in the saturation current between the electrodes. This stage can easily be attained with the non-penetrating secondary Röntgen rays or with the α rays. The number of ions generated by the α rays when totally absorbed by different gases has been determined by Rutherford,* and a similar investigation of the total ionization produced by the non-penetrating secondary Röntgen rays has been made by Langevin.† The results are as follows:

Gas.	Total Ionization produced by α rays.	Total Ionization produced by the non-penetrating Röntgen rays.
Air	100	100
Hydrogen	95	87
Coal gas	—	106
Oxygen	106	—
Carbonic acid	96	107
Hydrochloric acid	102	—
Ammonia	101	—
Sulphuretted Hydrogen	—	138

46. Lenard rays. Another kind of radiation, similar in its nature to the β rays from radio-active substances and the non-penetrating secondary Röntgen rays, known as the Lenard rays, also renders the air highly conducting. These rays, discovered in 1894 by Lenard, are cathode rays or negatively charged particles travelling with a high velocity which can be obtained outside a vacuum tube when the cathode rays fall on a small aperture covered with a thin sheet of aluminium. A study ‡ of their properties showed that they had the power of ionizing the air through which they passed, and a charged metal disc connected to an electroscope was discharged, whether its charge was positive or negative, when placed anywhere within a distance of 30 centimetres from the aluminium window in the vacuum tube. The rays themselves do not extend that distance, but only to about 8 or 10 centimetres; the ions are produced

* E. Rutherford, Phil. Mag. (5) **47**, p. 137, 1899.
† P. Langevin, Annales de Chimie et de Physique, (7) **28**, p. 289, 1903.
‡ P. Lenard, Wied. Ann. **63**, p. 253, 1897.

by the rays in the field of force of the charged disc, and those charged with electricity of opposite sign are attracted to the disc. It was found possible to prevent the disc being discharged by sending a strong stream of air from the disc towards the space traversed by the rays. Experiments also showed that the air remained a conductor for some time after it had been acted on by the rays.

Compared with other forms of radiations these rays are by no means a convenient method of obtaining a conducting gas, but as far as it is possible to judge from the limited number of experiments that have been made on ions generated in this way, it would appear that they have the same properties as the ions generated by Röntgen rays. The particular interest attached to the Lenard rays is that they pass through thin sheets of aluminium and afford a means of studying the properties of cathode rays in gases at various pressures.

47. Action of ultra-violet light on a metal surface. Action of ordinary light on alkali metals. The effect of ultra-violet light in facilitating the conductivity of a gas, generally known as the photo-electric effect, was first observed by Hertz* in 1887 when making experiments on the sparking between electrodes in circuits acted on by electro-magnetic waves. He noticed that a discharge between terminals in air takes place more easily when the electrodes are illuminated by the light from another spark, the effect disappearing when a plate of glass was interposed between the two spark gaps, while a plate of rock-crystal did not stop the action of the light. The action of the light takes place principally at the negative electrode. The phenomenon was at the same time investigated by Hallwachs † and Stoletow,‡ and they found that the effect of the light takes place with small electric forces, and currents can be obtained without the aid of potentials that produce discharges. Currents may thus be obtained from all metals commonly used for

* H. Hertz, Wied. Ann. **31**, p. 983, 1887.
† W. Hallwachs, Wied. Ann. **33**, p. 301, 1888.
‡ A. Stoletow, Comptes rendus, **107**, p. 91, 1888.

electrodes when ultra-violet light falls on them, but Elster and Geitel * discovered that the alkali metals possess this property in a remarkable degree, and with ordinary sunlight comparatively large currents may be obtained.

Using a petroleum lamp as the source of light, Elster and Geitel found that the metals may be arranged in the following order of decreasing photo-electric activity:

Potassium pure.
Alloy of Potassium and Sodium.
Sodium pure.
Amalgams of { Rubidium, Potassium, Sodium, Lithium, Magnesium, Thallium, Zinc.

The current obtained from a potassium surface, when the lamp is at a distance of 6 metres, may be easily detected with an ordinary electrometer.

The effective rays, traversing glass, extend from the ultra-violet to the red, the maximum effect being in the blue.

The wave-length of the most effective radiation varies with each metal, according to its position in the above list, potassium and sodium reacting to the long waves and zinc to the short waves. The effects were from the beginning attributed by Arrhenius † to the conduction of electricity by negative ions emitted from the metal on which the light fell.

48. Properties of ultra-violet light observed under ordinary conditions. Ionization of gases by strong ultra-violet light. When the light falls on a negatively charged body, negative ions are set free and are repelled by the electric force, and the body loses its charge; but positive electricity is not set free from the surface, so that no similar effect is

* J. Elster and H. Geitel, Wied. Ann. **43**, p. 225, 1891.
† S. Arrhenius, Wied. Ann. **32**, p. 565, 1887, and **33**, p. 638, 1888.

obtained when light falls on a body positively charged. Some experiments seemed to show that there is some corresponding effect in the latter case, but Elster and Geitel * found that the loss of the positive charge was due to negative ions set free by reflected light falling on other bodies, which are attracted to the body that is positively charged. When a beam of ultra-violet light passes between two parallel plates without touching a metal surface, no conductivity is observed under ordinary circumstances, the effect being certainly very small compared with that produced by light falling on a negatively charged plate.

Lenard † has, however, found that gases become conductors when exposed to strong ultra-violet light, the effect being noticeable up to a distance of ten centimetres from the spark of a Leyden-jar discharge. Both positive and negative ions were detected in air by allowing a stream of the gas to flow along a glass tube provided with an opening covered by a plate of quartz fixed to the glass with air-tight cement. The air then passed to the space between a metal tube and a coaxial cylinder, the metal tube being connected to earth and the inner cylinder to the leaves of an electroscope. When light from a spark gap in a Leyden-jar discharge circuit passed through the quartz plate, the charge on the electroscope gradually disappeared; the rate being the same whether the charge on the cylinder was positive or negative. The effect ceased to be noticeable when the origin of the light was withdrawn to a distance of more than 10 centimetres from the quartz window. In order to obtain as large an effect as possible Lenard made experiments with spark terminals of various metals, the metal finally adopted being aluminium. The wavelength of the light which gave the maximum effect was estimated at about $\lambda = 1{,}800$, when the light passed through a thin quartz plate. Measurements of the velocity under an electric force of the ions generated in air by ultra-violet

* J. Elster and H. Geitel, Wied. Ann. **57**, p. 24, 1896.

† P. Lenard, Annalen der Physik, (4) **1**, p. 486, 1900, and (4) **3**, p. 298, 1900.

light showed that they are more complex than those produced by Röntgen or Becquerel rays. Lenard found, for the velocities of the ions produced in air by the light, 3·13 centimetres per second for negative ions and ·0015 centimetre per second for positive ions under a force of one volt per centimetre. The former number is about double the velocity of the negative ions produced in air by Röntgen rays, and the latter is about 1/1000 of the velocity of the positive ions. The high mobility obtained for the negative ions may possibly be due to experimental error, as the method adopted is not very accurate, and has been abandoned by Lenard in his later researches.

The small value of the mobility of the positive ions shows that they consist of charged bodies of considerable mass.

As it is not certain whether dust was entirely absent during the experiment, it has been conjectured that the positive carriers consist of dust particles which have lost a negative electron by virtue of the ordinary photo-electric effect on solids.

Some support was lent to this view by the experiments of Bloch.* Strong ultra-violet light was passed through a quartz window into a box containing two parallel electrodes at a large distance apart, so that the light should not fall on the electrodes. A certain conductivity was observed when unfiltered air was in the box, but the conductivity diminished gradually to 1/60th of its original value. When a fresh supply of ordinary air was admitted the current between the plates was restored, while if the air was admitted through a cotton-wool filter the conductivity was not restored.

On the other hand, Lenard † found that in his experiments it made practically no difference whether the air was filtered or not.

49. Lenard and Ramsauer's experiments with strong ultra-violet light. Action of the longer-waved ultra-violet

* E. Bloch, Comptes rendus, **146**, p. 892, 1908.

† P. Lenard, loc. cit.

light on ordinary air. A new investigation of the subject has recently been made by Lenard and Ramsauer,* using a very strong source of ultra-violet light, obtained by means of an induction coil specially designed to give a large secondary current. A Wehnelt interrupter was used and the primary current amounted sometimes to 60 ampères. Experiments were made mainly under two different conditions in order to give special prominence respectively to the longer-waved ultra-violet light and to the highly absorbable type of short-waved light investigated by Schumann.

In the first case the light from the aluminium spark gap traversed 4 centimetres of air and a quartz plate before entering the testing vessel, and the wave-length of the effective rays was estimated to exceed 1,800. The standard atmospheric air employed was conducted through glass tubes from a neighbouring open space and after being dried over calcium chloride and phosphorus pentoxide was filtered through cotton-wool. Both positive and negative ions were obtained, the latter being somewhat more numerous on account of the photo-electric effect on the walls of the testing vessel.

Further investigations showed that the formation of ions in dust-free air was dependent on the presence of condensable vapour. Thus a great reduction in the number of positive ions was caused by passing the air through a glass spiral immersed in a freezing mixture at $-78°$ C., while air passed through rubber tubing showed a greatly increased effect. These facts rendered it undesirable to attempt to reduce the photo-electric effect on the walls of the vessel by coating it with a soap solution, as was done in the earlier experiments, and it was further necessary to preserve the air from contact with grease or other substances likely to emit small quantities of vapour. The magnitude of the volume effect on the gas may be approximately measured by the ratio of the number of positive ions to that

* P. Lenard and C. Ramsauer, Sitzungsberichte der Heidelberger Akademie, 5 parts, Aug. 1910–Aug. 1911.

of negative ions. The following table shows the results obtained by successive purifications:

	Ratio of the number of positive to the number of negative ions.
Unfiltered air	1 (approximately)
Standard filtered air	·113
Air purified by freezing to −78° C.	·041
Air evaporated from liquid air	less than ·001

The last results render it extremely likely that ultra-violet light of the longer wave-lengths has no volume effect on the pure gases of the atmosphere.

50. Masses associated with ions generated by ultra-violet light in filtered air and in purified air. The mobilities of the ions produced were found by the method due in principle to McClelland,* by determining the number of ions in a stream of air which are collected on the central electrode of a cylindrical condenser for various potential differences between the two cylinders. In the case when ions of widely different magnitudes are present, Langevin † has shown that the number of ions whose mobilities lie between given limits can be determined from the current-potential curve.

The size of an ion of given mobility may be estimated by a formula to be given later.‡ In the standard filtered air, large ions of both signs were found whose radius exceeded $1{\cdot}7 \times 10^{-6}$ centimetre, the positive ions being present in somewhat greater numbers than the negative ions. An increase in the intensity of the light produced an increase in the sizes of the ions, and the same effect could also be obtained by using a slower stream of air.

In air purified by freezing, the strongest light produced only small ions, the largest having a radius of $1{\cdot}4 \times 10^{-7}$ centimetre.

The presence of large ions is closely bound up with that of condensation nuclei,§ consisting of finely divided

* See Sections 95 and 110.

† P. Langevin, Comptes rendus, **140**, p. 232, 1905.

‡ See Section 124.

§ See Section 158.

products formed by the light and not originally carrying an electric charge. The number of such nuclei present, as revealed by the steam jet, varied with the purity of air in the same way as that of the large ions. It thus appears that the presence of the large ions may be due to the coalescence of a normal ion with an uncharged nucleus generated by the light.

The effect of short-waved ultra-violet light (Schumann rays) was examined by bringing the spark gap up to a distance of a few millimetres from the apparatus and replacing the quartz plate by one of fluor-spar. The results are, in many respects, less complicated than before. On account of the strong absorption of the air, the effect on the gas predominates over that on the walls of the containing vessel and approximately equal numbers of positive and negative ions are formed. Traces of impurity also played a much smaller part than in the case of longer wave-lengths.

In the purest air obtainable, Lenard and Ramsauer found that the ions generated by Schumann rays were small, 90 per cent. having a radius less than 7×10^{-8} centimetre. The influence of traces of impurity on the size of the ions was not very great, the chief exception occurring in ammonia in the presence of water vapour, when numbers of ions of radius 5×10^{-7} centimetre were formed.

51. Comparison of effects obtained with ultra-violet light and Röntgen rays. It is of interest to consider those properties of the photo-electric effect which may be contrasted with corresponding properties of other radiations. The setting free of negative ions from a metal surface by the light bears some resemblance to the action of Röntgen rays in producing non-penetrating secondary rays that consist of negative corpuscles travelling with a high velocity. The only difference between the two effects is that the ions set free by the ultra-violet light are projected from the surface with a velocity which is small compared with the initial velocity of the particles composing the secondary

Röntgen rays. The latter traverse about 5 millimetres of air at atmospheric pressure before their velocity is reduced by collisions with molecules to such a low value that they cease to produce ions in the gas. The ions set free by ultra-violet light, on the other hand, are not projected with a sufficiently high velocity to produce others by collisions with the molecules of the gas, so that there are no positive ions in the gas near the surface of the electrode. A stream of air passing by the surface would in this case contain only negative ions, and if it passed a charged conductor the charge would disappear only when the conductor was positively charged.

52. Velocity of electrons emitted from metal surfaces by the action of ultra-violet light. An upper limit to the velocity with which negative ions are projected from a metal surface has been determined by Lenard. It may be estimated by finding the maximum positive potential to which an insulated conductor will rise in a good vacuum above a surrounding conductor of the same metal when ultra-violet light falls on the insulated conductor. If v_0 is the initial velocity of the ion normal to the surface, E the potential of the inner surface (the outer surface being at zero potential), m the mass, and e the electric charge of the ion, then the velocity of the ion will be reduced to zero before it reaches the outer surface and will return to the inner surface if $eE > \frac{1}{2} mv_0^2$. If the lines of force between the two surfaces which terminate on the part of the inner surface on which the light falls are straight lines, then the trajectories of the ions that start in a direction normal to the inner surface will also be straight lines and the ions will come to rest before they touch the outer surface. If, however, the lines of force are curved, all the ions may traverse curved paths and not come to rest at any point, and the observed value of E would be $m(v_0^2 - v_1^2)/2e$, where v_1 is the minimum velocity of the ions on arriving at the outer boundary. In fact, when the ions collide with the outer boundary they may have a considerable velocity in the tangent plane.

Lenard* found that a positive potential of 2·1 volts was sufficient to prevent ions from escaping from an aluminium surface, and the potential,† while varying with the wave-length of the light, was not affected by the intensity. The latter quality can be altered without changing the nature of the light by using the same source at different distances from the aluminium surface.

53. Velocity independent of the temperature. Experiments of a similar kind on various metals in a highly exhausted enclosure were made by Millikan and Winchester,‡ in order to investigate the influence of temperature on the positive potentials to which metals rise, and also the effect of temperature on the total number of ions set free from a surface. They found that neither of these quantities was affected by variations of temperature. The positive potentials which they obtained for the various metals are as follows: silver 1·340, iron 1·225, gold 1·215, brass 1·174, copper 1·135, nickel 1·126, magnesium ·830, aluminium ·738, antimony ·394, zinc ·197, and lead ·0, and they remained constant, within the experimental error, over a range of temperature from 26° C. to 95° C.; also there were no changes in the potentials when the intensity of the light was altered. The results thus obtained are interesting, as they show that the ions are set free with the same velocity under different physical conditions. As to the absolute values of the potentials, it is difficult to reconcile the number ·738 obtained for aluminium with the number 2·1 obtained by Lenard; it may be that in Millikan and Winchester's experiments the lines of force from the metal disc on which the light fell did not run straight to the surrounding gauze, or that the field of force round the disc was affected by electromotive forces of contact due to the presence of different metals. What is important, however, is that there is no

* P. Lenard, Wien. Ber., Abt. **2**[a], p. 1649, Oct. 19, 1899, and Annalen der Physik, (4) **2**, p. 359, 1900.

† P. Lenard, Annalen der Physik, (4) **8**, p. 149, 1902.

‡ R. A. Millikan and G. Winchester, Phil. Mag. (6) **14**, p. 188, 1907.

temperature effect, and in this respect the action of ultra-violet light in generating ions resembles that of Röntgen rays.

It should be mentioned that the influence of temperature on the number of ions set free had previously been investigated by other physicists, but there was no agreement between their results, and it would be difficult to decide from them what effect, if any, was produced by altering the temperature. As the experiments of Millikan and Winchester were conducted in a good vacuum they are probably not subject to so many indirect influences, and it is reasonable to suppose that their conclusions are the most reliable. One of the earlier investigations made by Elster and Geitel* on the influence of temperature led to the conclusion that the photo-electric effect at the surface of potassium in a good vacuum increased by about 60 per cent. when the temperature was raised from 20·3° C. to 50·3° C. With zinc, Elster and Geitel found that an increase of temperature did not alter the photo-electric effect.

54. Photo-electric effect as a function of the distance from the origin. The law connecting the number of ions liberated from a surface with the intensity of the light has been investigated by Lenard,† and more recently by Griffith.‡ Lenard concluded that the number of ions set free from a surface is proportional to the intensity of the light. The numbers he found experimentally, however, indicate a more rapid variation in the photo-electric effect than in the intensity of illumination. Thus, if E is the number of ions set free by the action of light of intensity I, the experiments gave $E = 231$ when $I = 23{\cdot}6$, and $E = 10{\cdot}6$ when $I = 1{\cdot}44$. In the experiments made by Griffith, the quantities were varied over large ranges, and by using a very sensitive apparatus to multiply the quantity E it was possible to measure accurately the photo-electric effect at distances from 10 centimetres to 138 centimetres

* J. Elster and H. Geitel, Wied. Ann. **48**, p. 625, 1893.
† P. Lenard, Ann. der Physik, (4) **8**, p. 149, 1902.
‡ I. O. Griffith, Phil. Mag. (6) **14**, p. 297, 1907.

from the origin of the light, which in this case was a spark gap between aluminium terminals in a circuit through which a Leyden-jar discharge took place. In the experiments at the shortest distance two quartz plates were placed between the origin and the detector, and at the longer distances the plates were fixed to the end of a long wide tube which was then exhausted. It was thus possible to secure that the absorption of light should be the same in each case, so that the reduction in the intensity for the different distances was accurately proportional to the inverse square of the distance. The numbers obtained in a series of experiments are given in the following table, r being the distance of the zinc surface from the origin, in centimetres:

r	10	37·6	110·2	138
I	126	8·63	1	·64
E	318	15·7	1	·43
E/I	2·52	1·82	1	·67

The diminution of the ratio E/I with decreasing values of I is to be expected on the ordinary electromagnetic theory of light, in which the intensity of the electric and magnetic forces from a vibrating system are supposed to be inversely proportional to the distance from the origin. It is obvious that negative electricity in a metal is not in a state at ordinary temperature that it can be made to move into a gas when acted on by a small force. If, therefore, negative ions are set free from a metal, the force applied to them must exceed a certain order of magnitude, so that when the intensity of the light diminishes, the value of E must become zero for certain small values of I. The above experiments support this view of the constitution of the waves of light, inasmuch as they show that the quantity E diminishes more rapidly than I, and may therefore become zero when the intensity I has certain small values.

55. Effect of Röntgen rays as a function of the distance from the origin. The effect thus obtained with ultra-violet light is different from what has been observed in the corresponding case of Röntgen rays, for the number of ions produced per unit length of a narrow beam was found to be constant, so that the ratio E/I would be constant if E denoted the number of ions produced per cubic centimetre by a beam of intensity I.

This property of the rays has been carefully examined by Gill,* and the number of ions generated by a narrow beam was found to be constant up to a distance of one metre from the origin.

Thus there appears to be a remarkable difference between a beam of Röntgen rays and a beam of ultra-violet light, but perhaps if the experiments with Röntgen rays had been carried out with less penetrating rays, the ratio E/I would have diminished. As far as the results hitherto obtained are concerned, it would be difficult to reconcile this property of the rays with the ordinary theory of electromagnetic waves or pulses, so that some of the supporters of that theory have suggested that the electric intensity is neither uniform nor varies continuously in the wave-front of a pulse, but is located on very small areas in which the intensity does not diminish as the beam of rays opens out. Many physicists, accustomed to believe in the theory of wave-motion as illustrated by the properties of light, find it impossible to accept this view, and indeed it is obvious from the results of Griffith's experiments that a beam of ultra-violet light cannot be constituted in this way.

56. Properties of ultra-violet rays. The action of ultra-violet light on a metal surface provides a method of obtaining negative ions which is very convenient for many purposes. Thus, for experiments at low pressures, it can easily be arranged to bring a beam of light through a quartz window into an exhausted chamber, and ions are set free from a metal surface in large numbers even with the lowest pressures obtainable. When, however, rays that ionize the molecules of the gas, such as Röntgen rays, are used, the number of ions diminishes as the pressure is reduced, and at very low pressures (of the order of one millimetre of

* E. W. B. Gill, Phil. Mag., (6) **23**, p. 114, 1912.

mercury) it becomes difficult to make accurate measurements of the small electrical effects they produce. A beam of ultra-violet light may also be focused by a quartz lens on a surface or converted into a parallel beam which can be used at a long distance from the origin, as there is no loss of intensity except that due to the absorption of the air. The intensity may be reduced to any required amount by passing the beam vertically through a shallow vessel containing a suitable depth of water.

The best sources of ultra-violet light are the arc-lamp, the mercury vapour-lamp in a quartz tube, and the spark across an air-gap through which a series of condenser discharges takes place. The most intense light is obtained by the first two methods, but the latter provides the most constant source and the light is sufficiently intense for most experimental purposes.

57. Photo-electric currents of different intensities. The rate of discharge of negative electricity from a metal undergoes large variations depending on the state of the surface. Thus a zinc surface immediately after it has been cleaned with sand-paper or turned in a lathe may be twenty times as sensitive to the action of the light as the same surface after it has been exposed to the air for two or three days. In the latter state it is not subject to large variations, and exposure to light, which in some cases produces a temporary diminution in the photo-electric effect, does not affect the less sensitive surface, so that there is an advantage in using a surface that has been exposed to the air for some time. Owing to these causes it is difficult to determine the relative values of the rates of discharge from various metals, and there is little agreement between the results obtained by different physicists as the surfaces were not prepared in the same way.

Ladenburg,* who investigated the subject, using metals that had been polished once with emery and oil, obtained the following relative values for the photo-electric effects for

* E. Ladenburg, Ann. der Phys. (4) **12**, p. 558, 1903.

metals in a high vacuum: platinum 2·81, copper 2·45, zinc 2·22, brass 2·01, silver 1·31, antimony 1·20, lead ·89, bismuth ·78, iron ·61, nickel ·52. It was found that the numbers were not constant, but showed large variations with the polishing of the surface.

In order to avoid any effect that might arise from a film of oil on the surface, Millikan and Winchester polished the surfaces with dry emery, and then washed them in alcohol and dried by heating to 400° C. The photo-electric currents in arbitrary units which they obtained when the metals were in a high vacuum are: copper 25·1, gold 24·7, nickel 24·0, brass 23·8, silver 17·16, iron 16·4, aluminium 14·9, magnesium 11·0, antimony 4·0, zinc 1·2, lead 0·9. The heating process that was used to dry the metals must have oxidized some of them more than others, so that it is not surprising that the results do not agree with those of other observers. For instance, zinc and aluminium that have not been heated are usually found to be more sensitive to the action of the light than is indicated by these numbers.

58. Early investigations of electrical properties of gases. In the other well-known methods of obtaining gases that conduct at ordinary temperatures, the ions as a rule move through the gas with velocities that are very small compared with those acquired under similar conditions by the ions generated by Röntgen rays. Some of the electrical phenomena that occur in these cases were discovered many years ago, and are mentioned among the earlier researches that were made at a time when the nature of the effects that were obtained was imperfectly understood. An interesting account is given by Volta * of some of these experiments, which he made in company with Lavoisier and Laplace. They found, for example, that electricity was obtained from the evaporation of water, from the combustion of coal, and from the effervescence of iron-filings in dilute sulphuric acid.

The effects they obtained were probably due to the fact

* A. Volta, Phil. Trans. **72**, p. 275, 1782.

that in many cases an excess of ions of one sign is set free in the gas, so that electrical charges were carried by air currents away from the insulated conductors.

With regard to the evaporation of water, the experiments made by Volta on this point are open to various interpretations. The method adopted in most cases, apparently, was to place some burning charcoal on an insulated plate, and to throw water on the charcoal. The alteration in the potential of the plate that ensued, which no doubt was considerable, was attributed to the evaporation of the water.

It may, however, be easily shown that if the vapour escaping from a charged surface of water carries with it any charge it must be very small. Many experimenters have failed to observe any loss of charge from a liquid due to evaporation. Pellat,* in his experiments on this subject, made observations extending over an hour and was then able to detect only a very small loss from a large surface of water, with a very sensitive apparatus. But considerable electrical effects may be obtained by splashing water, as were first noticed by Lenard,† who observed that the air in the neighbourhood of waterfalls is negatively electrified. Also, Kelvin ‡ found that air becomes charged with negative electricity when it is bubbled through pure water, the density of the electrification being diminished when small quantities of acids or salts are dissolved in the water.

The properties of these conducting gases have recently been studied more systematically, and some information as to the nature of the electrical properties of the gases has been obtained.

It is intended here to mention only the investigations which show how the general character of the conductivity can be completely explained on the hypothesis that positive and negative ions are formed in the gas, or that ions are set free from the surface of incandescent solids, which move through the gas under the action of an electric force.

* H. Pellat, Comptes rendus, **128**, p. 169, 1899.
† P. Lenard, Wied. Ann. **46**, p. 584, 1892.
‡ Lord Kelvin, Proc. Roy. Soc. **57**, p. 335, 1894.

59. Conductivity produced by flames. Velocity of ions in flame gases. Among the causes of this kind whereby the conductivity of a gas is increased, the action of flames is of special interest, as it was a study of the electrical property of gases rising from flames that led Giese* to propose the theory of ionization that has been found to account for many of the characteristic features of the conductivity of gases.

In order to show that the conductivity is not due to the

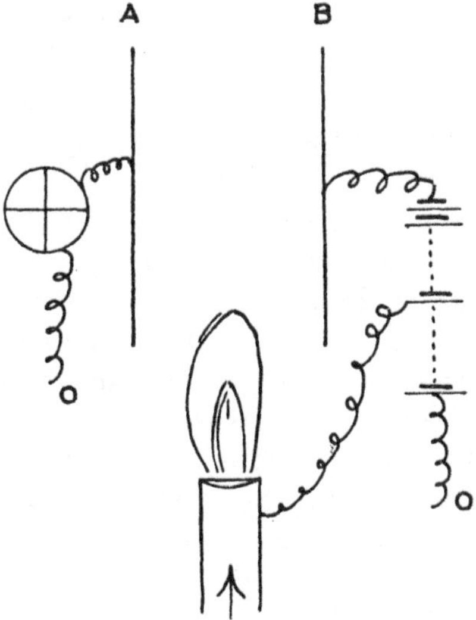

Figure 9.

escape of electricity from the surface of the metal electrodes, these may be placed at a distance from the flame where they remain cool. The explanation of the conductivity between the electrodes *A* and *B* when placed in the position shown in figure 9 would, according to Giese's theory, be as follows:

The positive and negative ions generated in the flame move in opposite directions from the flame under the electric force, and after passing through the air reach the electrodes.

* W. Giese, Wied. Ann. **17**, p. 519, 1882.

When the gases above the flame are examined in a similar manner the currents are much smaller, which is due to the recombination of positive and negative ions that takes place in the time that elapses between the formation of the ions in the flame and the arrival of gas in the field of force between the electrodes. Giese points out that the rate of recombination is greatest when there are a large number of positive and negative ions in the gas, and when there are a small number the chance of positive and negative ions coming into contact is very small, so that the gas retains a small conductivity for a considerable time after it has left the flame.

In many respects the currents between the electrodes for different forces are of the same kind as those obtained when Röntgen rays are passed through parallel plates. The currents increase with the force, and with certain large forces a maximum current is obtained. The maximum current is attained, in the case where the conducting gas moves between the electrodes, when the force is sufficient to collect on the electrodes all the ions that are brought between them before they are carried out of the field of force by the stream, or before the effect of recombination has had time to produce an appreciable reduction in the conductivity when the electrodes are long in the direction of the stream and the stream moves slowly. The force necessary to obtain the maximum current thus obviously depends on the mobility of the ions. McClelland* investigated in detail the mobilities, and found that in the gas above the flame the velocities of the ions under the electric force were much smaller than those of ions generated by Röntgen rays, and he also observed that the velocity diminished as the distance from the flame to the electrodes increased. He attributed this effect to the condensation of water vapour and other products of combustion on the ions as the temperature of the gas diminished.

60. Incandescent metals. Another case of interest in which the phenomena are very complicated is the ionization

* J. A. McClelland, Phil. Mag. (5) **46**, p. 29, 1898.

due to incandescent metals. Metal surfaces when cool do not, when charged to the highest potentials that can be applied, give off positive or negative ions, but when the temperature is raised, ions are produced at the surface which move through the air under an electric force. At first only positive ions are set free; so that if the metal were positively charged it would lose its charge, but if negatively charged the charge does not escape into the air. When the temperature is raised to a higher point, both positive and negative ions are generated at the surface, and electricity of either kind escapes from the metal.

McClelland examined the stream of air which passed an incandescent wire at one point of its course, and found that positive electricity could be collected on an electrode when the wire began to glow. As the temperature increased, negative ions were also found in the stream, and when the temperature exceeded 400° C. and the metal became bright yellow, the air contained positive and negative ions in equal quantities. The conducting air drawn from the neighbourhood of the wire has all the properties of air containing a large number of positively and negatively charged carriers which move slowly under an electric force. A maximum current can be obtained when large electric forces are used, and the stream is then deprived of its conductivity.

61. Chemical action. Mass associated with the ions. An example of a process in which ions are generated in air at ordinary temperatures is to be found in the case of the conductivity produced in the air surrounding a stick of moist phosphorus. Bloch* examined the stream of air which had passed over phosphorus, and it was found to have the same properties as air that had passed over an incandescent wire. The mobility of the ions was also examined in this case, and they were found to move very slowly under an electric force of one volt per centimetre, the velocity ranging from ·003 to ·0003 centimetre per second when the gas was dry.

* E. Bloch, Thèse de Doctorat, Paris, June 1904.

The effects obtained with air drawn from the neighbourhood of phosphorus resembles in many respects those obtained in newly prepared gases, as for instance the oxygen and hydrogen given off by electrolysis of alkaline or acid solutions.

The ions in these gases move very slowly under an electric force,* and there is generally an excess of electricity of one kind in the gas. This can easily be observed by passing the gas into an insulated vessel connected with an electrometer. In this case, as in all others in which the carriers of the atomic charge are very large compared with atomic dimensions, the gas may be sent through long lengths of narrow tubing or kept in enclosed spaces for several minutes without losing their conductivity or their charge. This is due to the small velocity of agitation of the ions of large mass, for the rate of recombination or diffusion to the sides of a tube depends on the velocity of agitation. This property must not be confused with a similar property of gases containing emanation. When a gas contains ions of large dimensions they can be removed by passing the gas through a strong field of force between two electrodes, and the gas emerges with its normal insulating power, which it retains. When a gas containing radio-active emanation is treated in a similar manner, the ions in the gas are removed, but the uncharged molecules of the radio-active gas are carried along with the stream and continue to generate ions in the gas which renew the conductivity.

* Phil. Mag. (5) **45**, p. 172, 1898.

CHAPTER III

THE MOTION OF IONS IN GASES

62. Phenomena due to the motion of ions. The determination of the currents between two electrodes in a conducting gas corresponding to different applied electromotive forces has formed the subject of many researches, the ions having been generated under different conditions in the various experiments. The simplest case to consider is that in which the gas is at rest in the space between two parallel-plate electrodes and the ions are generated in the gas by a beam of Röntgen rays or Becquerel rays. When no electromotive force is applied, the ions generated by the action of the rays increase in number until the rate at which they are lost by recombination or by diffusion to the electrodes is equal to the rate at which they are generated. The number that recombine per second in unit volume is proportional to the product of the numbers of positive and negative ions present, and the number lost per second by coming into contact with the boundary (which in this case consists of the two plate electrodes) is proportional to the number per cubic centimetre in the neighbourhood of the boundary.

When an electric force is applied, the positive and negative ions move in opposite directions towards the electrodes. The greater the velocity of the ions the smaller the number left in the gas, so that the number that recombine diminishes as the force increases. Also, since the positive ions move away from the positive electrode, the number that come into contact with that electrode by diffusion diminishes as the force increases; similarly with the negative ions. With sufficiently high forces, a point is reached when the length of time the ions are in the gas is so short that the number that recombine is inappreciable, and practically all the ions of one kind generated by the rays are collected on the electrode charged with electricity of the opposite sign. The current obtained with large forces, therefore, has a maximum value, known as the saturation current, which is equal to the charge on all the ions of one sign generated per second in the gas.

63. Variation of current with electric force. General character of experimental results. The relation between the current and the potential difference between the electrodes was determined experimentally with different gases and for electrodes of various shapes when the first investigations of the properties of the rays were being made. In all cases the current was found to be proportional to the force when small forces were used. The rate of increase of the current with the force diminished as the force was increased, and finally for large forces a maximum current was obtained. The main features of all the curves exhibiting the connection between the current and the electromotive force were the same, but they differed in points of detail according to the various conditions under which the experiments were made.

64. Potential gradient between parallel-plate electrodes. It is difficult to calculate mathematically the equation of the current-electric-force curve in the general case for parallel-plate electrodes, when recombination and diffusion are in progress, as the problem becomes very complicated owing to the fact that the number of ions per cubic centimetre varies at different points of the gas. Thus in the space adjacent to an electrode the number of ions approaching the electrode exceeds the number receding, since the former number contains ions generated at all parts of the field, whereas the latter only contains ions that are generated in the narrow layer near the electrode. The charges on the excess of negative ions in the gas near the positive electrode, and on the excess of positive ions near the negative electrode, disturb the electric field so that the force is increased near the surfaces of the electrodes and diminished in the space midway between the electrodes. This effect of polarization due to charges in the gas is appreciable when the ratio of the current to the electric force is above a certain value.

The electric force at different points in a current between two parallel plates has been investigated experimentally by Zeleny * and Child.† The potential gradient as deter-

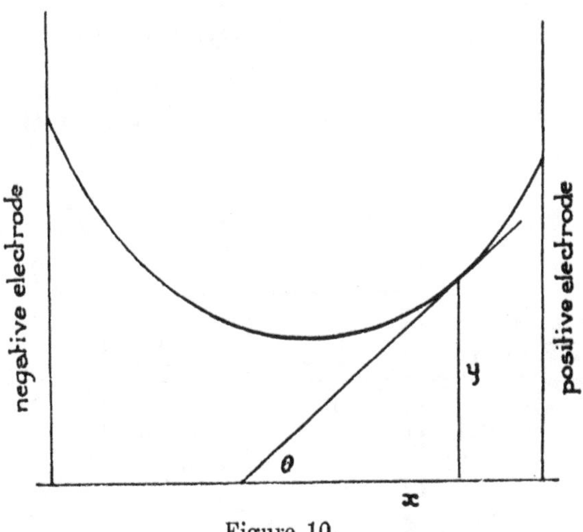

Figure 10.

mined by the potential assumed by an insulated wire parallel to the electrodes, or by a metallic point, showed that the electric force was greatest near the electrodes and nearly constant for some distance about midway between the plates.

It was also found that the field was not symmetrical about the central plane, the force being greater at the negative electrode than at the positive electrode. This, as has been shown by Zeleny, is due to the fact that the negative

* J. Zeleny, Phil. Mag. (5) **46**, p. 120, 1898.
† C. D. Child, Wied. Ann. **65**, p. 152, 1898.

ions move faster than the positive ions when in the same field of force, so that there is an excess of positive electricity in the gas which increases the force at the negative electrode and diminishes it at the positive electrode. The experimental results may be represented by the curve in figure 10, the ordinate y being the force at any point in the gas, and the abscissa x the distance of the point from the negative electrode. If q be the charge per unit volume at a point x, the difference dY between the forces at the points x and $x+dx$ is given by the equation $-4\pi q = \frac{dY}{dx} = \tan\theta$, and the curve therefore shows that there is a positive charge in the neighbourhood of the negative electrode and a negative charge near the positive electrode.

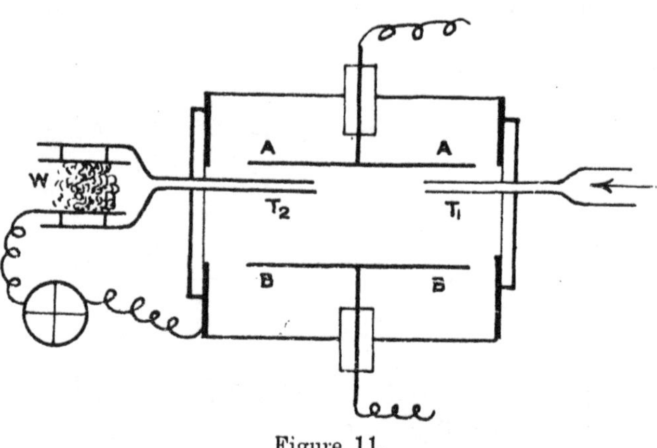

Figure 11.

65. Charge in the gas near the electrodes. This result is in agreement with the experiments made by Zeleny on the distribution of charge in the air between two parallel-plate electrodes when a current is flowing. The principle of the method that was used is shown in figure 11.

The air between the two plates A and B is ionized by Röntgen rays and a current is maintained between them by a battery of cells the centre of which is connected to the air-tight case which contains the electrodes. The tubes T_1 and T_2 project through the sides of the case, and are movable in a vertical direction so that they may be placed at various distances from the electrodes. When a current of air is blown into the tube T_1 it escapes from the box through the opposite tube T_2 and carries with it some of the ions moving in the gas under the electric force. The ions in the stream of air are discharged by the wool in the insulated tube W and the electrometer gives a deflection proportional to the charge in the air. When the tubes T_1 and T_2 were near the positive electrode the electrometer acquired a negative charge as the stream of air passed through the wool. When the tubes were moved towards the negative electrode the charge in the air diminished and was practically negligible at points near the centre of the field. As the tubes approached the negative electrode the charge became positive and was a maximum when the tubes were near the negative electrode.

These experiments are interesting, as they afford direct evidence of charges in the gas in the neighbourhood of the electrodes, and also show that the charges vary from point to point in agreement with the theory, but it would obviously be impossible to deduce from the results the exact value of the charge at a point in the gas between the electrodes when the current is flowing.

66. Potential assumed by an insulated wire. A difficulty also arises in drawing accurate conclusions from the potentials assumed by a wire in the conducting air, as there is good reason to believe that the potential at points in the gas along a line parallel to a plane electrode is not the same as the potential assumed by a wire when its axis is along those points.

Thus in the case of a wire placed near the positive electrode, its potential being initially the same as that of the gas, two streams of ions move in opposite directions past the sides of the wire, one containing a large number of negative ions and the other a smaller number of positive ions. It intercepts more negative ions than positive ions, so that its potential falls below the potential of the gas. The charge thus acquired by the wire increases until the effect which it produces in attracting positive and repelling negative ions causes them to come into contact with the wire in equal numbers. The final value of the potential assumed by the wire is therefore less than the gas, and the force near the electrodes as determined experimentally is too high by an amount which is difficult to estimate.

67. Mathematical investigations of the conductivity. Several mathematical investigations have been made by different physicists of the distribution of force between parallel-plate electrodes and of the currents obtained with various forces. Those made by Walker* and Robb† are principally of interest from a theoretical point of view, and deal with cases in which the differential equations representing the processes that take place in the gas are rendered more manageable by supposing that the gas is at particular pressures, which give convenient values to some of the constants involved.

The problem of finding the ratio of the electric force X_1 at an electrode to the force X_2 midway between the plates has been investigated by Thomson,‡ and up to a certain point the investigation is interesting, but the conclusion obtained, namely that the ratio X_1/X_2 is equal to 2·51 and independent of the current between the plates and of the intensity of the ionization produced by the rays, is consistent only when the current is a small fraction of the saturation current. The ratio 2·51 corresponds to the conductivity of air at atmospheric pressure, and at low pressures the ratio X_1/X_2 is said to vary inversely as the square root of the pressure, a result which is not true in general, as it may be seen from elementary considerations that the polarization diminishes as the pressure is reduced,

* G. W. Walker, Phil. Mag. (6) **8**, p. 650, 1904.

† A. A. Robb, Phil. Mag. (6) **10**, pp. 237, 664, 1905.

‡ J. J. Thomson, 'Conduction of Electricity through Gases', 1906 edition, pp. 84–88.

and it is possible to have comparatively large currents flowing between parallel plates in a gas at low pressure, which do not disturb appreciably the uniformity of the field.

68. Mie's investigation of the conductivity. The conductivity of a gas between parallel plates has also been investigated by Mie * for the general case in which the

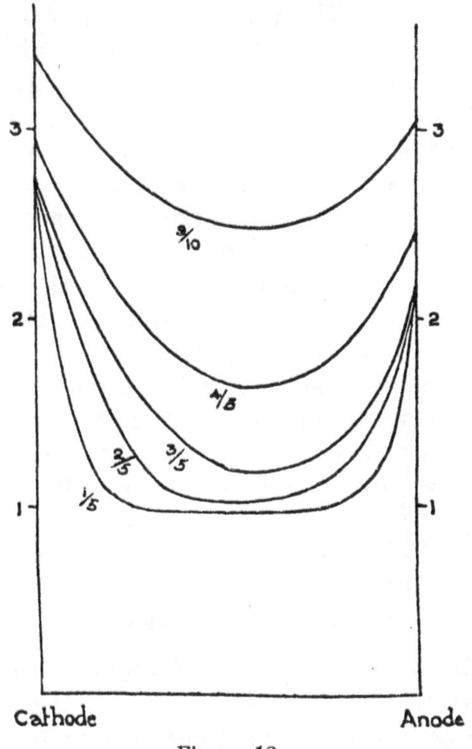

Figure 12.

difference of the velocities of the positive and negative ions is taken into consideration, the effect of diffusion being neglected. The air between the plates is supposed to be ionized uniformly by the action of rays, and the effect of recombination in reducing the currents corresponding to

* G. Mie, Ann. der Physik, (4) **13**, p. 857, 1904.

low forces is investigated, as well as the variation of the force in the gas at different points between the plates. Solutions of the general differential equation were obtained by a series of successive approximations, which are too lengthy to be reproduced here, but the results of the calculations for air at atmospheric pressure have been given in a convenient form by means of curves, corresponding to the cases in which the currents are various fractions of the saturation current. The curves given by Mie for currents which are 1/5th, 2/5ths, 3/5ths, 4/5ths, and 9/10ths of the saturation current are represented in figure 12. The ordinate y of the curve is proportional to the force at any point distant x from the negative electrode. In each case the force is greatest at the cathode: it has a minimum value at a point near the middle of the field and again assumes a large value at the anode. The curves are not symmetrical about the centre of the field, as the velocities of the positive and negative ions are different. The negative ions move fastest, so that there is an excess of positive ions in the space between the plates and the polarization at the cathode is therefore greater than that at the anode. The ratio of the force at the cathode to the minimum force has its greatest value 2·7 for a current of 1/5th the saturation value, and is nearly constant for the smaller currents; as the currents get large the effect of the polarization diminishes, and when the current is 9/10ths of the maximum the ratio of the force at the cathode to the minimum force is 1·39.

69. Conditions under which the field of force is approximately uniform. The investigation of the conductivity obtained with forces that give currents which are slightly less than the maximum current is of interest for many reasons, and it is possible to obtain a simple solution of the differential equations, when the effect of diffusion is neglected, which can easily be applied to gases at various pressures.*

When the potential difference between the electrodes is

* Proc. Roy. Soc. A, **86**, p. 72, 1911.

sufficiently large, the polarization may be neglected and the field of force between parallel plates may be considered uniform. The conductivity in a uniform field is therefore of a kind that can be realized in practice, and from the solution obtained for that case it is easy to see how the charge in the gas increases as the force is reduced, and to find exactly the percentage by which the current falls below the maximum current before the field of force is disturbed to such an extent as to introduce a serious error in the determination of the current.

In order to find an upper limit to the difference between the maximum force X_1 to the minimum force X_2 in a conducting gas, it is necessary to find the charge per cubic centimetre of the gas between the two points at which the forces are acting. Let ν_1 and ν_2 be the numbers of positive and negative ions per cubic centimetre, when the steady state is reached and a current i is flowing. If e be the charge on the ion, the current i per unit area of the electrodes is $e(u_1\nu_1+u_2\nu_2)$, where u_1 and u_2 are the velocities of the positive and negative ions, and if $e\nu_1 = n_1$ and $e\nu_2 = n_2$, then $i = n_1u_1+n_2u_2$. The velocities of the ions are proportional to the electric force, so that $u_1 = k_1X$ and $u_2 = k_2X$, where k_1 and k_2 are the velocities under unit electric force. Let a be the distance between the plates, and $x = 0$ at the positive electrode.

The difference between the forces at two points is

$$X_1-X_2 = 4\pi\int(n_1-n_2)\,dx$$

$$= 4\pi\int\left[\frac{n_1u_1+n_2u_2}{u_1} - \frac{n_2(u_1+u_2)}{u_1}\right]dx,$$

so that $X_1-X_2 < 4\pi\int\frac{i}{u_1}\,dx$ where the integral is taken over the distance between the two points at which the maximum and minimum forces are acting, which is less than the distance a.

Hence $$X_1-X_2 < \frac{4\pi ia}{u_1'},$$

where u_1' is the minimum velocity, so that

$$\frac{X_1 - X_2}{X_2} < \frac{4\pi i a}{k_1 X_2^2}.$$

If the force X_2 be increased to a large value, the current i remains constant when the saturation point is reached, and $X_1 - X_2$ becomes negligible in comparison with X_2.

70. Effect of recombination on the currents obtained with large forces. Let the space between the plates be ionized uniformly by rays producing q/e positive or negative ions per cubic centimetre per second, and let the rate of recombination be αe, so that the number of ions of either kind that recombine per second is $\alpha e \nu_1 \nu_2$.

When the steady state is reached, the rate at which ions are being generated per unit volume must be equal to the loss by recombination together with the loss by the motion of the ions through the boundary of the volume. When the diffusion of the ions is neglected, the steady state is represented by the equation

$$\frac{q}{e} = \alpha e \nu_1 \nu_2 + \frac{d}{dx}(\nu_1 u_1),$$

or

$$\frac{d}{dx}(n_1 u_1) = q - \alpha n_1 n_2, \quad . \quad . \quad . \quad . \quad . \quad (1)$$

also

$$n_1 u_1 + n_2 u_2 = i, \quad . \quad . \quad . \quad . \quad . \quad . \quad (2)$$

i being the current, which is the same at all points in the gas.

Hence $$\frac{d}{dx}(n_1 u_1) = \frac{\alpha}{u_1 u_2}\left[n_1^2 u_1^2 - i n_1 u_1 + \frac{q u_1 u_2}{\alpha}\right].$$

The velocities u_1 and u_2 are constant as the field of force is uniform, so that on integration the above equation gives

$$\left(\frac{q u_1 u_2}{\alpha} - \frac{i^2}{4}\right)^{-\frac{1}{2}} \tan^{-1}\left\{\left(n_1 u_1 - \frac{i}{2}\right)\left(\frac{q u_1 u_2}{\alpha} - \frac{i^2}{4}\right)^{-\frac{1}{2}}\right\}$$
$$= \frac{\alpha x}{u_1 u_2} + C,$$

or

$$n_1 u_1 - \frac{i}{2} = \sqrt{\frac{q u_1 u_2}{\alpha} - \frac{i^2}{4}} \tan\left\{\frac{\alpha}{u_1 u_2}\left(x - \frac{a}{2}\right)\sqrt{\frac{q u_1 u_2}{\alpha} - \frac{i^2}{4}}\right\}, \quad (3)$$

the constant C being determined by the conditions $n_1 u_1 = i$ when $x = a$, and $n_1 u_1 = 0$ when $x = 0$. The latter condition also gives

$$\sqrt{\frac{4\, q k_1 k_2 X^2}{\alpha i^2} - 1}\, \tan \left\{ \frac{\alpha a i}{4\, k_1 k_2 X^2} \sqrt{\frac{4\, q k_1 k_2 X^2}{\alpha i^2} - 1} \right\} = 1, \qquad (4)$$

which shows the connection between the current and the electric force.

In order to see over what range of forces the equation may be considered to hold, it is necessary to find to what extent the field becomes disturbed by the charge $(n_1 - n_2)$ per cubic centimetre of the gas. For this purpose it will be sufficient to consider that the ions move with equal velocities, so that the force will have the same value X_1 at each electrode and the value X_2 at the centre of the field. If u, the mean velocity of the positive and negative ions, be substituted for u_1 and u_2 in the expression for the current i, equation (3) becomes

$$u\,(n_1 - n_2) = i\, \sqrt{p^2 - 1}\, \tan \left\{ \frac{2\, q}{i p^2} \left(x - \frac{a}{2} \right) \sqrt{p^2 - 1} \right\},$$

where

$$p^2 = \frac{4\, q k_1 k_2 X^2}{\alpha i^2}.$$

Hence

$$X_1 - X_2 = -4\, \pi \int_0^{\frac{a}{2}} (n_1 - n_2)\, dx$$

$$= -\frac{4\, \pi i^2 p^2}{2\, q u} \log \cos \left(\frac{a q}{i} \frac{\sqrt{p^2 - 1}}{p^2} \right),$$

or

$$\frac{X_1 - X_2}{X} = -\frac{8\, \pi k_1 k_2}{\alpha k} \log \cos \left(\frac{a q}{i} \frac{\sqrt{p^2 - 1}}{p^2} \right), \qquad (5)$$

where k is the mean velocity of an ion under unit electric force, and X the mean force between the electrodes.

71. Currents in air at atmospheric pressure. As an example of the application of equations (4) and (5) the conductivity produced in air at atmospheric pressure may be considered. In this case the mean velocity of the ions may be taken as 450 centimetres per second under a force of one electrostatic unit and $\alpha = 3400$.

Equation (5) therefore becomes

$$\frac{X_1 - X_2}{X} = -\ 3{\cdot}3 \log \cos \phi,$$

where $$\phi = \frac{aq}{i} \frac{\sqrt{p^2 - 1}}{p^2}. \qquad (6)$$

Since the investigation only applies to cases where $(X_1 - X_2)/X$ is small, the angle ϕ must be small, and the above equation becomes

$$\frac{X_1 - X_2}{X} = 3{\cdot}3\, \frac{\phi^2}{2}. \qquad (5a)$$

Equation (4) becomes $p^2 - 1 = \cot^2 \phi = \phi^{-2} - \frac{2}{3}$ when ϕ is small, and (6) reduces to

$$i = \frac{aq}{1 + \frac{2}{3}\phi^2}, \qquad (6\,a)$$

which shows that the current must be nearly saturated, aq being the maximum value of i.

The value of the mean force X in terms of the current is obtained by substituting for p and q their values in terms of ϕ in the equation $\frac{4\,qk_1k_2X^2}{\alpha i^2} = p^2$, which becomes $\frac{X^2}{ai} = \frac{\alpha}{4\,k_1k_2} \frac{1}{\phi^2}$. When the values of k_1, k_2 and α for air at atmospheric pressure are substituted, the equation reduces to

$$X = \frac{\sqrt{ai}}{15{\cdot}5} \frac{1}{\phi} \qquad (7\,a)$$

The equations (5 a), (6 a), and (7 a) give the corresponding values of $\frac{X}{\sqrt{ai}}$, $\frac{i}{aq}$, and $\frac{X_1 - X_2}{X}$ in terms of the parameter ϕ.

Thus when

$$\phi = {\cdot}1; \quad i = \frac{aq}{1{\cdot}006}; \quad \frac{X_1 - X_2}{X} = {\cdot}016; \quad X = \frac{\sqrt{ai}}{1{\cdot}55},$$

$$\phi = {\cdot}141; \quad i = \frac{aq}{1{\cdot}012}; \quad \frac{X_1 - X_2}{X} = {\cdot}033; \quad X = \frac{\sqrt{ai}}{2{\cdot}17},$$

$$\phi = {\cdot}3; \quad i = \frac{aq}{1{\cdot}06}; \quad \frac{X_1 - X_2}{X} = {\cdot}148; \quad X = \frac{\sqrt{ai}}{4{\cdot}65}.$$

The figures in the first row show that when the current is ·6 per cent. less than the maximum there is a variation of 1·6 per cent. in the electric force, and the potential difference between the plates required to produce a current that is ·6 per cent. less than the maximum current is $\frac{a\sqrt{ai}}{1{\cdot}55}$ or $\frac{a^2\sqrt{q}}{1{\cdot}55}$ since aq is approximately equal to i. For example, if $a = 1$ centimetre and $q = 10^{-3}$ in electrostatic units, the potential required to give a current $\cdot 994 \times 10^{-3}$ is 6·1 volts approximately, since 300 volts are equal to one electrostatic unit of potential.

In order to see to what extent the method is accurate when applied to currents several per cent. below the saturation current, the figures in the third row may be considered. In that case the force at the electrodes is greater by 14 per cent. than the force at a point midway between the plates. The rate of recombination is greatest in the latter region because the product $n_1 n_2$ is greatest there. The mean velocity of the ions in the centre of the field where recombination is taking place is therefore about 7 per cent. less than in the undisturbed field, and the numbers of positive and negative ions present are also greater than the computed numbers by the same percentage, owing to the reduction in the velocity of the ions. Consequently the amount of recombination is under-estimated to the extent of about 14 per cent., so that the value of i should be $\frac{aq}{1{\cdot}07}$ instead of $\frac{aq}{1{\cdot}06}$. Hence the true value of the current is about 1 per cent. less than the estimated value.

72. Effect of reducing the pressure. The effect of reducing the pressure may be found by substituting for the numerical constants in the above equations their values for low pressures. The velocities k_1 and k_2 vary inversely as the pressure, and according to the most reliable determinations of the variation of the rate of recombination with pressure, which have been made by Langevin, the quantity

α varies approximately in direct proportion to the pressure. If the pressure be reduced to $1/n^{\text{th}}$ of an atmosphere, the equations for determining the variable quantities become

$$X = \frac{\sqrt{ai}}{15{\cdot}5}\,\frac{1}{n^{\frac{3}{2}}\phi}, \quad \frac{i}{aq} = \frac{1}{1+\frac{2}{3}\phi^2}, \quad \frac{X_1 - X_2}{X} = \frac{3{\cdot}3\,n^2\phi^2}{2}.$$

For a pressure one-hundredth of an atmosphere, when the current and electric force are the same as at one atmosphere, these equations give on substituting for ϕ the value 10^{-4} instead of $\cdot 1$,

$$X = \frac{\sqrt{ai}}{1{\cdot}55}, \quad \frac{X_1 - X_2}{X} = {\cdot}00016, \quad \frac{i}{aq} = \frac{1}{1+6\times 10^{-9}}.$$

Hence the polarization is 1/100th of its original value, the effect of recombination becomes negligible, and the current almost attains the saturation value. In practice, however, the current would not be completely saturated, for the effect of diffusion increases as the pressure is reduced in the same proportion as the velocities of the ions, and the difficulty of obtaining saturation at low pressures is generally due to the loss of ions by diffusion.

73. Effect of recombination under various conditions. The above determination of the force necessary to produce a certain degree of saturation may be used to explain some of the variations of the current-electric-force curves that have been observed experimentally for different densities of ionization and for different gases. Thus the force necessary to obtain saturation is much greater for large densities of ionization than for small. Also saturation is much more easily obtained with the gases of low atomic weight, the velocities of the ions being greater in these cases.

Since it would be impossible to avoid all recombination and arrive at the maximum current, it is necessary in a theoretical discussion to suppose that saturation is reached when the current is less than the maximum current by a small fraction specified by the quantity ϕ, where $i = aq/(1+\frac{2}{3}\phi^2)$. In order that recombination may be the predominating effect that causes the currents to fall below the maximum when the force is reduced, it will be necessary to suppose that the currents and the pressures are not too low.

It will be seen from the equation $X = \sqrt{\frac{\alpha}{4k_1k_2}}\frac{\sqrt{ai}}{\phi}$ that the force required to produce a given degree of saturation, corresponding to a given value of ϕ, is proportional to the square root of the current or the intensity of ionization, since $i = aq$ approximately. The current-electric-force curve may be indicated by the curve A, figure 13, for

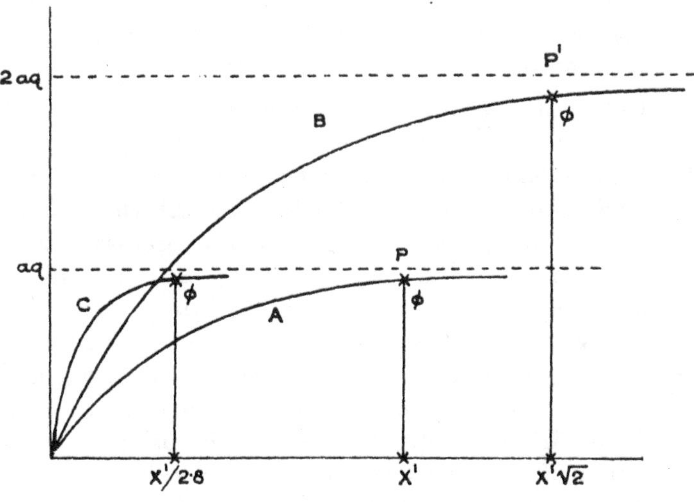

Figure 13.

a given intensity of ionization between plates at a fixed distance apart. The force corresponding to the point P at which $i = aq/(1+\frac{2}{3}\phi^2)$ is X'. When the intensity of ionization becomes $2q$ the corresponding curve may be represented by B, and at the point P', for which the force is $X'\sqrt{2}$, the given degree of saturation ϕ will be obtained.

The curve corresponding to a pressure of half an atmosphere will be given by the curve C, when the intensity of ionization is q, as at atmospheric pressure. In this case the force corresponding to the specified degree of saturation ϕ would be $X'/2{\cdot}8$. It is to be noticed that with this force the polarization would be four times as great as that at atmospheric pressure when the force is X', since

$$\frac{X_1 - X_2}{X} = \frac{8\pi k}{\alpha} \cdot \frac{\phi^2}{2},$$

and k/α is inversely proportional to the square of the pressure.

The forces necessary to obtain saturation in different gases may easily be found by substituting for k_1 and k_2 the velocities of the ions under unit electric force, the value of α being practically the same for the different gases.

74. Effect of diffusion. When the effect of recombination is negligible, the variations in the currents between the electrodes depend on the rate of diffusion and the velocity under the electric force. It is usual to take as the starting-point for these investigations the general equations established by Maxwell for the interdiffusion of gases, but in order to get an insight into the processes taking place it is necessary to begin further back and to consider the actual motions of the ions, each ion moving along free paths between collisions with molecules with a velocity which is large compared with the velocity acquired under the electric force. It is only in such cases that the theory of diffusion in its simplest form can be used to explain the motion of an ion, for when the force becomes large the interval between collisions is affected appreciably by the velocity due to the force. When an ion is immersed in a gas the velocity of agitation which it acquires owing to the collisions with molecules varies about a certain mean value, which is such that the kinetic energy of translation of the ion is equal to the mean kinetic energy of a molecule of the gas. The motion of an ion is thus quite independent of the presence of other ions, provided that the total number per c.c. is small and the charge on the ions produces no appreciable electric force.

When an ion moves in an electric field it acquires a mean velocity along the direction of the force but does not travel exactly along the line of force, and for some parts of its path it may travel in a direction opposite to the direction of the force. If a number of ions are generated at a point P* on a line of force between two conductors A and B they diffuse outwards from the point and a spherical distribution is formed, the centre of which moves along the line of force. If the ions are generated continuously at P they move in a stream which gradually opens out owing to the motion of agitation, the central line of the stream coinciding with the line of force.

When there is no force acting and the point P is near the conductor A, most of the ions reach the conductor A and are discharged by it, but if a force is applied so as to repel the ions from A, then the chance of an ion reaching the conductor A is diminished and a current in the direction of the force is established between A and B. The current increases with the force and approaches a maximum value which would be attained if all the ions were collected on B. Theoretically some ions will always reach A, but the number is so small when the force is very high that in practice a saturation current is generally obtained with moderate forces which depend on the distance between the plates. It is obvious that for small distances between the plates, or when the ions are generated near the surface of one of the plates, the force required to produce saturation must be greater than in other cases.

75. Application of the kinetic theory of gases to the motion of ions. The fundamental principles on which the investigation of the motion of the ions is based are the same as those which are used in the kinetic theory to explain the properties of gases. It is only when an electric force is applied that it is necessary to take into consideration the charge on an ion, and the kinetic theory of gases may be applied to the diffusion of ions as if they were uncharged particles.

The particles on this theory acquire a motion of agitation which is determined by the mass of the particle and the

* See figure 15.

temperature of the gas in which they are immersed. Let m be the mass of an ion, V the mean velocity of agitation, n the number of ions per cubic centimetre of the gas.

The kinetic energy of translation $\frac{1}{2}mnV^2$ is the same as that of an equal number of molecules of the gas, so that the quantity $\frac{1}{3}mnV^2$, which is the partial pressure of the ions, is the same as the pressure that would be exerted on a boundary by any gas when the number of molecules per cubic centimetre is equal to n.

If N be the number of molecules per unit volume of a gas at 760 mm. pressure and temperature 15° C., $\frac{1}{3}mNV^2$ expressed in dynes per square centimetre is approximately 10^6; and when the density mN is known the velocity of agitation V^2 may be estimated. The determinations of the rate of diffusion have shown that the mass associated with an ion is about 30 times the mass of a molecule of oxygen, so that the velocity of agitation of an ion is about one-fifth of that of a molecule of oxygen. The latter is about 480 metres per second at 15° C., and the value of V is therefore about 9×10^3 centimetres per second. The velocity of an ion in air at atmospheric pressure under an electric force of one volt per centimetre is about 1·5 centimetres per second, so that the velocity due to an electric force is small compared with the velocity of agitation even when the force is as great as 100 volts per centimetre. The velocity of agitation is independent of the pressure, and the velocity under an electric force is inversely proportional to the pressure, and therefore the ordinary theory of diffusion becomes inaccurate when the ratio of the force to the pressure exceeds a certain limit. In what follows it will be supposed that the forces are such that they do not produce velocities of as high an order as V.

76. Langevin's formulae for the coefficient of diffusion and velocity under an electric force. The determination of the rate of diffusion of ions, and the velocity in the direction of an electric force, in terms of the mean velocity of agitation, the masses, and the linear dimensions of the ions and molecules is very complicated in the general case when the variations in the velocities of agitation are taken into consideration. A complete investigation of the coefficient of interdiffusion K of two gases has been made by Langevin,* on the hypothesis that the collisions between the molecules are of the same type as those between smooth elastic spheres. The value of K which was obtained is

$$\dagger K = \frac{3}{16\,\sigma^2 N} \cdot \sqrt{\frac{m+m'}{\pi h m m'}},$$

m and m' being the masses of the molecules of the two gases, σ the sum of their radii, N the total number of molecules per cubic centimetre, and $2h = N/\Pi$, where Π is the pressure of a gas containing N molecules per cubic centimetre.

In the same notation it may be seen that if e is the charge on an ion the velocity U in the direction of an electric force X is given by the formula

$$U = \frac{3\,e X}{8\,\sigma^2 N} \cdot \sqrt{\frac{h(m+m')}{\pi m m'}}.$$

Since the ions are few in number compared with the molecules of the gas, the quantity N may be taken as the number of molecules per cubic centimetre, so that the rate of diffusion and velocity under the electric force are inversely proportional to the pressure. The quantity h is inversely proportional to the temperature and the formulae show that when N is constant the rate of diffusion increases and the velocity under the electric force diminishes as the temperature is increased.

Several simple methods ‡ of investigating the motion of

* P. Langevin, Ann. de Chim. et de Phys. (8) **5**, p. 245, 1905.

† This formula differs by the numerical factor 3/4 from the formula for the coefficient of interdiffusion of two gases, given by Maxwell (Collected Papers, vol. ii, p. 345).

‡ The following formula, involving m' the mass of a molecule of the gas, has been given by Lenard (Ann. der Physik, [4] **3**, p. 312, 1900) for the velocity of an ion due to an electric force:

$$U = \frac{eXl}{V}\left[\frac{1}{m'} + \frac{1}{2m}\right],$$

l being the mean free path, m the mass, and V the velocity of agitation of the ion.

the ions have been suggested, which give approximate values of the quantities U and K. The expressions, which are usually given in terms of the mean free path of an ion between collisions with molecules, may be obtained by assuming that the velocity of agitation of the ions is constant. According to Maxwell's law of the distribution of velocity, the velocities of a particle along various free paths do not frequently differ by a large factor from the mean velocity as given by the equation $\frac{1}{3}mnV^2 = p$, p being the pressure corresponding to the number of molecules n in unit volume. The quantity V^2, which is the mean of the squares of the velocities of agitation, is somewhat greater than the square of the mean velocity, the latter being about $\frac{12}{13}V$. For simplicity the velocities may all be taken as being equal to V.

77. Lengths of the free paths: mean free path. The particles move along free paths of different lengths between the collisions, and the mean length of a large number of consecutive free paths is called the mean free path. The lengths of the various paths differ considerably, and it is desirable to distinguish between the mean of the squares of the free paths and the square of the mean free path. This may be done by finding the number of paths which exceed a given length x.

Let a number of particles F_0 start normally with the same velocity from the plane $x = 0$, and let F be the number that arrive at a distance x without colliding with molecules of the gas. The number of collisions that occur in the distance dx will be $F\alpha dx$, where α is a constant, since the number that collide with the molecules must be proportional to the distance dx and the number that enter the space between the two planes x and $x + dx$.

Hence $$\frac{dF}{dx} = -\alpha F,$$

and therefore $$F = F_0 e^{-\alpha x}.$$

The number of paths whose lengths lie between x and $x + dx$ is $F_0 e^{-\alpha x}\alpha dx$, and the mean free path l is

$$\frac{1}{F_0}\int_0^\infty F_0 e^{-\alpha x}\alpha x dx = \frac{1}{\alpha}\cdot$$

Hence the number of paths that exceed the distance x is

$$F_0 e^{-x/l}.$$

The mean of the squares of the paths is

$$\frac{1}{F_0}\int_0^\infty F_0 e^{-x/l}\frac{x^2}{l}dx = 2l^2,$$

or twice the square of the mean free path.

78. Approximate formulae for the velocity under an electric force. There are two cases of particular interest in connection with the motion of ions which may be investigated by the simpler methods. In some experiments in gases at low pressures the electrons move freely and the mass of a negative ion is very small compared with the mass of a molecule, whereas at ordinary pressures the mass associated with an ion is usually large compared with the mass of a molecule of the gas. In the first case, when the electron collides with a molecule it may be deflected in any direction, and all directions of motion may be considered equally probable. In the second case, the direction of motion of the large ion after a collision is inclined at a small angle to the direction of motion before collision and the ion continues to move with a velocity having a large component in the original direction of motion.

Considering the case in which the mass m of the ion is small and all directions of motion after a collision are equally probable, then the ion does not retain any of the momentum Xet acquired while moving along a free path l in the time $t = l/V$, but on an average the ions may be considered to travel the distance $\frac{1}{2}\frac{Xe}{m}\cdot t^2$ in the direction of the electric force in the time t.

The sum of the distances $s_1 + s_2 + \dots s_n$, traversed in the direction of the electric force while the ion traverses the paths $l_1, l_2, \dots l_n$, with the constant velocity V is

$$\sum_{r=1}^{r=n} s_r = \tfrac{1}{2} \cdot \frac{Xe}{m} \cdot \frac{l_1^2 + l_2^2 + \dots l_n^2}{V^2}$$

$$= \tfrac{1}{2} \cdot \frac{Xe}{m} \cdot \frac{2\,nl^2}{V^2},$$

where l is the mean free path.

Dividing this distance by the time nl/V, the velocity U in the direction of the electric force is obtained,

$$U = \frac{Xe}{m} \cdot \frac{l}{V},$$

$$= \frac{Xe}{m} \cdot T,$$

where T is the mean interval between collisions.*

This approximate formula gives too high a value of the velocity U. When V is not constant, but represents the mean velocity of agitation, the accurate value of U may be deduced from the expression given in section 76, as in this case the mean free path of the ion is $1/\pi\sigma^2 N$. If m', the mass of a molecule, be large compared with m, the mass of the electron, the ratio $\frac{m+m'}{mm'}$ becomes $\frac{1}{m}$ and the value of U reduces to

$$U = \frac{Xe}{m} \cdot \frac{1}{\pi\sigma^2 N V} \times \cdot 815$$

$$= \frac{Xe}{m} \cdot \frac{l}{V} \times \cdot 815.$$

79. Definition of the coefficient of diffusion K. It is usual to express the coefficient of diffusion K in terms of the flow produced when there is a variation of the number of particles n in unit volume of a medium. The number of particles passing out through a geometrical boundary is proportional at each point to the number of particles per cubic centimetre at an adjacent point inside the boundary. The number entering is proportional to the number per cubic centimetre outside, so that if there is a variation in the density the numbers crossing the boundary

* See Langevin, Ann. de Chim. et de Phys. **28**, p. 336, 1903.

in the two directions will be different. The difference, referred to unit area of the plane perpendicular to the axis of x, may be expressed as $K\frac{dn}{dx}$ per second, the constant K being the coefficient of diffusion. This quantity may also be expressed by the product nu, where u is a velocity, defined by the equation

$$nu = -K\frac{dn}{dx}.$$

The velocity u is therefore the velocity that must be attributed to all the particles, otherwise supposed to be at rest, in order that the number that cross the area should be equal to the difference between the numbers that actually cross in opposite directions.

Similarly there are two other equations for the velocities v and w corresponding to the axes of y and z, namely

$$nv = -K\frac{dn}{dy},$$

and

$$nw = -K\frac{dn}{dz}.$$

80. Rate of change of the mean square of the distances of particles from any point. The coefficient of diffusion K may be found in terms of the rate of change of the mean square of the distances of the particles from any point. For simplicity* let the distribution be symmetrical about the centre O, so that the number of particles n per cubic centimetre is a function of the distance r. The number of particles between two spheres of radii r and $r+dr$ is $4\pi r^2 n dr$ and the mean square of the distance of all the particles from the centre is

$$\frac{\int_0^{r'} 4\pi n r^4 dr}{\int_0^{r'} 4\pi n r^2 dr} = R^2.$$

* The investigation applies also to an unsymmetrical distribution, for as far as the motion along the radius from O is concerned, it is immaterial whether the particles between two concentric spheres of radii r and $r+dr$ are distributed irregularly or uniformly.

The radius r' is taken so large that the integration extends beyond the space in which the particles are included. Hence $n = 0$ and $\frac{dn}{dr} = 0$ when $r = r'$. The denominator, which represents the total number of particles under consideration, is constant, so that when differentiated with respect to the time, the above equation gives

$$\int_0^{r'} \frac{dn}{dt} r^4 dr = \frac{d(R^2)}{dt} \int_0^{r'} nr^2 dr.$$

The rate at which the number of particles in the space between the two spheres of radii r and $r + dr$ is increasing, namely $4\pi r^2 \frac{dn}{dt} dr$, has to be accounted for by the difference between the rates at which particles are crossing the two spheres. Hence

$$4\pi r^2 \frac{dn}{dt} dr = 4\pi r^2 nu_r - \left\{4\pi r^2 nu_r + \frac{d}{dr}(4\pi r^2 nu_r) dr\right\}$$

or
$$r^2 \frac{dn}{dt} = -\frac{d}{dr}(r^2 nu_r) = \frac{d}{dr}\left(r^2 K \frac{dn}{dr}\right).$$

Hence
$$\int_0^{r'} r^2 \frac{d}{dr}\left(r^2 K \frac{dn}{dr}\right) dr = \frac{d(R^2)}{dt} \int_0^{r'} nr^2 dr.$$

The values of n and $\frac{dn}{dr}$ being zero at the limit $r = r'$, the above equation, on integration by parts, gives

$$2K \int_0^{r'} r^3 \frac{dn}{dr} dr = 6K \int_0^{r'} nr^2 dr = \frac{d(R^2)}{dt} \int_0^{r'} nr^2 dr.$$

Hence the rate of change of the mean square of the distance of any distribution from any point is $6K$.

This result also follows from the expression given by Fourier * for the temperature θ at any point of an infinite solid initially at zero temperature throughout, except at the origin of co-ordinates, where the temperature exceeds that of the surrounding material. After a time t the tem-

* J. B. J. Fourier, Théorie analytique de la chaleur.

perature at a point at the distance r from the origin is proportional to $t^{-\frac{3}{2}} e^{-\frac{r^2}{4Kt}}$. Since the equation for the diffusion of heat from a point $\frac{d\theta}{dt} = \frac{K}{r^2}\frac{d}{dr}\left(r^2 \frac{d\theta}{dr}\right)$ is the same as the equation for the diffusion of particles, it follows that if a number of particles are placed at any point in a gas after a time t the number n per cubic centimetre is proportional to $t^{-\frac{3}{2}} e^{-\frac{r^2}{4Kt}}$. When this value is substituted for n the volume integral $\int_{r=0}^{r=\infty} r^2 n \, . \, r^2 dr$ becomes $6\,Kt \int_{r=0}^{r=\infty} nr^2 dr$, which shows that the mean square of the distance of the particles from the origin is $6\,Kt$. Similar results have been found by Einstein and Smoluchowski * in their investigations of the theory of the Brownian movement.

81. Rate of change of the mean square of the distance in terms of the mean free path. The rate of change of the

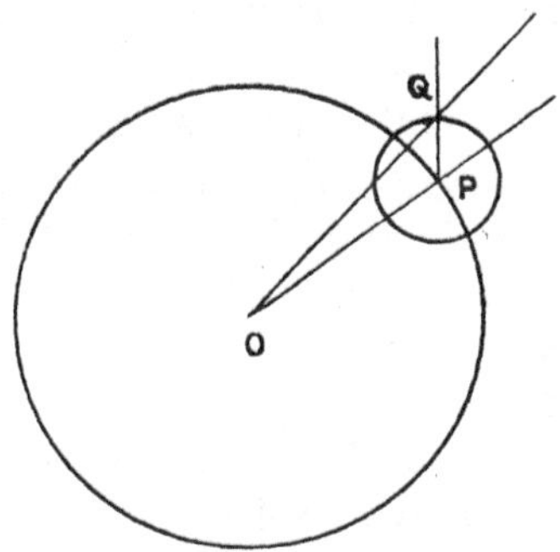

Figure 14.

mean square of the distance of a distribution from any point may easily be found when the motion is of the type considered in section 78, that is, when the particles move with a constant velocity between collisions, and all directions of motion after collision are equally probable.

If Q, figure 14, be any point on the surface of the small

* See sections 175 and 176.

sphere of radius x, the mean square of the distance of the points on the surface of the sphere from O is

$$\frac{1}{4\pi x^2}\int_0^{\pi}(r^2+x^2+2\,rx\cos\theta)\,2\,\pi x^2\sin\theta\,d\theta = r^2+x^2,$$

where r is the distance of P from the point O and θ the angle between OP and PQ. Hence the mean square of the distances of the ions from O is increased by x^2 after traversing the free paths of length x. When several paths are considered, the change in the mean square of the distances will be Σx^2 in the time $\frac{\Sigma x}{V}\cdot$ Hence the rate of change of the mean square of the distances from O is $V\frac{\Sigma x^2}{\Sigma x} = 2\,l V$, which is independent of r, and therefore the same for all ions in the field.

Hence $K = \frac{1}{3}\,l\,V$.

82. Motion of large particles in a gas. In a large number of cases the rate of diffusion of ions in gases is much smaller than the rate of interdiffusion of ordinary gases, so that the mass associated with an ion is large compared with the mass of a molecule. The formulae given in the preceding sections for U and K in terms of the mean free path of an ion do not apply in these cases, since the effect of a collision only causes the ion to deviate from its original direction by a small angle.

Let m and m', V and V' be the masses and velocities of agitation of the ion and a molecule of the gas respectively. If the collisions are such that all subsequent directions of motion of the smaller mass m' are equally probable, the greatest transfer of momentum that occurs at any collision is $2m'V'$, which happens when the direction of motion of the molecule is reversed. Since $mV^2 = m'V'^2$, the ratio of the momentum of the ion to that of the molecule is $\sqrt{m/m'}$; hence the transfer of momentum required to reverse the motion of the ion is greater than that which could be imparted to it by a single collision with a molecule. The collisions, therefore, which may have a large effect in altering the direction of motion of the smaller mass will have a comparatively small effect on the larger mass, and it is only after several collisions with molecules that the direction of motion of the ion becomes independent of its original motion.

83. Coefficient of diffusion when the particles are large compared with the molecules of the gas. Let the ion of large mass m travel along straight paths of lengths $l_1, l_2, \dots l_n$ between collisions, in the intervals of time $t_1, t_2, \dots t_n$ with the constant velocity V. Let the components of the velocity of the ion along the axes of x, y, and z be u_1, v_1, w_1 during the first interval, u_2, v_2, w_2 during the second interval, etc. After a time $t = t_1 + t_2 + \dots t_n$ the ion will have travelled the distances x, y, z along the axes given by the formulae

$$x = u_1 t_1 + u_2 t_2 + \dots u_n t_n,$$
$$y = v_1 t_1 + v_2 t_2 + \dots v_n t_n,$$
$$z = w_1 t_1 + w_2 t_2 + \dots w_n t_n,$$

and the average rate of change $\frac{R^2 - R_0^2}{t}$ in the square of the distance of the ion from any point is $\frac{x^2 + y^2 + z^2}{t}$, which may be written in the form

$$\frac{(u_1^2 + v_1^2 + w_1^2)\, t_1^2 + \dots + 2\,(u_1 u_2 + v_1 v_2 + w_1 w_2)\, t_1 t_2 + \dots}{t_1 + t_2 + \dots t_n}$$
$$= \frac{l_1^2 + l_2^2 + \dots + 2\, l_1 l_2 \cos\phi_{12} + \dots}{t_1 + t_2 + \dots t_n},$$

where $\phi_{rr'}$ is the angle between the paths l_r and $l_{r'}$.

Also, since

$$l_1^2 + l_2^2 + \dots l_n^2 = 2\,nl^2 = 2\,l\,(l_1 + l_2 + \dots l_n),$$

where l is the mean free path, the numerator of the preceding fraction may be written as the sum of n terms, the first two being

$$2\,l_1\,(l + l_2 \cos\phi_{12} + l_3 \cos\phi_{13} + \dots),$$

and

$$2\,l_2\,(l + l_3 \cos\phi_{23} + l_4 \cos\phi_{24} + \dots),$$

the others being similar expressions.

The quantity $l_2 \cos\phi_{12} + l_3 \cos\phi_{13} + \dots$ is the sum of the projections of the second, third, and subsequent paths on the direction of the first path. The sum of the projections has a mean value λ greater than zero, since the second path and those that immediately follow it are inclined at small angles to the direction of the first path. After a certain average number of collisions s, the angle ϕ_{1s} becomes of the order $\frac{\pi}{2}$, and then the sum of the subsequent terms $l_s \cos \phi_{1s} + \text{etc.}$ becomes on the average equal to zero. The quantity $(l + l_2 \cos \phi_{12} + \dots l_n \cos \phi_{1n}) = l + \lambda$ represents the distance that an ion travels in the direction of its initial motion before all directions of motion become equally probable, or in other words, if a large number of ions start in a direction normal to a plane their mean distance from the plane attains a definite value $l + \lambda$ in a short time, which includes several intervals between collisions. When the total number of collisions n is large compared with s, the quantity $l_1^2 + l_2^2 \dots l_n^2 + 2 l_1 l_2 \cos \phi_{12} + \dots$ becomes equal to $2(l + \lambda)(l_1 + l_2 + \dots l_n)$ or $2nl(l + \lambda)$.

Hence $\frac{d(R^2)}{dt} = \frac{2n(l + \lambda) l V}{l_1 + l_2 + \dots l_n} = 2(l + \lambda) V.$

Since $\frac{d(R^2)}{dt}$ is equal to $6K$,

$$K = \tfrac{1}{3}(l + \lambda) V.$$

84. Velocity of large particles under an electric force. The velocity of the ions under an electric force may also be found under the above conditions.* It is necessary in this case to take into consideration the drift in the direction of the force after a collision that arises from the velocity of the ion before the collision. It will be seen, from the investigation in the preceding section, that if an ion starts along any direction with a velocity V it reaches, on an average, the distance $l + \lambda$ in that direction before the effect of the momentum disappears, l being the average distance traversed before the first collision occurs. If u, v, w are the components of V before collision, then the average distances

* Proc. Roy. Soc. A, **86**, p. 197, 1912.

travelled along the axes of co-ordinates after the collision will be $\frac{u\lambda}{V}$, $\frac{v\lambda}{V}$, and $\frac{w\lambda}{V}$.

When an ion moves under the electric force X, the acceleration is Xe/m and the ion drifts a distance $\frac{1}{2}ft_1^2$ while moving along the free path l_1 with the velocity V, and immediately before the collision the velocity of the ion in the direction of the force is ft_1. The effect of this velocity is not lost by the collision, but the ion, after the collision, drifts on for a distance $\frac{ft_1\lambda}{V}$ in a time which is short compared with the sum of the intervals between all the collisions under consideration. Hence the effect of the action of the force X for the time t_1 is to make the ion travel a distance

$$\frac{1}{2}ft_1^2 + ft_1 \cdot \frac{\lambda}{V} = \frac{\frac{1}{2}fl_1^2 + fl_1\lambda}{V^2}.$$

The total distance S traversed in the time $t = \frac{nl}{V}$ in which n collisions occur is

$$\frac{f}{V^2}\left[\frac{1}{2}(l_1^2 + l_2^2 + \ldots l_n^2) + \lambda(l_1 + l_2 + \ldots l_n)\right],$$

or

$$S = \frac{nfl}{V^2}(l + \lambda),$$

so that the velocity under the electric force U is

$$\frac{SV}{nl} = \frac{f(l+\lambda)}{V}.$$

Hence

$$U = \frac{Xe}{m} \cdot \frac{l+\lambda}{V}.$$

Both the quantities l and λ are inversely proportional to the pressure of the gas in which the ions move, so that U and K are both inversely proportional to the pressure, and their ratio is

$$\frac{U}{K} = \frac{3Xe}{mV^2} = \frac{NeX}{\Pi}.$$

Here $\Pi = \frac{1}{3}mNV^2$ is the pressure of a gas in which there are N molecules per cubic centimetre.

85. Approximate formulae for U and K for large particles. The quantity λ will obviously be large compared with the mean free path l if the mass of the ion is large compared with that of a molecule, and in order to obtain definite information from the above formulae it is necessary to find how these two lengths are related to each other. A simple method of procedure is to consider the interdiffusion of two gases in two special cases. From elementary considerations it may be shown that the rate of diffusion K of a gas A into a gas B or of B into A depends on the total pressure of the two gases, and is independent of the relative numbers of molecules of each that are present. When there are a small number ν of molecules of A per cubic centimetre and a large number N of molecules of B, the rate of diffusion of A into B will be the same as when the number of molecules per cubic centimetre of B is ν and the number of molecules of A is N.

In the first case, if the mass m of the molecule of A is large compared with m', the mass of a molecule of B, the value of K is $\frac{1}{3}(l+\lambda)V$. Similarly in the second case the rate of diffusion of B into A is $\frac{1}{3}l'V'$, it being supposed that when the smaller mass m' collides with the mass m all directions of motion of the former become equally probable.

Hence $$(l+\lambda)V = l'V'.$$

The total number of collisions C that occur in unit volume per second is proportional to the product $N\nu$, so that the total number of free paths of the molecules of A in the first case is equal to the total number of free paths of B in the second case.

Hence $$Cl = V\nu \text{ and } Cl' = V'\nu,$$

so that $$l' = \frac{lV'}{V}.$$

Hence $$\frac{l+\lambda}{l} = \frac{V'^2}{V^2} = \frac{m}{m'}.$$

Thus when m, the mass of an ion, is large compared with m', the mass of a molecule of the gas, U and K are given by the formulae

$$U = \frac{Xe}{m} \cdot \frac{L}{V}, \text{ and } K = \frac{LV}{3},$$

or
$$U = \frac{Xe}{m'} \cdot \frac{l}{V}, \text{ and } K = \frac{mlV}{3m'},$$

l being the mean free path of an ion and $L = l + \lambda = ml/m'$.

86. Maxwell's equations of motion. The velocities u, v, w of the ions along the directions of the axes, due to the combined effect of diffusion and electric force, may now be written in the form

$$u = -\frac{K}{n}\frac{dn}{dx} + \frac{Xe}{m}\frac{L}{V},$$

or
$$\frac{1}{K}(nu) = -\frac{dn}{dx} + \frac{3n\,Xe}{mV^2}.$$

If p be the partial pressure due to a number of particles n per cubic centimetre, $p = \frac{1}{3}mnV^2$, and the above equation reduces to

$$\frac{1}{K}(pu) = -\frac{dp}{dx} + nXe.$$

Similarly
$$\frac{1}{K}(pv) = -\frac{dp}{dy} + nYe,$$

and
$$\frac{1}{K}(pw) = -\frac{dp}{dz} + nZe.$$

These three equations are the same as the equations which may be obtained from the equations for the interdiffusion of gases given by Maxwell,* when the molecules of one of the gases are treated as charged particles, and are few in number compared with the number of molecules of the uncharged gas.

The application of these equations to special cases, arranged for the measurement of K, and the ratio of U/K from which the value of Ne may be deduced, will be explained in another chapter, but the general effect of diffusion on the currents between electrodes may be examined without solving the equations for any particular case.

* J. C. Maxwell, Phil. Trans., vol. **157**, 1866.

87. Rate of diffusion independent of the concentration. When ions are generated at a point P, figure 15, on a line of force between two electrodes, they diffuse outwards and form a spherical distribution round a point which moves with a velocity $U = \frac{Xe}{m}\frac{L}{V}$ along the line of force. The number that arrive at any conductor in the field will depend on the ratio of the rate of diffusion to the velocity due to the electric force. When the force is very large, the diffusion effect is small and the ions generated at P are practically all collected on the electrode at the point P' at which the line of force through P terminates. When the electric force is small and ions are produced in the space between the two electrodes A and B, A being charged positively and B provided with a guard ring C, some of the positive ions generated at P will arrive on the electrode A and on the guard ring instead of on B.

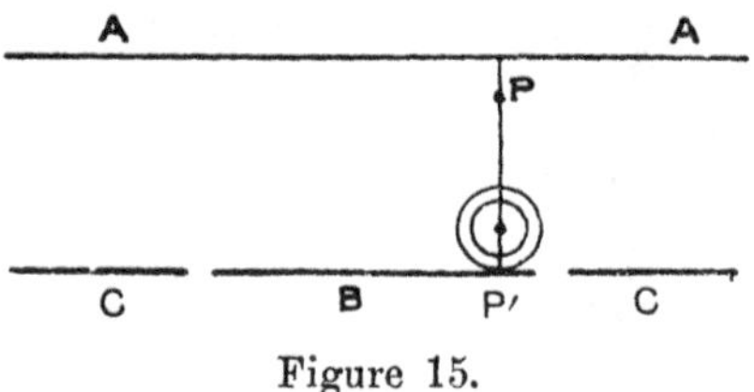

Figure 15.

It is to be noticed that in all cases in which a certain quantity of ions are generated by rays in a limited space between two electrodes, and the currents are measured by the charge acquired by an electrode, fluctuations in the rate at which the ions are generated by the rays do not affect the result, provided that the force is not altered during the time the rays are acting, and is kept on for a sufficient time to remove all the ions from the gas. In such cases the ions may be supposed to be generated at a uniform rate, and the currents may be investigated as if a constant stream were flowing through the space between the plates, and the number of ions per cubic centimetre at any point did not vary with the time. Thus the current-electric-force curves do not vary with the intensity of the current and are the same whether the ions are generated continuously or at irregular intervals. In all cases the proportion of the whole which reach an electrode depends only on the force.

In this respect these current-electric-force curves differ from those obtained when the effect of recombination predominates, for then the loss of ions is much greater when they are produced in a short interval of time than when the same number are produced continuously in a longer time. When the currents are nearly saturated, the force X required to produce a given degree of saturation is proportional to the square root of the current.

88. Currents independent of the pressure. When the diffusion and velocity under an electric force are the principal processes that determine the proportion of the ions that reach an electrode, the current to the electrode is not affected by altering the pressure of the gas. If the pressure be lowered the mean free path of the ions is increased, and the velocity under the electric force $\frac{Xe}{m} \cdot \frac{L}{V}$ and the rate of diffusion $\frac{LV}{3}$ are both increased in the same proportion. Thus in the case in which a number of ions are generated at P, when the mean square of their distance from the centre of the distribution has attained a value R^2, the centre of the distribution has travelled a distance S given by the equation

$$\frac{R^2}{S} = 2LV \cdot \frac{mV}{XeL} = \frac{2mV^2}{Xe}.$$

The ratio R^2/S is thus independent of the pressure, and of the nature of the gas in which the ions are moving, for although the mass of the ion m varies in different gases, the quantity mV^2 depends only on the temperature. It is obviously the ratio of R^2 to S that determines the number that arrive on any electrode, hence the current-electric-force curve depends only on the temperature of the gas, and for a given initial distribution of ions is independent of the pressure and nature of the gas.

89. Motion in a magnetic field. When the motion of ions in a gas takes place in a magnetic field the rates of diffusion and the velocities due to an electric force may be determined by the methods similar to those given in the preceding sections.* The effect of the magnetic field may be determined by considering the motion of each ion between collisions with molecules. The magnetic force causes the ions to be deflected in their free paths, and when no electric forces are acting the paths are spirals, the axes being along the direction of the magnetic force. If H be the intensity of the magnetic field, e the charge, and m the mass of the ion, then the radius r of the spiral is mv/He, v being the velocity in the direction perpendicular to H. The distance that the ion travels in the interval between two collisions in a direction normal to the magnetic force is a chord of the circle of radius r. The distances the ion travels in those directions are thus reduced, and the rate of diffusion of the ions in directions perpendicular to the magnetic force is less than the rate of diffusion in the direction of the force.

It has already been shown that the coefficient of diffusion K is a measure of the rate of increase of the mean square R^2 of the distance of any distribution from any point, $\frac{dR^2}{dt}$ being equal to $6K$. Since $R^2 = X^2 + Y^2 + Z^2$, it follows that $2K = \frac{dX^2}{dt}$ and $4K = \frac{d\rho^2}{dt}$, where ρ^2 is the mean of the squares of the distances of the ions from the axis of Z.

When a magnetic force H is applied in the direction of the axis of Z, the rate of diffusion K_h in the directions perpendicular to Z is less than the rate of diffusion K along Z which is unaltered by the magnetic field. As the phenomena are of special interest in connection with negative ions in the electronic state when magnetic fields pro-

* Proc. Roy. Soc. A, **86**, p. 571, 1912.

duce large effects, the collisions with molecules may be supposed to affect the motion of the ions in the same way as the motion of small bodies is affected by colliding with comparatively large particles. All directions of motion of an ion after collision with molecules may therefore be supposed to be equally probable, so that the mean values of the velocities of the ions after colliding will be zero. Since the magnetic field only alters the direction of motion, the number of collisions that an ion makes per second is not affected, and the intervals between the collisions will be distributed as in the ordinary case and will have the same mean value T. The number of intervals that exceed the time t, out of a total number N, will be $N\epsilon^{-\frac{t}{T}}$.

In the following investigations two expressions occur which may be found in terms of the mean time between collisions. These are the series of cosines, $\cos \omega t_1 + \cos \omega t_2 + \dots$, and the series of sines $\sin \omega t_1 + \sin \omega t_2 + \dots$, $t_1, t_2 \dots$ being intervals between a large number N of consecutive collisions that an ion makes with molecules. The number of intervals that lie between the values t and $t + dt$ is $\frac{N}{T}\epsilon^{-\frac{t}{T}}dt$, so that the series of cosines may be expressed as the integral
$$\frac{N}{T}\int_0^\infty \cos \omega t\, \epsilon^{-\frac{t}{T}}\,dt.$$

This expression may be integrated by parts and its value is $\frac{N}{1+\omega^2T^2}$. Similarly the series of sines will be found to be equal to $\frac{N\omega T}{1+\omega^2 T^2}$.

90. Rate of change of the mean distance from an axis. The rate of increase of the mean square of the distance of any distribution from the axis of Z, $d\rho^2/dt$, may be determined by considering the motion of an ion parallel to the plane of xy.

Let x and y be the ordinates of an ion at any time t after a collision. The equations of motion are

$$m\frac{d^2x}{dt^2} = -He\frac{dy}{dt},$$

$$m\frac{d^2y}{dt^2} = He\frac{dx}{dt},$$

and the velocities $\dot{x}$ and $\dot{y}$ are given by equations of the form

$$m^2\frac{d^2\dot{x}}{dt^2} + H^2e^2\dot{x} = 0,$$

so that $\dot{x} = A\sin(\omega t+\alpha)$ and $\dot{y} = -A\cos(\omega t+\alpha)$, where $\omega = He/m$.

Let the ion travel the distances δx_1 and δy_1 in the first interval t_1 between collisions; δx_2 and δy_2 in the second interval t_2, etc.

In the first interval the distances δx_1 and δy_1 traversed by the ions are

$$\delta x_1 = \frac{A}{\omega}[\cos\alpha_1 - \cos(\omega t_1+\alpha_1)]$$

and

$$\delta y_1 = \frac{A}{\omega}[\sin\alpha_1 - \sin(\omega t_1+\alpha_1)].$$

If $x_0 y_0$ be the initial position of the ion, then after the time $NT = t_1+t_2+\ldots$, the position will be

$$x_0+\Sigma\delta x_s, \text{ and } y_0+\Sigma\delta y_s,$$

and the square of the distance from the axis will have increased by the amount

$$2x_0\Sigma\delta x_s + 2y_0\Sigma\delta y_s + \Sigma(\delta x_s)^2 + \Sigma(\delta y_s)^2 + 2\Sigma(\delta x_s\delta x_{s'} + \delta y_s\delta y_{s'}).$$

The latter terms are zero, since on the average the sum of the cosines of the angles which any path makes with the consecutive path vanishes. Hence for any ion the square of the distance changes by the amount

$$2x_0\Sigma\delta x_s + 2y_0\Sigma\delta y_s + \Sigma(\delta x_2)^2 + \Sigma(\delta y_s)^2.$$

For a large number of ions starting from the point x_0y_0 the average values of $\Sigma\delta x_s$ and $\Sigma\delta y_s$ vanish, since the ions are free to move in all directions. Hence the rate of change of the mean square of the distance of the distribution from the axis is $\Sigma[(\delta x_s)^2+(\delta y_s)^2]/NT$.

Substituting for δx_s and δy_s their values, the quantity $\frac{d\rho^2}{dt}$ becomes $\frac{2}{NT\omega^2}\Sigma A_s^2(1-\cos\omega t_s)$.

The velocities A_s are independent of the times t_s, so that the mean value of A^2 may be substituted for A_s^2 in this expression. Also since A_s is the velocity in the plane xy the mean value of A_s^2 is $2\,V^2/3$, V being the mean velocity of agitation of the ions, and since the series of cosines $\Sigma\cos\omega t_s$ is equal to $N/(1+\omega^2T^2)$ the above expression reduces to

$$\frac{d\rho^2}{dt}=\frac{2}{NT\omega^2}\cdot\frac{2\,V^2}{3}\cdot\frac{N\omega^2T^2}{1+\omega^2T^2}=\frac{4\,V^2T}{3\,(1+\omega^2T^2)}.$$

The rate of diffusion K along the direction of the magnetic force is $\frac{lV}{3}$ or $\frac{V^2T}{3}$.

Hence
$$\frac{d\rho^2}{dt}=\frac{4\,K}{1+\omega^2T^2}.$$

91. Rate of diffusion in directions perpendicular to the magnetic force. The motion of the ions may therefore be expressed in the usual form by the equations

$$\frac{nu}{K_h}=-\frac{dn}{dx},$$

$$\frac{nu}{K_h}=-\frac{dn}{dy},$$

$$\frac{n\omega}{K}=-\frac{dn}{dz}.$$

The change of the number of ions n per cubic centimetre at any point is

$$-\frac{dn}{dt}=\frac{d\,(nu)}{dx}+\frac{d\,(nv)}{dy}+\frac{d\,(nw)}{dz}$$

$$=K_h\left(\frac{d^2n}{dx^2}+\frac{d^2n}{dy^2}\right)+K\frac{d^2n}{dz^2}.$$

In this notation $\frac{d\rho^2}{dt}$ is given by the equation

$$\frac{d\rho^2}{dt}\cdot\iiint ndxdydz=\iiint(x^2+y^2)\,\frac{dn}{dt}\,dxdydz,$$

and it is easy to see that when the above value is substituted for $\frac{dn}{dt}$ the integral on the right of this equation reduces to

$$4\,K_h\iiint n\,dx\,dy\,dz.$$

Hence
$$\frac{1}{4}\frac{d\rho^2}{dt}=K_h=\frac{K}{1+\omega^2T^2}.$$

92. Motion in electric field with transverse magnetic field. When ions are moving in an electric field and a magnetic force H is applied along the axis of z, the motion in that direction is unchanged, but in the directions x and y the velocities are altered. Taking the axis of x as the direction of the electric force X, the equations of motion become

$$m\frac{d^2x}{dt^2}=Xe-He\frac{dy}{dt},$$

$$m\frac{d^2y}{dt^2}=He\frac{dx}{dt},$$

so that the velocities $\dot{x}$ and $\dot{y}$ are given by the equations

$$\dot{x}=A\sin(\omega t+\alpha)\text{ and }\dot{y}=\frac{X}{H}-A\cos(\omega t+\alpha),$$

the velocities at the beginning of the path being $A\sin\alpha$ and $\frac{X}{H}-A\cos\alpha$. If $\dot{x}_1, \dot{x}_2, \ldots, \dot{y}_1, \dot{y}_2, \ldots$ be the velocities after a number of consecutive collisions of an ion with molecules of the gas, the sum of the velocities $\dot{x}_1+\dot{x}_2+\ldots$ differs for different ions and has an average value equal to zero for a group of ions.

Hence $\Sigma A_1\sin\alpha_1=0$ and $\Sigma A_1\cos\alpha_1=\frac{NX}{H}$.

The distances that an ion travels in the first interval t_1 are

$$\delta x_1=\frac{A_1}{\omega}(\cos\alpha_1-\cos(\omega t_1+\alpha_1)),$$

$$\delta y_1=\frac{X}{H}t_1+\frac{A_1}{\omega}(\sin\alpha_1-\sin(\omega t_1+\alpha_1)).$$

The distances traversed in the time $NT = t_1 + t_2 + \dots t_n$ are

$$\Sigma\delta x_s = \frac{NX}{H\omega} - \frac{1}{\omega}\Sigma\left(\frac{X}{H} - \dot{y}_s\right)\cos\omega t_s + \frac{1}{\omega}\Sigma\dot{x}_s\sin\omega t_s,$$

$$\Sigma\delta y_s = \frac{NXT}{H} - \frac{1}{\omega}\Sigma\left(\frac{X}{H} - \dot{y}_s\right)\sin\omega t_2 - \frac{1}{\omega}\Sigma\dot{x}_s\cos\omega t_s.$$

The terms independent of X obviously vanish in estimating the mean displacement of a group of ions starting from the origin, and the velocities U_h and V_h along the axes become

$$U_h = \frac{\Sigma\delta x_s}{NT} = \frac{X}{H\omega T}\cdot\frac{\omega^2 T^2}{1+\omega^2 T^2},$$

$$V_h = \frac{\Sigma\delta y_s}{NT} = \frac{X}{H}\cdot\frac{\omega^2 T^2}{1+\omega^2 T^2},$$

and since $\omega = \frac{He}{m}$, the expressions for the velocities reduce to

$$U_h = \frac{Xe}{m}\cdot\frac{T}{1+\omega^2 T^2}; \text{ and } V_h = \frac{Xe}{m}\cdot\frac{\omega T^2}{1+\omega^2 T^2}.$$

The velocity U when the magnetic force is zero is $\frac{Xe}{m}\cdot T$,

hence $$U_h = \frac{U}{1+\omega^2 T^2}; \text{ and } V_h = U_h \cdot \frac{He}{m}\cdot T.$$

Also the following relation holds, $\frac{U_h}{K_h} = \frac{U}{K}$, which shows that the velocity in the direction of the electric force is reduced by the magnetic field in the same proportion as the rate of diffusion.

The equations of motion from which the number of ions per cubic centimetre at any point in the gas may be determined, when electric and magnetic forces are acting, are

$$\frac{pu}{K_h} = -\frac{dp}{dx} + nXe,$$

$$\frac{pv}{K_h} = -\frac{dp}{dy} + nXe\cdot\frac{He}{m}\cdot T,$$

$$\frac{pw}{K} = -\frac{dp}{dz} + nZe,$$

the component of the electric force along the axis of y being zero.

93. Deflection of a stream of ions. When a stream of ions is moving under the action of an electric force the deflection of the stream from the direction of the electric force may be measured by the ratio

$$\frac{V_h}{U_h} = \frac{He}{m} \cdot T = \tan \theta.$$

The deflection varies with the pressure since T, the mean interval between collisions, is inversely proportional to the pressure of the gas. Also when e/m is constant the deflection is independent of the electric force.

The value of $\tan \theta$ may be written in various forms, thus

$$\tan \theta = \frac{H}{X} \cdot \frac{Xe}{m} \cdot T = \frac{HU}{X} \cdot$$

Hence the velocity U in the direction of the electric force may be determined by observing the deflection θ produced by the magnetic force H.

When the deflection is small the free paths are only slightly curved by the action of the magnetic force, since the radius r of the circle described by the ion in the plane $Z = 0$ is $\frac{mv}{He}$ and the mean length s of the arc of the circle described between collisions is $v \,.\, T$.

Hence $$\frac{s}{r} = \frac{HeT}{m} = \tan \theta.$$

The length of the arc s described by the ion is therefore small compared with the radius of the circle.

CHAPTER IV

THE VELOCITY OF IONS IN AN ELECTRIC FIELD

94. Velocity of ions in gases at high pressures. The velocity acquired by ions under the action of an electric force was first determined by Rutherford in the cases in which the ions were generated in gases at atmospheric pressure by Röntgen rays and Becquerel rays. He found that in air the sum of the velocities of the positive and negative ions produced by Röntgen rays was 3·2 centimetres per second under a force of one volt per centimetre.* Afterwards, by a method similar to that used by McClelland to investigate the velocities of the ions from flames, it was shown that the mobility of the ions produced by uranium oxide † was the same as that of the ions produced by Röntgen rays. In this method a stream of gas is blown between two concentric cylinders and the potential difference between the cylinders necessary to remove a certain proportion of the ions from the gas is found. The principle was subsequently applied by Zeleny to determine accurately the velocities of the positive and negative ions.

Rutherford also determined the velocity of the negative ions set free from a metal surface by ultra-violet light, by measuring the distance the ions travelled from the plate under the action of an alternating force.‡ The value of the velocity thus found for negative ions in air at atmospheric pressure was 1·4 centimetres a second under an electric force of 1 volt per centimetre, which is of the same order as the velocity of the ions produced by Röntgen rays, but

* E. Rutherford, Phil. Mag. (5) **44**, p. 429, 1897.
† Id., Phil. Mag. (5) **47**, p. 109, 1899.
‡ Id., Proc. Camb. Phil. Soc. **9**, p. 410, 1898.

is smaller than the velocity of negative ions as found by the more accurate investigations of Zeleny and Langevin.

Zeleny, who was the first to make accurate determinations of the velocities of ions generated by Röntgen rays, found that there was a considerable difference between the velocities of the positive and negative ions.

Langevin, using a different method, obtained results in good accordance with Zeleny's, and in addition investigated the effect of reducing the pressure. It was found that the velocity of the positive ions was inversely proportional to the pressure, the same law holding for the negative ions except at the lower pressures, when it was observed that the velocity increases more rapidly than the inverse of the pressure.

The effects obtained at low pressures were subsequently investigated more fully by Lattey, who found that the velocity of the negative ions was proportional to the force X and inversely proportional to the pressure p for small values of the ratio X/p. For large values of X/p the velocity increases more rapidly than the ratio X/p, and eventually the velocity undergoes large changes when small variations are made in the force or in the pressure.

The smaller velocities may be determined by these methods, which are all founded on the principle of measuring the interval of time occupied by the ions in travelling a given distance in the direction of the electric force. It has, however, been found impracticable to measure thus directly velocities of an order greater than 5×10^3 centimetres per second.

In order to investigate the conditions under which higher velocities are obtained, a method depending on the effect of a magnetic force on a stream of ions moving in an electric field has recently been used.* It is applicable to high velocities of any order, provided that the motion of the ions is controlled by the viscous resistance of the gas

* J. S. Townsend and H. T. Tizard, Proc. Roy. Soc. A, **87**, p. 357, 1912; **88**, p. 336, 1913.

to such an extent that the velocity under the electric force assumes a constant value.

95. Velocity in gases at atmospheric pressure. Zeleny's investigations. The principle of the method used by Zeleny,* for gases at atmospheric pressure, is shown by the diagram figure 16, which represents the essential part of the apparatus.

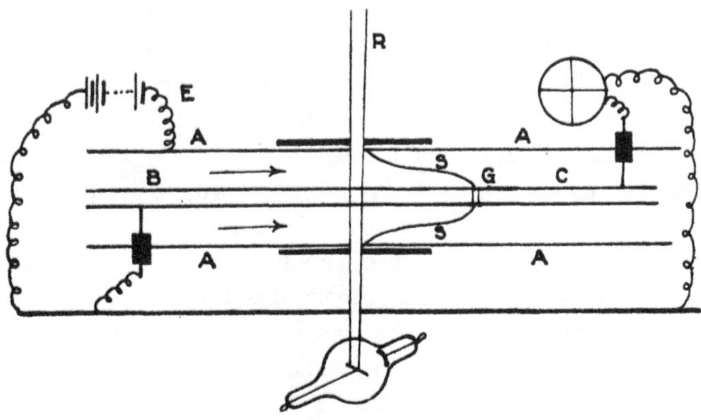

Figure 16.

The gas is passed from one gasometer to another through a long wide tube A, provided with two coaxal cylinders B and C insulated from each other by a narrow air-gap G. The potential E of the outer cylinder A is adjustable and is either positive or negative according as it is desired to collect the positive or negative ions on the inner cylinders. The cylinder B is connected to earth, and the cylinder C is insulated and connected to a pair of quadrants of a sensitive electrometer. The electrometer deflections being always small, the potentials of the two cylinders B and C are approximately the same. The stream of gas is run at a uniform rate along the space between the outer and inner cylinders and is ionized by a beam of Röntgen rays R which is limited by lead screens to a narrow space between two planes perpendicular to the axis of the tube.

* J. Zeleny, Phil. Trans. A, **195**, p. 193, 1900.

The ions are carried by the stream in a direction parallel to the axis of the tubes, and move along directions perpendicular to the axis under the action of the electric force. When the potential E is sufficiently high, the ions that move inwards are all collected on the electrode B, and as the potential is reduced the distance they are carried by the stream increases, and eventually some of the ions reach the electrode C and an electrometer deflection is obtained. The ions which first reach the electrode C are those generated at the surface of the outer tube, and their paths are in the form of the curves S when the potential E is adjusted so that the electrometer just begins to be affected. The velocity of the ions under unit electric force may be calculated in terms of this potential and the quantity of gas passing through the tube per second.

Let a be the distance from the beam R to a point on the gap G midway between the electrodes B and C, r_1 the radius of the large tube and r_2 the radius of the inner electrode. The ions generated at the surface of A travel the distance $r_1 - r_2$ under the electric force while they are carried a distance a by the stream. Let r be the distance of an ion at any point P of its path S from the axis of the cylinders, x the distance of the point P from the plane of R. The electric force X at any point is inversely proportional to r, and if E is the potential difference between the outer and inner electrodes then $X = \dfrac{E}{r \log r_1/r_2}$; so that if k be the mobility, or the velocity of an ion under unit electric force, the velocity towards the axis at P is $\dfrac{kE}{r \log r_1/r_2}$. Thus in the time dt the ion travels a distance dr towards the axis given by the equation $dr = \dfrac{kEdt}{r \log r_1/r_2}$ and a distance dx along the axis equal to udt, where u is the velocity of the stream. Hence $urdr = \dfrac{kE}{\log r_1/r_2} \cdot dx$.

Multiplying this equation by 2π and integrating along

the whole length of the curve S the equation to determine k is as follows:

$$\int_{r_2}^{r_1} 2\pi u r dr = \frac{2\pi k E a}{\log r_1/r_2}.$$

The expression on the left of this equation represents the volume of gas Q passing through the tube per second, hence

$$k = \frac{Q \log r_1/r_2}{2\pi a E}.$$

In order to obtain accurate results it is necessary that the secondary radiation should produce inappreciable effects between the primary beam and the electrode C. Any ions thus produced near the surface of the larger cylinder would reach the electrode C when E is adjusted so that the ions produced by the beam R travel along the path S. In Zeleny's experiments the tubes were made of aluminium which gives out a small secondary radiation, so that probably no serious error is thus introduced.

96. Principle of Langevin's method. In the method used by Langevin * to determine the velocities of ions, the gas was at rest in the space between two parallel-plate electrodes and was ionized by rays given out by a single discharge (that lasts for a very short period) through a Röntgen-ray tube. The electric field was established between the electrodes before the discharge was passed, and was reversed after a short interval of time t. Thus ions of one sign were collected on one electrode for a time t, and afterwards all the ions of opposite sign that were in the gas at the instant the force was reversed were collected on the same electrode. The total effect was measured by having the insulated plate connected to an electrometer. The experimental investigation consisted in finding the charges acquired by the insulated electrode for different values of the time t.

When the space between the plates AB and CD, figure 17,

* P. Langevin, Comptes rendus, **134**, p. 646, 1902; Annales de Chimie et de Physique, (7) **28**, p. 289, 1903.

is ionized uniformly throughout the volume by a beam of rays, it is easy to calculate the charge acquired by one of the electrodes in terms of the velocities k_1 and k_2 of the positive and negative ions under unit electric force.

Let X be the electric force, which is applied for the time t before reversal, and let the positive ions move towards AB. During the interval t the negative ions generated in the space between the electrode CD and a parallel plane at a distance $k_2 Xt$ are collected on the electrode CD, and the positive ions generated in the space of thickness $k_1 Xt$ are collected on the electrode AB. At the instant of reversal there remain in the gas the negative ions that were produced in the layer of thickness $l - k_2 Xt$, which have their direction of motion reversed and are collected by the plate AB. Thus the total charge q acquired by AB is pro-

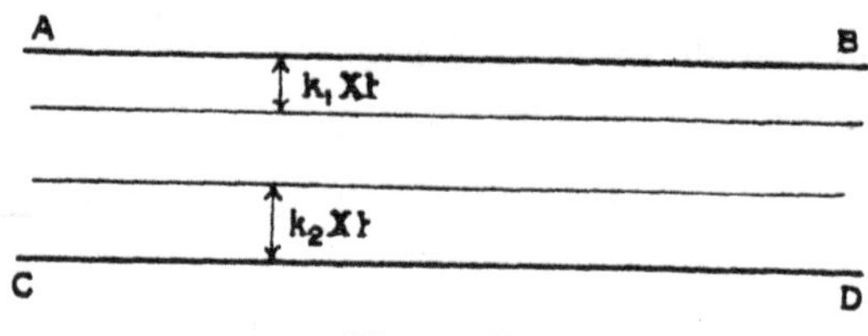

Figure 17.

portional to $-l + (k_1 + k_2)Xt$ when t is small and $k_2 Xt$ does not exceed l.

Thus q increases uniformly with the time t, having a value $-Q$ when $t = 0$. When t exceeds $l/k_2 X$ all the negative ions are collected on the electrode CD before the force is reversed, and q is then proportional to $k_1 Xt$, so that at the point $t_2 = l/k_2 X$ the rate of increase of q with the time t changes abruptly from $(k_1 + k_2) X$ to $k_1 X$. As the period t is further increased, the time $t_1 = l/k_1 X$ is reached when all the positive ions are collected on the electrode AB and q remains constant for larger values of t. The value of q in terms of t is thus represented by the curve, figure 18.

The curve found experimentally agrees with the theoretical curve and has two points corresponding to times t_1

and t_2 at which there is an abrupt change in the slope of the curve.

When there is intense secondary radiation at one of the plates, as for instance when the electrode AB is of lead, then when t is zero the charge acquired by that electrode is negative as before, but as t increases the number of positive ions acquired by AB increases rapidly. This is obvious, since in the interval t before the potential is reversed the positive ions generated in the layer of thickness k_1Xt near the electrode are collected on the plate AB. Also when t approaches the value $t_2 = l/k_2X$ the negative ions generated near AB begin to reach the electrode CD, and the charge q

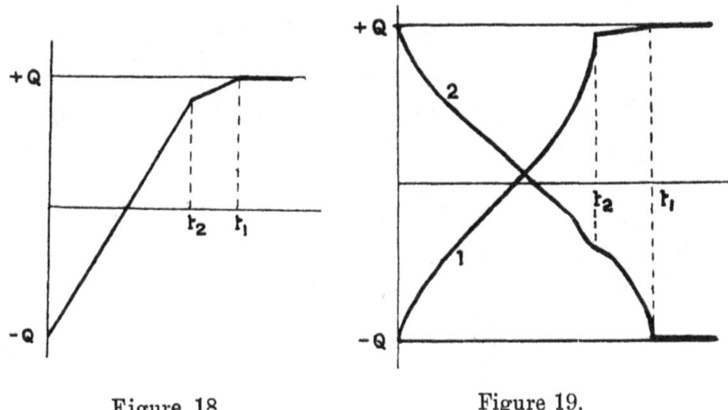

Figure 18. Figure 19.

acquired by AB approaches rapidly the value $+Q$. When t attains the value t_2 all the negative ions are collected on the electrode CD, and the charge acquired by AB is less than the maximum $+Q$ by a small quantity corresponding to the positive ions generated near the electrode CD which do not reach AB before the force is reversed. The value of q in terms of t is then given by the curve 1, figure 19.

When the electric force is applied in the opposite direction initially, the negative ions are collected during the period t on the plate AB and q is given by the curve 2, figure 19. There is a small change in the slope of the curve when $t = t_2$ and a large change when $t = t_1$, which is the time required for the large number of positive ions that are generated near the surface of AB to reach the electrode CD, and at that time the electrode AB has acquired the charge $-Q$.

97. Apparatus used by Langevin for determinations at different pressures. It is necessary in these experiments to have a special device for reversing the potential of one of the plates when a short interval of time has elapsed after the discharge has passed through the Röntgen-ray tube. Also, in order to obtain accurate results, it is neces-

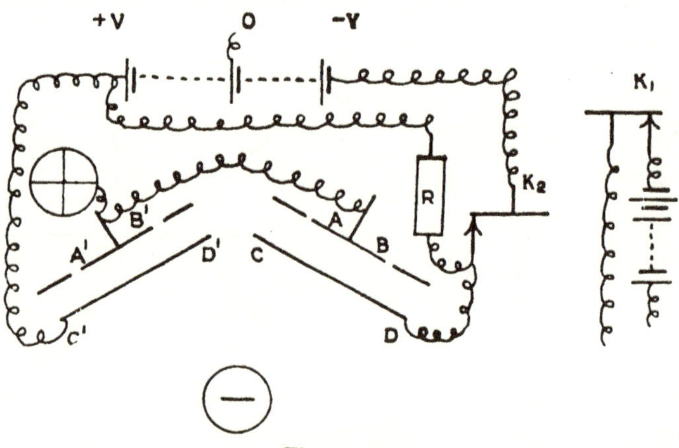

Figure 20.

sary to avoid errors that may arise from variations in the intensity of the rays. The principle of the arrangement of the apparatus that was used is shown in figure 20. A heavy weight falling through a known height operated two keys at different distances from the point at which it starts. The first key, K_1, breaks the primary circuit of the coil working the discharge tube, and the second key, K_2, reverses the electric force between AB and CD, R being a large resistance.

The compensating charge acquired by $A'B'$ was adjusted to be equal and opposite to the maximum charge acquired by AB by ionizing the gas between the two sets of electrodes with rays from the same bulb, and limiting the beams that fell on CD and $C'D'$ so that the electrometer gave no deflection when constant forces of opposite sign were established between the two pairs of electrodes.

If in a time t after a single discharge through the bulbs the force between AB and CD is reversed, then the total charge acquired by AB and $A'B'$ will continue to be zero for large values of t. The times t_1 and t_2 are determined as the minimum times that elapse between the discharge and the reversal of the field when the electrometer is unaffected.

98. Values of the velocity at ordinary pressures. The velocities obtained by the various methods are given in the following tables. The numbers given by Zeleny for gases at atmospheric pressure were obtained from experiments in which various forces were used, and the effect of moisture on the velocity was also examined. The velocity was found to be proportional to the force. The following are the velocities k_1 and k_2 of the positive and negative ions, in centimetres per second, for an electric force of one volt per centimetre in dry and moist gases. The temperature in degrees centigrade of the gas is given in the column T.

	Dry gases.			Moist gases.		
	k_1	k_2	T	k_1	k_2	T
Air	1·36	1·87	13·5	1·37	1·51	14·0
Oxygen	1·36	1·80	17·0	1·29	1·52	16·0
Hydrogen	6·70	7·95	20·0	5·30	5·60	20·0
Carbon dioxide	·76	·81	17·5	·82	·75	17·0

The ions in these experiments, as also in Langevin's experiments, were generated by the action of Röntgen rays.

The effect of pressure on the velocities of the ions is shown by the following numbers given by Langevin for the velocities in dry air at various pressures p, k_1 and k_2 being the velocities of the positive and negative ions in a field of one volt per centimetre.

p	k_1	k_2	$\frac{pk_1}{76}$	$\frac{pk_2}{76}$
7·5	14·8	21·9	1·46	2·16
20·0	5·45	7·35	1·43	1·97
41·5	2·61	3·31	1·42	1·77
76·0	1·40	1·70	1·40	1·70
143·5	·75	—	1·42	—
142·0	—	·90	—	1·68

The numbers in the fourth column show that the velocity of the positive ions is inversely proportional to the pressure, and those in the last column show that for negative ions the product of the velocity and the pressure increases as the pressure diminishes.

99. Velocities in different gases and vapours. Effect of oxygen on the negative ions in argon and nitrogen. The velocities of ions in a large number of gases and vapours have been determined by Wellisch,* adopting the method used by Langevin.

The following table gives the velocities k_1 and k_2 of the positive and negative ions for one volt per centimetre, at a pressure of 760 mm. The molecular weights and absolute critical temperatures are given in the first and second columns of numbers.

	Formula.	Molecular weight.	Absolute critical temperature.	Velocities in centimetres per second.	
				k_1	k_2
Air	—	—	—	1·54	1·78
Carbon monoxide	CO	28	137	1·10	1·14
Carbon dioxide	CO_2	44	304	·81	·85
Nitrous oxide	N_2O	44	310	·82	·90
Ammonia	NH_3	17	404	·74	·80
Aldehyde	C_2H_4O	44	454	·31	·30
Ethyl alcohol	C_2H_6O	46	513	·34	·27
Acetone	C_3H_6O	58	511	·31	·29
Sulphur dioxide	SO_2	64	429	·44	·41
Ethyl chloride	C_2H_5Cl	64·5	458	·33	·31
Pentane	C_5H_{12}	72	470	·36	·35
Methyl acetate	$C_3H_6O_2$	74	507	·33	·36
Ethyl formate	$C_3H_6O_2$	74	507	·30	·31
Ethyl ether	$C_4H_{10}O$	74	467	·29	·31
Ethyl acetate	$C_4H_8O_2$	88	522	·31	·28
Methyl bromide	CH_3Br	95	467	·29	·28
Methyl iodide	CH_3I	142	528	·21	·22
Carbon tetrachloride	CCl_4	154	557	·30	·31
Ethyl iodide	C_2H_5I	156	554	·17	·16

* E. M. Wellisch, Phil. Trans. A, **209**, p. 249, 1909.

The values of k_1 and k_2 for helium were determined by Franck and Pohl * by a method similar to that used by Rutherford to determine the velocities of the ions set free by ultra-violet light. The numbers they obtained are 5·09 centimetres per second for the positive ions and 6·31 centimetres per second for negative ions, at normal pressure and temperature.

The velocities in argon and nitrogen that had been carefully purified were subsequently investigated by Franck.†

Very little effect was produced by impurities on the motion of the positive ions, and the values of k_1 were found to be 1·37 and 1·27 centimetres per second in argon and nitrogen respectively.

With negative ions the velocities were very high in the pure gases, k_2 being 206 centimetres per second in pure argon. When 1·2 per cent. of oxygen was admitted, the velocity fell to 1·7 centimetres per second and remained at the value when the amount of oxygen was raised to 10 per cent. Similarly in nitrogen, when the gas is pure the value of k_2 was found to be 144 centimetres per second, but with a small percentage of oxygen the value fell to 1·84 centimetres per second.

These results are interesting, as they show that the group formed by oxygen molecules round a negative ion is larger and more stable than that formed by the molecules of nitrogen or argon.

As the velocities were determined by an alternating force, the value of the force to which these results correspond can only be obtained within certain limits. Other investigations show that the effect of impurities on the velocity of the negative ions depends on the pressure of the gas and the electric force that is applied. At low pressures the high velocity of the negative ions is not affected by comparatively large percentages of oxygen. Thus, in dry air at 20 milli-

* J. Franck and R. Pohl, Deutsch. Phys. Gesellsch., Verh. **9**, p. 195, 1907.

† J. Franck, Deutsch. Phys. Gesellsch., Verh. **12**, p. 291, 1910, and Verh. **12**, p. 613, 1910.

metres pressure, Lattey found that the velocities of the positive and negative ions were 108 and 3,500 centimetres per second respectively, under an electric force of two volts per centimetre. If a small percentage of oxygen had the same effect here as at higher pressures, the latter velocity would have been about 140 centimetres per centimetre.

Water vapour behaves in a similar manner, but is more effective than oxygen. When a small percentage of water vapour is present in air at 20 millimetres pressure, the velocity of the negative ions under a force of 2 volts per centimetre is of the same order as the velocity of the positive ions. This effect ceases as the pressure is reduced and the force is increased, as has been found in many experiments. For example, when the ratio X/p is of the order required to obtain new ions by the collisions of negative ions with molecules, the mass of the negative ions is not influenced by the presence of water vapour. The negative ions under these conditions move as free electrons and generate new ions by collisions in water vapour as in dry gases.

100. Velocity of negative ions in terms of the ratio X/p. The methods of determining the velocities of ions described in the preceding sections are applicable to cases in which the ratio of the force to the pressure is small. In these cases the velocity is proportional to the ratio X/p both for positive and negative ions. When the pressure is reduced and X/p is of the order ·05 or ·1 (X being in volts per centimetre and p in millimetres of mercury) the velocity of the negative ions in air increases rapidly with the force. Under these conditions the motion of the ions in the direction of the electric force may be measured by the method used by Lattey, when the velocity does not exceed 3×10^3 centimetres per second. With larger values of the ratio X/p, when velocities of the order 10^6 or 10^7 centimetres per second are obtained, it is necessary to use a method depending on the action of a magnetic force on the motion of a stream of ions.

In the method used by Lattey* an electric force X is applied to the space between two parallel gauzes G and G' for a short time t and is then replaced during an equal interval of time by a somewhat larger force in the opposite direction. Ions are supplied to the gauze G from the side remote from G', so that some of the ions that come through the gauze G will reach G' if $Ut = l$, U being the velocity of the ions and l the distance between the two gauzes. When t is less than l/U none of the ions reach G', and when the force is reversed they are all brought back again to G. The arrangement of the apparatus for measuring the

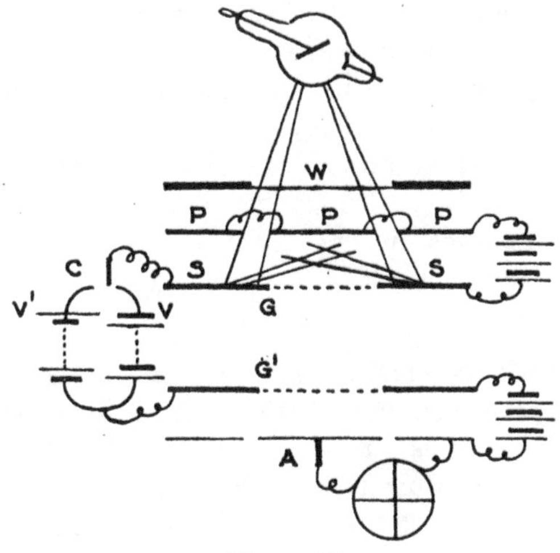

Figure 21.

velocities on this principle is shown in figure 21. The gas is contained in a brass cylinder with an aluminium window W through which the Röntgen rays enter the apparatus and fall on the metal surface S, which gives off secondary rays that ionize the gas in the space between the plate P and the gauze G. A fixed difference of potential is established between P and G by means of an insulated battery. The second gauze G' is raised to a fixed potential so that the force X from G' to the insulated electrode A should be in the same direction as the force from P to G.

* R. T. Lattey, Proc. Roy. Soc. A, 84, p. 173, 1910.

By means of a rotating commutator at c, the alternate segments of which were at potentials V and V' with respect to G', a field of force $X = V/l$ was created between G and G' for a time t and then changed for an equal time to a larger force V'/l in the opposite direction. The electrode A is connected to an electrometer, which is deflected when any of the ions that pass through G reach the second gauze G', since some of them will pass through the openings in the gauze and come into contact with the plate A.*

The Röntgen-ray tube giving out the primary rays is excited by an induction-coil with an ordinary interrupter in the primary coil, and the rays may be allowed to act for any convenient period while the commutator is rotating. The electrometer receives no charge when the commutator rotates with a high velocity, but as the speed is reduced and t reaches the value l/U the electrometer begins to be deflected.

Small variations in the intensity of the rays do not affect the accuracy of this method of measuring the velocities.

101. Lattey's determinations with air at low pressures. Effect of water vapour. With positive ions Lattey found that the velocity is proportional to the ratio X/p, with pressures varying from 14·3 to 28·8 millimetres, and electric forces from 1·6 to 2·7 volts per centimetre. The mean value of Up/X for a large number of determinations was 1120, which corresponds to a velocity of 1·475 centimetres per second at atmospheric pressure, under a force of one volt per centimetre.

At these pressures the velocity of the negative ions increases more rapidly than the electric force, so that it is necessary to record both the force and pressure at which

* The object of the second gauze G' is to screen the insulated plate A from the electric effects of the alternating field. If G' were removed and the plate A substituted, the apparatus would work as before, but there would be difficulty in determining accurately the force from G to A when the potential of G varies from $+V$ to $-V'$, and also in maintaining a uniform force X from G to A. The introduction of the second gauze removes these difficulties.

the velocity is determined. The following table gives the velocities of negative ions in centimetres per second for air at pressures p, when different forces X are acting. The numbers at the head of each column represent the ratio X/p, X being in volts per centimetre and p in millimetres of mercury:

p	$X/p = 0{\cdot}04$	0·05	0·06	0·07	0·08	0·09
14·3	107	175·5	310	580	1126	(2200)
18·4	103	163	279	514·5	1106	(2050)
24·5	(119·5)	172·5	286	509	936·5	1799
28·8	116	180	298·5	519·5	926	(1652)
Mean	112·5	173·2	287·8	510·5	953	1845

The numbers in brackets were obtained by extrapolation, and are only approximate values.

The table shows that the velocity is increased in the proportion 8·5 : 1 when X/p is increased from ·04 to ·08.

Over the above range of pressures the velocity is approximately a function X/p only, but the numbers obtained with the higher forces show that the velocity rises more rapidly by reducing the pressure than by making proportional increases in the force.

102. Velocities in hydrogen and carbon dioxide at low pressures. Similar experiments were made with hydrogen and carbon dioxide at low pressures by Lattey and Tizard.* With forces from 1·26 to 3 volts per centimetre the velocities of positive ions in dry carbonic acid were proportional to the ratio X/p for pressures from 3·68 to 13 millimetres, the mean value of pU/X being 642. This corresponds to a velocity of ·84 centimetres per second at atmospheric pressure under a force of one volt per centimetre.

The velocity of positive ions in hydrogen was approximately proportional to the ratio X/p for pressures between 72 millimetres and 5 millimetres when forces of the order of 2 volts per centimetre were used. The mean value of the quantity pU/X was 4058, but the numbers obtained at the lower pressures were a few per cent. greater than those

* R. T. Lattey and H. T. Tizard, Proc. Roy. Soc. A, **86**, p. 349, 1912.

found at the higher pressures. This number corresponds to a velocity of 5·34 centimetres per second for one volt per centimetre at atmospheric pressure.

With negative ions the results were similar to those obtained with air. In carbonic acid, with forces of the order of 2 volts per centimetre, and pressures ranging from 20 to 10 millimetres, the following are the velocities that were obtained:

X/p	·08	·09	·10	·11	·12	·13	·14	·15	·16	·17
U	64	73	82	92	106	132	182·5	277	468	977

The corresponding numbers for hydrogen with similar forces, and pressures ranging from 197 to 67 millimetres, were as follows:

X/p	·010	·012	·014	·016	·018	·020	·022	·024	·026	·027
U	68	95	127	174	236	336	489	730	1100	1347

The velocities of the negative ions for the above values of X/p are very much reduced when the gas is slightly moist. Thus with a very small percentage of water vapour the velocity in air was 1,302 centimetres per second when $X/p = \cdot 11$, the velocity in dry air being probably about 7500. When a larger percentage of water vapour is present, 2·1 millimetres in a total of 18 millimetres, the velocity was found to be 113 centimetres per second when $X/p = \cdot 127$, which is of the order of one hundredth of the velocity in dry air for the same force and pressure. This is a large effect compared with that obtained at atmospheric pressure with small forces, for under those conditions, as appears from Zeleny's determinations, the velocity under a force of one volt per centimetre is 1·87 centimetres per second in dry air, and 1·51 in air saturated with water vapour.

103. Transition from ionic to electronic state as X/p increases. The velocities of positive and negative ions in dry air are given by the curves, figure 22, as a function of X/p.

For positive ions, the quantity pU/X is constant and the curve is a straight line, the value of the constant being about 1080, which is the mean of the values obtained by Zeleny, Langevin, and Lattey.

The curve representing the velocity of the negative ions is a straight line for the smaller values of X/p, and the equation of the curve near the origin is $pU/X = 1350$. For the larger values of X/p the velocity increases rapidly and

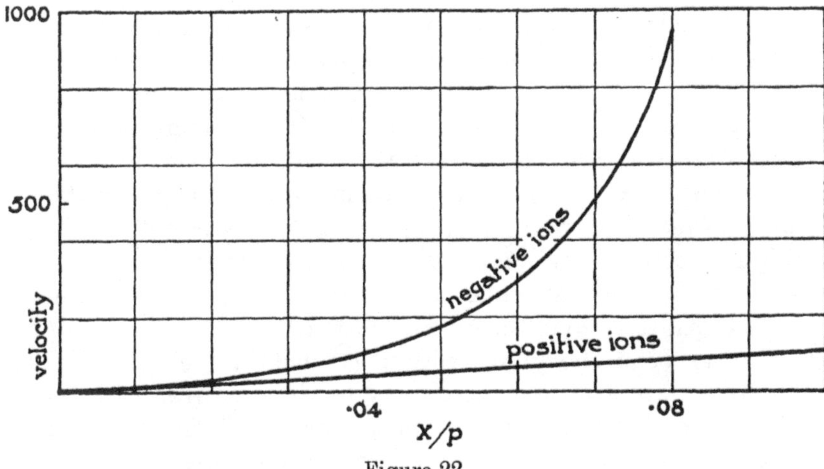

Figure 22.

the curve becomes inclined at a very small angle to the axis of y when $X/p = \cdot 10$.

From the results of the experiments on the mobility and the rate of diffusion of ions into gases the dimensions of the ions may be estimated in terms of the dimensions of molecules of gases. It may be shown that when the velocity is proportional to X/p, the mass associated with the ion is a constant, hence for a large range of pressures and forces the groups of molecules that move with the positive and negative ions are unaltered by the force or the pressure. The above experiments show that the velocity of the negative ions increases rapidly as the value of X/p increases above certain values, which indicates that the mass associated with the negative ions diminishes as the higher values of X/p are attained. Since there is no corresponding effect with positive ions, it appears that the group of molecules that move with the negative ions when X/p is small is more unstable than that which moves with the positive ions, and that the stability is greatly increased when there is a small proportion of water vapour in the gas.

104. Velocity of electrons under an electric force. This increased mobility of the negative ions may be investigated by determining the motion of the ions in a magnetic field, and from the results obtained it appears that under forces of one or two volts per centimetre, negative ions move along some of their free paths in air at 10 millimetres pressure as if the mass associated with the negative ion were very small compared with the mass of a molecule of the gas. Thus even when the pressure of the gas is sufficiently high to control the motion of the ions so that they attain a constant velocity under an electric force, the negative ions begin to assume the electronic state.

It has been shown in section 93, that the velocity U of an ion in an electric field X may be deduced from the deflection of a stream of ions in a magnetic field. Thus if the direction of motion of an ion, in a magnetic field of intensity H, makes an angle θ with the direction of the electric force, the deflection is given by the formula

$$\tan\theta = \frac{He}{m} \cdot T,$$

where T is the interval between the collisions of the ion with the molecule of the gas.

The velocity due to the electric force is

$$U = \frac{Xe}{m} \cdot T,$$

so that

$$U = \frac{X\tan\theta}{H}.$$

This equation may also be obtained from elementary principles if it be assumed that the ions move with a constant velocity U in a resisting medium. In this case the forces acting on the moving charge are Xe in the direction of the electric force, and HUe at right angles to the directions of X and H. The direction of motion will be along the resultant of the two forces, so that $HU/X = \tan\theta$, and when θ is small U is the velocity under the force X.

For a certain range of forces and pressures, there is a transition stage in which the average mass of a negative ion diminishes from that of a group of molecules to that of an electron. Under a given electric force the electron moves freely along some of its free paths, and the velocity varies, but the mean velocity U as deduced from the magnetic deflection is not then the same as the mean velocity as obtained by a direct method. The experiments on the magnetic deflection show that in these cases the deflection is not proportional to H. In dry air the negative ions are in the transition stage when the ratio X/p lies between ·01 and ·2.

For the larger values of X/p the electron moves freely along all the paths between collisions. The phenomena then becomes more simple, and the velocity given by the equation $U = \frac{X}{H}\tan\theta$ is the same as the velocity that would be obtained by a direct method. A characteristic feature of the motion of electrons is the increase of the velocity of agitation with the value of X/p, so that when the pressure is constant the interval between collisions, T, diminishes as the force increases and the velocity in an electric field is not proportional to the force.*

105. Deflection of a stream of electrons by a transverse magnetic force. Values of the velocity in terms of the ratio X/p. The apparatus used for measuring the deflection of a stream of ions by a magnetic force is shown in figure 23. A beam of ultra-violet light entered the space enclosed by an air-tight cover through the quartz plate W and fell upon a metal plate A. The ions that were set free moved

* A more complete discussion of the motion that takes place, as X/p increases, is given in sections 132 and 133.

under an electric force towards plate B, and some of them passed through a narrow slit in the centre of B.

The plates were charged by means of a battery of cells to negative potentials, the potential of A being greater than that of B, so that the electric force is in the same direction on the two sides of the aperture in B, and the ions that come through continue their motion towards the three electrodes C. Six metal rings, R, were arranged at intervals of one centimetre from R_0 to B and were connected in series by five equal resistances, the upper ring being joined to the plate B and the lower to the ring R_0 by similar

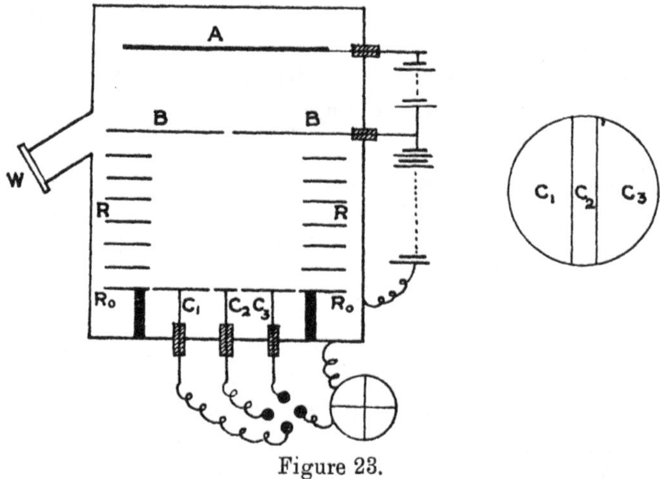

Figure 23.

resistances. The ring R_0 was connected to earth through the case of the apparatus, so that when the plate B was raised to a potential V, by means of a battery, the potential at any point of the electric field bounded by the rings was proportional to the distance of the point from the electrodes C. The ions, therefore, move in a field of uniform force after passing through the aperture in B. The electrodes C were parts of a disc seven centimetres in diameter, which was divided by two narrow air-gaps into two equal segments C_1 and C_3 and a narrow central segment C_2. The gaps between the electrodes were parallel to the slit in B.

The stream of ions that comes through the slit in B diffuses laterally, and some of the ions are received on the plates C_1 and C_3. When no magnetic field is acting, the centre of the stream falls on the centre of C_2 and the charges received by the electrodes C_1 and C_3 are equal. In very dry gases at low pressures the rate of diffusion is abnormally great, so that the proportion of the total charge received by C_2 diminishes as the pressure is reduced, when the electric force is constant. Thus the charges n_1, n_2, and n_3 received by the electrodes were in the proportion of 1 : 4 : 1 when the air inside the apparatus was at a pressure of 10 millimetres, and in the proportion 1 : 1·8 : 1 when the pressure was 2·8 millimetres.

When the magnetic force H is applied in a direction parallel to the slit in the metal plate B, the stream of ions is deflected so that the centre is displaced from the centre of the electrode C_2, and for a certain value of the magnetic force the centre of the stream is deflected to the centre of the slit between the electrodes C_1 and C_2. In that case the charge n_1 received by the electrode C_1 is equal to the charge $n_2 + n_3$ received by the two electrodes C_2 and C_3.

Experiments were made at different pressures with a constant electric force of one volt per centimetre in the field between the plate B and the electrodes C. The electrodes C_2 and C_3 were joined and the charge, $n_2 + n_3$, which they received was compared with the charge n_1 received by the electrode C_1 when various magnetic forces were acting. The results are represented by the curves, figure 24.* It will be seen that at the higher pressures the ratio $(n_2 + n_3)/n_1$ at first diminishes as the force H increases, and attains a minimum value. At the lower pressures the deflection increases continuously with the force H, and the ratio $(n_2 + n_3)/n_1$ diminishes until the charge received by the plates C_2 and C_3 becomes very small.

When the charge n_1 is equal to $n_2 + n_3$ the centre of the stream is deflected through a distance $a/2$, where a is the

* See paper by the author and H. T. Tizard, Proc. Roy. Soc. A, 87, p. 357, 1912.

width of the electrode C_2, and the velocity of the ions is given by the equation

$$\frac{H_1 U}{X} = \frac{a}{2l},$$

l being the distance from B to C, which is large compared with a, and H_1 the force for which the charge $n_2 + n_3$ becomes equal to n_1.

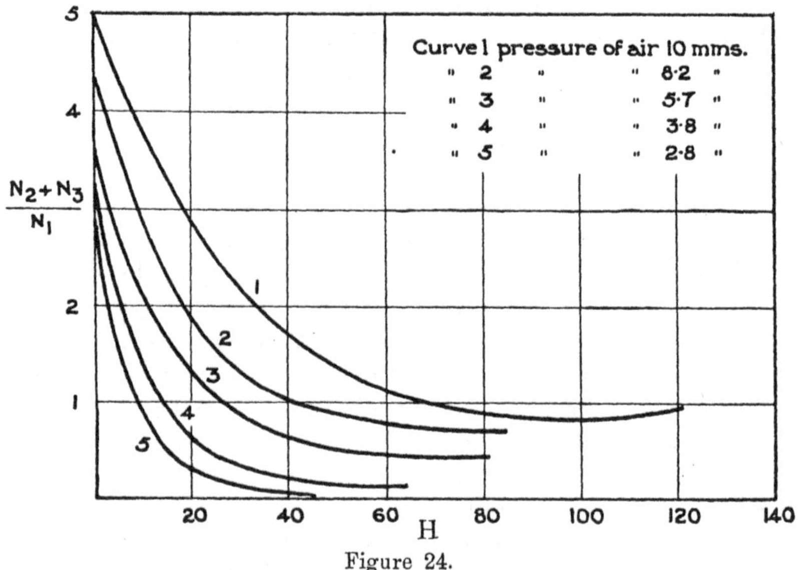

Figure 24.

In a series of determinations * of velocities that were made on this principle, with dry air at pressures between ·25 millimetre and 18·5 millimetres, and electric forces from 2 volts per centimetre to 50 volts per centimetre, it was found that the velocity was a function of the ratio X/p only.

The velocities obtained are as follows:

X/p	0·2	0·5	1	2	5	10	20	50	100	150	200
$U \times 10^{-6}$	·5	·9	1·25	1·75	3·0	5·2	9·0	17·3	27	35	44

The table shows that when the pressure p is constant, the velocity does not increase in proportion to the force X.

* J. S. Townsend and H. T. Tizard, Proc. Roy. Soc. **88**, p. 336, 1913.

106. Effect of temperature on the velocity of ions in air, at atmospheric pressure. The velocities of ions in air at high and low temperatures have been determined by Phillips,* who used Langevin's method. The results of the experiments are given in the following table:

Absolute temperature.	k_1	k_2
411	2·00	2·495
399	1·95	2·4
383	1·85	2·3
373	1·81	2·21
348	1·67	2·125
333	1·60	2·00
285	1·39	1·785
209	·945	1·23
94	·235	·235

The exact pressure of the air was not recorded, but presumably it was constant and approximately 760 mm.

It is remarkable that the velocity is practically directly proportional to the temperature, except at the lowest temperature.

The expression for the velocity $U = XeT/m$ shows that U is proportional to the interval between collisions T. When the temperature is raised from 209 to 411 degrees the free paths are increased in the ratio 2 : 1, p being constant, and since the velocity of agitation of the molecules and the ions are both increased in the proportion $\sqrt{2} : 1$, the value of T is increased by the factor 1·41 : 1 if the mass m is constant. The results show that the velocities are increased in the ratio 2 : 1 in this case, which indicates a diminution in the value of m as the temperature increases.

107. Velocity of ions in point discharges. The mobility of ions in a point discharge through air has been investigated on a different principle by Chattock.† The method consists in finding the change of pressure at different points in the gas, arising from the motion of the charge that is

* P. Phillips, Proc. Roy. Soc. **78**, p. 167, 1907.

† A. P. Chattock, Phil. Mag. (5) **48**, p. 401, 1899.

repelled by the electric force from the point. The simplest form of apparatus for measuring the variation of pressure is illustrated in figure 25. The gas is contained in a glass tube E, and the current flows from the point A to the ring B. The wire that terminates in the point A is surrounded by a capillary tube coaxial with the tube E, and the ring is supported by a glass tube G, electrical connection being made with the ring by means of a wire passing through the tube. The distance between the point and the ring may be adjusted to any required length by means of a micrometer screw which moves the tube G in a direction parallel to the axis of the outer tube.

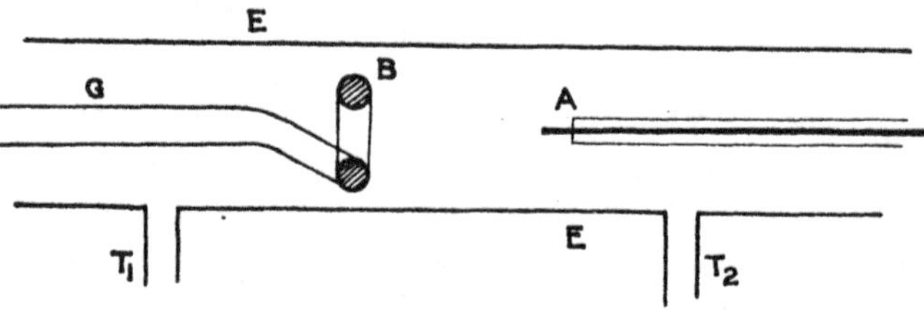

Figure 25.

The difference of pressure produced by the current is measured by a manometer connected to the side tubes T_1 and T_2.

Let C be the current flowing from the point, S the sectional area of the tube E, ρ the density of electrification at any point, and ω the velocity of the ions parallel to the axis of the tube, then

$$C = \iint \rho\omega \, dS,$$

the integration being taken over the whole area S. Assuming that the velocity of the ions is proportional to the electric force Z, then

$$\frac{C}{k} = \iint \rho Z dS,$$

k being the velocity corresponding to unit force.

If z and $z+dz$ be the distances of two sections of the

tube from the plane of the ring, the force acting on the charge between the two sections is

$$dz \iint \rho Z dS = \frac{C}{k} dz,$$

and since C is constant along the tube, the force acting on the gas between two planes at a distance apart a, due to the presence of the charge, is Ca/k.

As this formula only applies to that part of the field containing ions of one sign moving with a velocity proportional to the electric force, it cannot be applied to the gas in the neighbourhood of the point. For just near the point there are both positive and negative ions in the gas, and when the point is negatively charged the negative ions that are repelled move in the strong field of force that extends to some distance from the point with velocities that exceed the value kZ.

When the ring and the point are sufficiently far apart, these effects take place within a fixed distance b from the point. The difference of pressure P between the gas in the tubes T_1 and T_2 may therefore be divided into two parts, one being a constant P' arising from the current in the immediate neighbourhood of the point, and the other proportional to the distance a, $a+b$ being the distance from the ring to the point. Hence

$$P = P' + \frac{Ca}{kS}.$$

When the current is constant, and the distance between the ring and the point is increased from $a_1 + b$ to $a_2 + b$, the difference of pressure in the tubes T_1 and T_2 increases from P_1 to P_2, where

$$P_2 - P_1 = \frac{C(a_2 - a_1)}{kS}.$$

This equation was verified by the experiments, which showed that for a given current the difference of pressure, as indicated by the manometer, increased in proportion to the increase of distance between the point and the ring.

The velocities corresponding to a volt per centimetre appear to depend on the arrangement of the apparatus, such as the relative size of the disc and the tube. The values deduced from the earlier experiments are 1·14 and 1·41 centimetres per second for the positive and negative ions respectively. Subsequently, with another apparatus which was considered to be more satisfactory, the numbers 1·38 and 1·80 were found for the positive and negative ions.

Experiments were also made to measure the pressure at different points of the surface of a plate when the current from a point discharge is received by the plate. The results are interesting, as they show how the pressure is distributed over the plate when the point is situated at various distances from it. It is also possible to deduce the values of the velocities from measurements of the total pressures of the air on the plate, but the method was not considered to be as satisfactory as that subsequently used, in which the current took place between a point and a ring inside a glass tube.

108. Comparison of results obtained under various conditions. It is to be remarked in connection with these determinations that the electric force acting on the ions is much larger than that used in the more direct methods of determining the mobilities. Neglecting the strong field in the neighbourhood of the point, the force acting on the ions in the more uniform part of the field was of the order of 3,000 volts per centimetre, and the calculations of the velocities which have been made involve the assumption that at atmospheric pressure the velocity of the ions is directly proportional to the force, over a range of forces up to that order of magnitude. The value of X/p (p being measured in millimetres) is therefore as great as 4, which is large compared with ·05, the value X/p at which the negative ions in dry air at low pressures were found to acquire velocities much greater than k_2X/p.

The difference may be due to the fact that, with point discharges, gases such as oxides of nitrogen are formed which would tend to condense on the ions, and if any water vapour were present a further increase of the mass of the ion would take place. This hypothesis is supported by the determinations of the rates of diffusion, as it has been found that positive and negative ions formed by point discharges in air have smaller rates of diffusion than the ions generated by Röntgen rays.

109. Velocity of ions associated with large masses. The ions generated by the methods that have been described are more mobile than those generated by other methods. The ions have the same atomic charges in all cases, so that the small velocity with which they move under certain conditions must be attributed to an increase of the mass associated with the ion. This, as has already been explained, may be brought about by the condensation of a gas or vapour on the ion. In most cases ions of large mass are produced while chemical actions are taking place, and gases are formed which condense more readily than the constituents of the air. It would appear that very small quantities of such impurities are sufficient to produce large effects, since in the case of the conductivity of air in the neighbourhood of an incandescent wire the velocities of the ions are of the order of one hundredth of the velocities of the ions generated by Röntgen rays. The mass associated with the charge is further increased when water vapour is present, and in some cases the particles form a visible cloud even when the gas is not completely saturated.

As might be expected, the mobility of the ions varies considerably in all these cases, and it is difficult to specify the exact conditions under which a particular velocity is acquired. It is not necessary, therefore, to measure the velocity to a high degree of accuracy, and the chief interest is in finding the effects produced by varying the conditions under which the experiments are made.

110. Conductivity of flame gases. A good example illustrating some of the properties of these ions is to be found in the experiments made by McClelland * to determine the velocity of the ions in the heated air above a gas flame. A

* J. A. McClelland, Phil. Mag. (6) **46**, p. 29, 1898.

form of apparatus which was used for some of these experiments is illustrated in figure 26. The air passes up a tube T provided with cylindrical electrodes A, B, and C at different distances from the flame. The electrodes are coaxial with the tube and are supported by rods projecting through ebonite plugs fixed in the tube. The upper end of the tube was connected to a water-pump and an arrangement was made for measuring the volume drawn through the apparatus per second. The mean temperature of the stream as it passed each electrode was obtained by removing the electrodes and inserting thermometers through the apertures that were made in the tube to hold the ebonite plugs for insulating the electrodes. It was thus possible to obtain the temperature of the air and to calculate the velocity of the stream at various points of the tube.

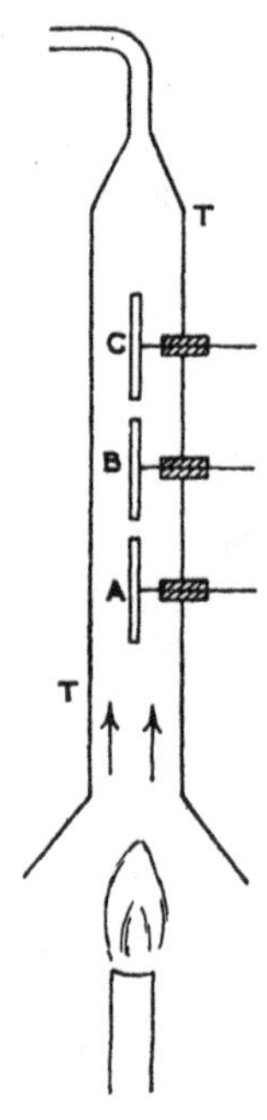

Figure 26.

The conductivity of the gas at different distances from the flame was found by connecting one of the electrodes to a pair of quadrants of an electrometer, both pairs being raised to a potential V by connecting them to a battery of accumulators. The tube T and the other two electrodes were connected to earth. A current flows between the tube and the electrode at high potential when a Bunsen flame is placed in the inverted funnel at the lower end of the tube and a stream of air is drawn up the tube. The conductivity is measured by the electrometer deflection obtained in a given time on insulating the quadrants connected to the electrode.

For small forces, the current increases rapidly with the force, but with large forces a point is reached for a certain value of the potential V' when the increases of the current corresponding to further increases of the potential are comparatively small.

It is easy to explain the effects thus obtained by considering the processes that contribute to the removal of the ions from the gas. The ions carried by the stream into the field of force disappear from the air, partly by recombination and partly by coming into contact with the electrode. The rest pass by the electrode and may be detected in the gas higher up the tube. As the potential difference between the tube and an electrode, B for example, is increased the conductivity in the upper parts of the tube diminishes, and when the potential V' is reached the gas above the electrode does not conduct. This may be shown by connecting the electrode C to the quadrants of the electrometer, and joining the electrode B to the battery of accumulators, so that the potential difference V is maintained between B and the outer tube while the conductivity at C is being measured. Under the circumstances it was found that the current between the tube and the electrode C attains a maximum value and then diminishes as the potential V is increased, and becomes zero when the value V' is attained.

For potentials greater than V' the current between the tube and the electrode B continues to increase with the potential, although the rate of increase is small. This arises from the fact that a smaller number of ions recombine as the force increases, and it is practically impossible in this case to obtain a saturation current, since any increase in the potential of one electrode must necessarily bring additional ions to that electrode from points lower down in the tube.

The velocity of the ions under unit force may be determined from the potential V'. Let r_1 be the radius of the tube T, r_2 the radius and a the length of the electrode. All the ions will be removed from the gas when they traverse the distance $r_1 - r_2$ under the electric force while they are carried a distance a by the stream. If k be the velocity of the ions under unit electric force, then, as has been shown in section 90,

$$k = \frac{Q \, . \log (r_1/r_2)}{2\,\pi a V'},$$

Q being the volume of gas passing the electrode per second.

111. Velocity of ions in gases rising from a flame. The following table of the values of the velocity k of the ions under a volt per centimetre is given by McClelland. They are deduced from measurements of the currents to the three electrodes at different distances from the flame. The temperature of the gas is given in the second column of figures. The recombination produces a large effect, as is shown by the diminution in the conductivity as the gas rises in the tube. This may be seen by the numbers given in the third column, which are proportional to the currents between the tube and the electrodes corresponding to the potentials V', and represent approximately the numbers of ions in the gas when it reaches the electrodes.

	Velocity of ions under a force of one volt per centimetre.	Temperature of air.	Conductivity.	Distance from the flame.
At Electrode A	·23 cm. per sec.	230° C.	26	5·5 cm.
" B	·21 " "	160° C.	3	10·0 "
" C	·04 " "	105° C.	1	14·5 "

There is thus a large fall in the value of the velocity of the ions when the temperature changes from 160° C. to 105° C., indicating an increase in the mass associated with the charge.

The gas used in the above experiments was the ordinary gas supply containing a large proportion of hydrogen, so that a corresponding amount of water vapour is formed when it is burnt in the air. The rapid growth of the size of the ions is not, however, due altogether to the water vapour that is included among the products of combustion. Similar experiments were made with a flame of carbon monoxide burning in air, and it was also found in this case that the ions increased in size as the air above the flame cooled, the only water vapour that was present being that contained in the air of the room.

112. Ions in the air surrounding a glowing wire. Similar experiments were made * to determine the velocities of ions that give rise to the conductivity of the air in the neighbourhood of a glowing wire, and of an arc discharge.

In these cases large volumes of air were used, and the velocities were determined at a distance from the source of ionization, so that the temperature of the air as it passed the electrodes was practically the same as that of the room.

The velocities were found to vary over large ranges, depending on the temperature of the glowing wire and on the current passing through the arc.

In the case of a platinum wire, the velocity of the ions under a volt per centimetre fell from ·01 centimetre per second to ·003 centimetre per second when the temperature of the wire was increased.

A similar effect was obtained with the arc. The velocity of the ions was found to be ·33 centimetres per second when a small current was passing through the arc, and on increasing the current the velocity fell to ·015 centimetre per second.

In all these experiments, including those with flames, the velocity of the negative ions was greater by about 20 per cent. than that of the positive ions.

113. Ions produced by chemical action in gases. In some cases when slowly moving ions are produced by chemical actions there is a large excess of ions of one sign in the gas, and it is possible under these conditions to estimate the mobility by a very simple method. When there is no applied electric force arising from charges on electrodes the gas loses its charge to the sides of the vessel in which it is contained in two ways. Owing to the charge in the gas there is an electric force that repels the ions towards the boundary of the vessel in which they are contained, and in addition the ions come into contact with the sides of the vessel by the ordinary process of diffusion. The rate at which ions are discharged, on account of their mutual

* J. A. McClelland, Proc. Camb. Phil. Soc. **10**, p. 241, Dec. 1899.

repulsion towards the sides, depends only on the charge per cubic centimetre, while the rate at which the ions are discharged by the process of diffusion depends on the dimensions* of the vessel. It is possible, therefore, to arrange an experiment in which the former process largely predominates, and to deduce a value of the mobility of the ions from observations of the rate at which a gas loses its charge when it is contained in a closed vessel.

Let u, v, w be the velocities of the ions along the axes of co-ordinates, and ϕ the potential of the electric force. Neglecting the diffusion, the velocities are given by the equations

$$u = -k\frac{d\phi}{dx}, \quad v = -k\frac{d\phi}{dy}, \quad w = -k\frac{d\phi}{dz}.$$

On substituting these values of u, v, and w in the equation of continuity,

$$\frac{1}{\rho}\frac{\delta\rho}{\delta t} + \frac{du}{dx} + \frac{dv}{dy} + \frac{dw}{dz} = 0,$$

the relation between ρ and ϕ is obtained, namely,

$$\frac{1}{\rho}\frac{\delta\rho}{\delta t} = k\nabla^2\phi,$$

ρ being the density of the charge at a point moving with the ions.

Since $\nabla^2\phi = -4\pi\rho$, the equation † for ρ becomes

$$\frac{1}{\rho^2}\frac{\delta\rho}{\delta t} = -4\pi k,$$

which on integration gives

$$\frac{1}{\rho} - \frac{1}{\rho_0} = 4\pi kt,$$

ρ_0 being the initial density, and ρ the density after a time t.

* Phil. Mag. (5) **45**, p. 471, 1898.

† It is to be noticed that if the motion of the ions is due to forces arising from external charges, then $\nabla^2\phi$ is zero, so that $\frac{\delta\rho}{\delta t} = 0$ at every point moving with the ions. Hence if a number of ions are distributed initially throughout a volume V_0 and move through the gas under the action of forces arising from charges, on electrodes of any shape, the motion will be such that at any subsequent time all the ions will be contained in a volume equal to V_0.

114. Ions of large mass in newly-prepared gases. Measurements of the density of electrification of a gas may be made by sending a stream of the gas through a metal cylinder, so as to fill it with a gas having a uniform electrification. The gas is then allowed to stand in the cylinder for a given time t_1, when it is displaced by a current of air. If ρ_1 be the charge per c.c. at the time t_1 and V the volume of the cylinder, the charge $\rho_1 V$ is carried out of the cylinder with the gas. The cylinder being insulated and connected to an electrometer the deflection obtained will be the same as if a charge $-\rho_1 V$ were given to the insulated system. The density of ρ_1 may thus be found. Repeating the experiment and leaving the gas in the cylinder for a longer time t_2, a smaller electrometer deflection proportional to $-\rho_2 V$ is obtained on displacing the charged gas. The mobility k is then given by the formula

$$k = \frac{\rho_1 - \rho_2}{t_2 - t_1} \cdot \frac{1}{4\pi\rho_1\rho_2}.$$

The mobilities of the ions in oxygen or hydrogen given off by the electrolysis of sulphuric acid may be obtained by this method, as the gases are given off with a constant density of electrification.*

In these cases the size of the particle associated with the charge depends to a great extent on the amount of water vapour that is present. When the oxygen was partially dried by bubbling it once through sulphuric acid the velocity of the ions (under a force of one volt per centimetre), as determined by the above method, was 2×10^{-4} centimetre per second. The mobility could doubtless be increased by more complete drying.

115. Ions produced during the oxidation of phosphorus. There are several other cases in which slowly moving ions are found in gases: in fact it is generally possible to detect some charged particles in newly prepared gases or in gases in which chemical actions are taking place. The very

* Phil. Mag. (5) **45**, p. 135, 1898.

high conductivity obtained by the oxidation of phosphorus in air is an interesting example of these phenomena.

The properties of the ions obtained by this method have been fully investigated by Bloch. In this case positive and negative ions appear in equal numbers in the air, and their mobility was determined by a method similar to that used by McClelland. Under a force of one volt per centimetre the velocity of the ions in dry air was found to be between ·001 and ·0003 centimetre per second.* As in the case of newly prepared gases, a cloud is formed when the air is bubbled through water, and when dried the cloud almost entirely disappears, but the small particles which are left give rise to a hazy appearance when the gas is in a strong light.

* These investigations and others of a similar kind are described in the Memoir by F. Bloch, Ann. de Chim. et de Phys. **4**, pp. 25–144, 1905.

CHAPTER V

DIFFUSION OF IONS

116. Diffusion towards the sides of a vessel. The ions, being small particles moving with high velocities along their free paths, tend, like the molecules of a gas, to distribute themselves uniformly throughout any space in which they are generated. When they come into contact with the sides of the vessel in which they are contained they lose their charge, and no charged particles return from the surface into the gas. The density ρ of ionization at the surface rapidly diminishes, and the mean velocity towards the surface becomes of an order equal to the velocity of agitation. The rate at which ions are removed from the gas is then ρV per unit area of the surface, where V is a large velocity of the order 10^4 centimetres per second. The value of ρ at the surface must therefore be very small, otherwise the ions would disappear almost instantaneously from a gas contained in a vessel of ordinary dimensions. The number of ions, and the partial pressure of the ions, at the surface may therefore be taken as zero.

The principle on which the coefficient of diffusion of ions into a gas has been determined may be illustrated by a simple example of a similar process, which depends on the rate of interdiffusion of gases. The ordinary method of removing impurities from gases—for instance, the method of drying gases by bubbling them through sulphuric acid—is exactly analogous.

If the water vapour did not diffuse through the gas, only a very small proportion of it would be removed, as very few of the molecules would come into contact with the acid. In practice it is found that a large percentage of the moisture is removed, particularly in the case of hydrogen, through which the moisture diffuses more rapidly than through other gases. This is due to the motion of agitation of the molecules of the vapour in virtue of which a large number come into contact with the boundary, the effect being increased when the free paths of the molecules are long, as happens with hydrogen or with gases at reduced pressure.*

A conducting gas would be affected in a similar manner: the ions would diffuse and become discharged at the surface.

117. Conductivity of a gas after passing through small tubes. In order to determine the coefficient of diffusion of ions it is necessary to arrange the experiments so that the number that are lost by recombination is small compared with the number that come into contact with the boundary. The method that has been adopted is to pass a conducting gas through fine metal tubes with a constant velocity, and to find experimentally the relative conductivities of the gas after passing through tubes of different lengths. The coefficient of diffusion may be deduced from the observed diminution in conductivity when the velocity of the gas and the dimensions of the tubes are known.†

Let a be the radius of the tube, W the velocity of the gas at any point at a distance r from the axis of the tube. The velocity W is given in terms of r by the equation $W = \frac{2V}{a^2}(a^2 - r^2)$, V being the mean velocity of the gas as measured by the volume Q that passes through in a time t, namely $Q = \pi a^2 Vt$.

The differential equations giving the velocities u, v, w, of the ions along the axes in terms of their partial pressure p, and coefficient of diffusion K, become in this case

$$\frac{1}{K}(pu) = -\frac{dp}{dx}; \quad \frac{1}{K}(pv) = -\frac{dp}{dy}; \quad \frac{1}{K}p(w - W) = -\frac{dp}{dz},$$

the axis of z being taken along the axis of the tube.

* Phil. Mag. (5) **45**, p. 471, 1898.

† Phil. Trans. A, **193**, p. 129, 1899.

The gas may be supposed to be uniformly ionized before entering the tubing, and when the steady state is reached the equation of continuity becomes

$$\frac{d}{dx}(pu) + \frac{d}{dy}(pv) + \frac{d}{dz}(pw) = 0.$$

In all the experiments the velocity W of the gas along the tube was large compared with the quantity $\frac{K}{p} \cdot \frac{dp}{dz}$, so this term may be neglected and the velocity w of the ions along the axis of the tube may be taken as being equal to W.

Substituting the values of the velocities in the equation of continuity, the differential equation for the partial pressure p of the ions at any point is obtained,

$$K\left(\frac{d^2p}{dx^2} + \frac{d^2p}{dy^2}\right) - W\frac{dp}{dz} = 0.$$

This equation may be reduced to one in two independent variables r and z by substituting for the first two terms their equivalent in terms of the cylindrical co-ordinate r. The equation for p thus becomes

$$\frac{K}{r}\frac{d}{dr}\left(r\frac{dp}{dr}\right) - \frac{2V}{a^2}(a^2 - r^2)\frac{dp}{dz} = 0 \quad . \quad . \quad . \quad (1)$$

It is necessary to find a solution of this equation satisfying the following conditions:

1. On entering the tube the gas is uniformly ionized. Thus if the origin of co-ordinates is taken on the axis of the tube at the end at which the gas enters, the partial pressure $p = p_0$, a constant, for all values of r when $z = 0$.

2. Since the ions are discharged when they come into contact with the surface of the tube, p is zero for all values of z when $r = a$.

The quantity p may be expressed as a series of terms of the form $\phi\epsilon^{-\frac{\theta a^2 K}{2V}z}$ where ϕ is a function of r, and θ a constant, the values of which depend on the surface conditions.

Substituting this expression for p in equation (1), the equation for ϕ becomes

$$\frac{d^2\phi}{dr^2} + \frac{1}{r}\frac{d\phi}{dr} - \theta\,(r^2 - a^2)\,\phi = 0 \quad . \quad . \quad . \quad . \quad (2)$$

The solution of this equation is $\phi = AM + BN$, M and N being the two independent solutions one of which, N, becomes infinite when $r = 0$. The multiplier B must therefore be zero, since p is finite along the axis of the tube.

The solution M is a series of powers of r,

$$M = 1 + B_1 r^2 + B_2 r^4 + \ldots,$$

the coefficients B_1, B_2, ... being determined so as to satisfy equation (2).

The first few terms in the expansion are

$$M = 1 - \frac{\theta a^2 r^2}{4} + \tfrac{1}{16}\left(\theta + \frac{\theta^2 a^4}{4}\right) r^4 + \ldots \quad . \quad . \quad . \quad (3)$$

The value of p thus becomes

$$p = c_1 M_1 \epsilon^{-\frac{\theta_1 a^2 K}{2V} z} + c_2 M_2 \epsilon^{-\frac{\theta_2 a^2 K}{2V} z} + \ldots \quad . \quad . \quad . \quad . \quad (4)$$

The boundary condition $p = 0$ when $r = a$ requires that such values of θ be chosen as will reduce the series M, equation (3), to zero when $r = a$.

Hence θ_1, θ_2, ... are the roots of the equation

$$1 - \frac{\theta a^4}{4} + \tfrac{1}{16}\left(\theta a^4 + \frac{\theta^2 a^8}{4}\right) + \ldots = 0. \quad . \quad . \quad . \quad (5)$$

It is easy to see that the roots of this equation are all positive, so that the terms in p involving the larger roots become so small in comparison with the others that they may be neglected.

The two smallest roots are

$$\theta_1 = 7{\cdot}313/a^4\,; \text{ and } \theta_2 = 44{\cdot}5/a^4.$$

The coefficients c_1, c_2, ... in equation (4) remain to be determined. These may be found by the same principles that are used to find the coefficients in a series of Bessel's functions which has a known value for a certain range of the variable.

Since M satisfies equation (2) it is easy to obtain the following relations by Green's theorem,

$$\iiint (U\nabla^2 V - V\nabla^2 U)\, dx\,dy\,dz = \iint \left(U\frac{dV}{dn} - V\frac{dU}{dn}\right) ds,$$

for transformation from volume integrals to surface integrals, U and V being any functions of the co-ordinates.

$$\int_0^a M_n \cdot M_{n'}(a^2 - r^2)\, r\,dr = 0, \quad . \quad . \quad . \quad . \quad . \quad . \quad . \quad . \quad (6)$$

$$\int_0^a M_n{}^2 (a^2 - r^2)\, r\,dr = a\left[\frac{dM_n}{d\theta_n} \cdot \frac{dM_n}{dr}\right]^{r=a}, \quad . \quad . \quad (7)$$

$$\int_0^a M_n (a^2 - r^2)\, r\,dr = -\frac{a}{\theta_n}\left[\frac{dM_n}{dr}\right]^{r=a} . \quad . \quad . \quad . \quad . \quad (8)$$

Thus equation (6) is obtained by substituting M_n and $M_{n'}$ for U and V in Green's equation, and integrating throughout the volume of a cylinder of radius a. The surface integral vanishes, since M_n and $M_{n'}$ are zero when $r = a$,

and $$\nabla^2 M_n = \frac{d^2 M_n}{dr^2} + \frac{1}{r}\frac{dM_n}{dr} = \theta_n (r^2 - a^2) M_n.$$

Hence $M_n \nabla^2 M_{n'} - M_{n'} \nabla^2 M_n = (\theta_{n'} - \theta_n) M_n M_{n'} (r^2 - a^2)$, and the volume integral reduces to

$$2\pi(\theta_{n'} - \theta_n)\int_0^a M_n M_{n'} (r^2 - a^2)\, r\,dr,$$

so that the integral must vanish, since $\theta_{n'}$ and θ_n are not the same.

Equation (7) is obtained by substituting in Green's equation

$$U = M_n \text{ and } V = M_n + \frac{dM_n}{d\theta}\, d\theta,$$

and equation (8) by substituting

$$U = 1 \text{ and } V = M_n.$$

The condition $p = p_0$ when $z = 0$ is satisfied if the coefficients c are determined to satisfy the equation

$$p_0 = c_1 M_1 + c_2 M_2 + \ldots,$$

for values of r between $r = 0$ and $r = a$.

Multiplying this equation by $M_n (a^2 - r^2)\, r\,dr$, and integrating from $r = 0$ to $r = a$, the integrals obtained are given by the equations (6), (7), and (8), and the expression for c_n becomes

$$-\frac{p_0 a}{\theta_n}\left[\frac{dM_n}{dr}\right]^{r=a} = c_n a\left[\frac{dM_n}{d\theta}\cdot\frac{dM_n}{dr}\right]^{r=a}.$$

Hence
$$c_n = -\frac{p_0}{\theta_n}\cdot\frac{1}{\left[\dfrac{dM_n}{d\theta_n}\right]^{r=a}}.$$

Thus equation (4) becomes

$$p = -\frac{p_0}{\theta_1\left[\dfrac{dM_1}{d\theta_1}\right]^{r=a}}\cdot M_1 \epsilon^{-\frac{\theta_1 a^2 K}{2V}z} - \frac{p_0}{\theta_2\left[\dfrac{dM_2}{d\theta_2}\right]^{r=a}}\cdot M_2 \epsilon^{-\frac{\theta_2 a^2 K}{2V}z} + \ldots \quad . \quad (9)$$

The ratio R of the number of ions that pass a section at a distance z from the origin to the number that enter the tube is

$$R = \frac{\int_0^a 2\pi r p\cdot\frac{2V}{a^2}(a^2 - r^2)\,dr}{p_0 \pi a^2 V}.$$

When the value of p, equation (9), is substituted, the integration of the numerator of this fraction may be obtained from equation (8), and the expression for R becomes

$$R = \frac{4}{a^3}\left\{\frac{1}{\theta_1^2}\left[\frac{dM_1/dr}{dM_1/d\theta_1}\right]^{r=a}\epsilon^{-\frac{\theta_1 a^2 Kz}{2V}} + \frac{1}{\theta_2^2}\left[\frac{dM_2/dr}{dM_2/d\theta_2}\right]^{r=a}\epsilon^{-\frac{\theta_2 a^2 Kz}{2V}} + \ldots\right\}.$$

The coefficients of the exponential terms corresponding to the two smallest roots of equation (5),

$$\theta_1 = \frac{7{\cdot}313}{a^4}, \text{ and } \theta_2 = \frac{44{\cdot}5}{a^4},$$

may be obtained from the following numbers that have been found for the values of dM/dr and $dM/d\theta$:

$$\frac{1}{\theta_1 a^3}\left[\frac{dM_1}{dr}\right]^{r=a} = \cdot 1321, \quad \frac{1}{\theta_2 a^3}\left[\frac{dM_2}{dr}\right]^{r=a} = \cdot 0302,$$

$$\frac{1}{a^4}\left[\frac{dM_1}{d\theta_1}\right]^{r=a} = \cdot 0926, \quad \frac{1}{a^4}\left[\frac{dM_2}{d\theta_2}\right]^{r=a} = \cdot 0279.$$

Hence $R = 4\left(\cdot 1952\,\epsilon^{-\frac{7\cdot 313\,Kz}{2a^2V}} + \cdot 0243\,\epsilon^{-\frac{44\cdot 5\,Kz}{2a^2V}} + \ldots\right)$.

In practice it is impossible to determine the number of

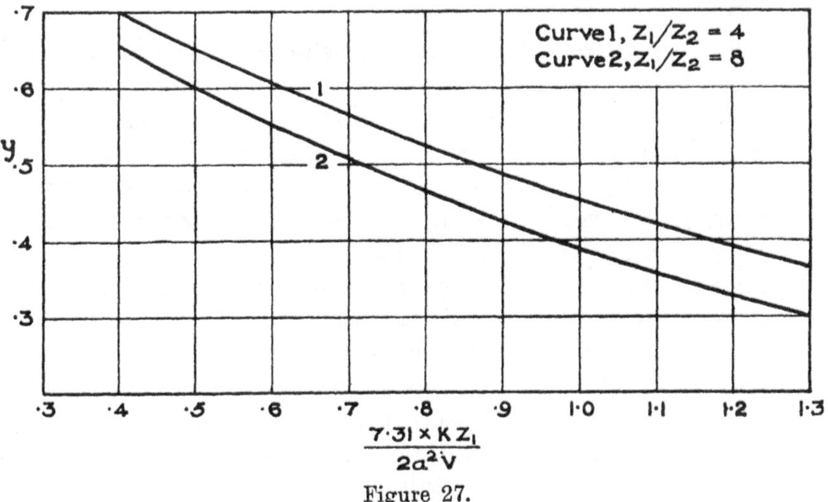

Figure 27.

ions that enter the tube, so that the ratio of the numbers of ions n_1 and n_2 emerging with a stream from tubes of different lengths z_1 and z_2 is determined experimentally, the same number of ions entering the tubes in each case.

The ratio n_1/n_2 is therefore

$$\frac{n_1}{n_2} = \frac{\cdot 195\,\epsilon^{-\frac{7\cdot 313\,Kz_1}{2a^2V}} + \cdot 0243\,\epsilon^{-\frac{44\cdot 5\,Kz_1}{2a^2V}}}{\cdot 195\,\epsilon^{-\frac{7\cdot 313\,Kz_2}{2a^2V}} + \cdot 0243\,\epsilon^{-\frac{44\cdot 5\,Kz_2}{2a^2V}}}.$$

In order to deduce K from the experimental results it is necessary to draw a curve whose equation is

$$y = \frac{\cdot 195\,\epsilon^{-x} + \cdot 0243\,\epsilon^{-\frac{44\cdot 5}{7\cdot 31}\cdot x}}{\cdot 195\,\epsilon^{-\frac{xz_2}{z_1}} + \cdot 0243\,\epsilon^{-\frac{44\cdot 5}{7\cdot 31}\cdot\frac{xz_2}{z_1}}},$$

so that when $y\,[= n_1/n_2]$ is given, the value of x which is $7\cdot 313\,Kz_1/2a^2V$ may be found.

The curves, figure 27, give the values of y in terms of the quantity $\frac{7\cdot 313\,Kz_1}{2\,a^2V}$, curve 1 being for the case in which $z_1/z_2 = 4$ and curve 2 for $z_1/z_2 = 8$.

118. Comparison of effects of diffusion and recombination on the conductivity of a stream. When a suitable ratio of the lengths z_1 and z_2 is adopted [such as 4 : 1 or 8 : 1], the ratio of the conductivities n_1/n_2 depends on the values of Kz_1/a^2V, and for accurate determinations it is necessary to adjust the velocity V so that n_1/n_2 is about ·5. The average time z_1/V that the gas is in the longer tube is then proportional to a^2. The reduction in the conductivity arising from recombination is proportional to the time z_1/V, so that it is advantageous to have tubing of small bore, and thus reduce as far as possible the correction for recombination. Also when the gas contains ions of one sign the self-repulsion of the ions gives rise to a motion towards the surface of the tubing which tends to diminish n_1. The correction for this effect, like the correction for recombination when ions of both signs are present, rises in importance in proportion to the product of the density of ionization and the time z_1/V.

The upper limit of the intensity of ionization that may be used without introducing a serious error arising from the process of recombination or self-repulsion may be easily found.

Let ρ be the mean value of the charges on the positive and negative ions per cubic centimetre in the longer tubes. The loss $\delta\rho$ due to recombination is given approximately by the formula

$$\frac{\delta\rho}{\rho} = \alpha\rho t,$$

α, the coefficient of recombination, being about 3300 for gases at atmospheric pressure. It is desirable, for accurate measurements, that the quantity $\alpha\rho t$ should not exceed ·01.

When the gas contains ions of one sign the diminution in density due to self-repulsion is given by the formula

$$\frac{\delta\rho}{\rho} = 4\,\pi k\rho t,$$

where k, the velocity of the ions under a force of an electrostatic unit, is about 540 centimetres per second for negative ions in dry air and 440 for positive ions.

It is interesting to observe that sometimes both effects take place. The negative ions diffuse much faster in dry gases than the positive ions, so that when equal numbers of positive and negative ions enter the tubes, an excess of positive ions emerges with the stream. The gas has a positive charge while it is in the longer tubes, and this tends to diminish the number of negative ions, and to increase the number of positive ions that are lost to the sides.

Thus after making the correction for recombination the values of K_+ and K_- for the positive and negative ions are still subject to a small additional error, K_- being too small and K_+ being too large.

It is therefore necessary to use tubing of very small bore, and density of ionization of a small order of magnitude, so as to reduce the errors that arise from these causes.

The quantity of electricity carried through a single tube would under these conditions be inconveniently small, and in order to avoid the necessity of making electrometer observations of several minutes duration, a number of small tubes were used in parallel.

In the first experiments twelve tubes, 3 millimetres in diameter, were used in parallel, and it was necessary to apply corrections to the quantities n_1 to compensate for the loss of conductivity arising from recombination. These corrections were of the order of one or two per cent. for the smaller conductivities, and rose to nine per cent. for the larger conductivities.

In the later experiments twenty-four tubes of one millimetre in diameter were used in parallel, and in this case it was not considered necessary to make corrections. The shorter tubes were ·5 centimetre in length, and the longer tubes 4 centimetres.

119. Description of apparatus used to determine the rate of diffusion of ions. The arrangement of the tubes in these

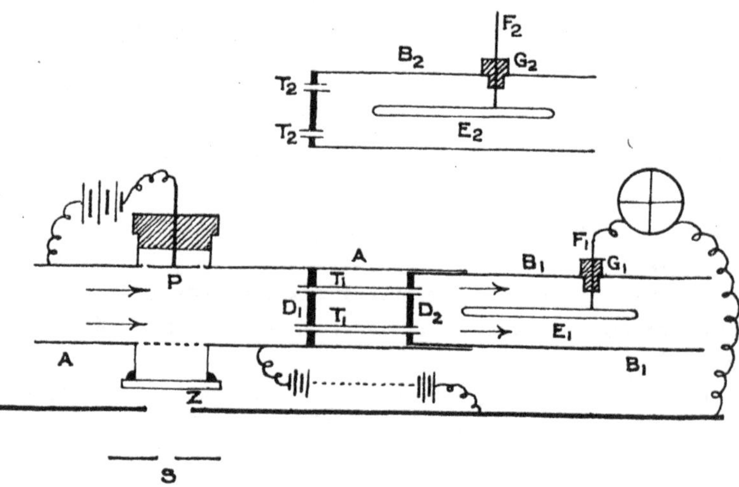

Figure 28.

experiments is shown in figure 28. The tubes T_1 in which diffusion takes place were passed through holes drilled in two brass discs D_1 and D_2 which fitted exactly into the large tube A. The holes in the discs lay on a circle whose centre coincided with that of the disc, so that the tubes T_1 were symmetrically placed in the large tube and equal streams of gas passed through each. Twenty-four tubes 4 centimetres long and 1 millimetre in internal diameter were thus arranged in parallel. The disc D_2 was soldered to the end of a tube B_1 provided with an electrode E_1 with which contact was made through the rod F_1. The latter was insulated and held in position by the ebonite plug G_1. The shorter tubes T_2, ·5 centimetre in length, were similarly arranged and were held in position by a single disc soldered to the end of the tube B_2, which was exactly similar to B_1. The tubes T_1 with the electrode E_1 could thus be easily removed from the tube A and the shorter tubes put in their place.

The stream of gas from a gasometer that passes down the large tube A may be ionized by any of the well-known methods before it reaches the small tubes. The figure shows the apparatus as arranged for generating ions by ultra-violet light. The light originates from the spark-gap S in the circuit of a Leyden-jar discharge, and enters the tube A through the quartz plate Z. The ions are produced at the surface of the metal plate P, which was made a little smaller than the opening in the tube A in order to insulate it from the tube. The positive terminal of a small battery was connected to the tube A and the negative terminal to the rod that carried the metal plate P, so that when negative ions are generated at the surface of the metal they are repelled into the stream of gas passing along A and are thus carried into the small tubes.

The same apparatus may be used to measure the rates of diffusion of the positive and negative ions generated by Röntgen rays or Becquerel rays coming from a source placed at a convenient distance from the opening in the tube A. In these cases the quartz window is removed, a thin aluminium disc put in its place, and the upper plate P is disconnected from the battery and connected by a wire to the tube A.

The ions that come through the tubes T_1 are collected on the electrode E_1 when the tube A is raised to a high potential of about 80 volts by means of a battery of accumulators, and after the charge n_1 acquired in a given time is determined, the tubes T_1 are removed and the shorter tubes T_2 are put in their place. The same density of ionization is again produced in the stream of gas and the charge n_2 coming through the shorter tubes is determined.

120. Values of the coefficients of diffusion of ions. The results of the experiments* made with different gases in

* Phil. Trans. A, **193**, p. 129, 1899.

L 2

which the conductivity was produced by Röntgen rays are given in the following tables. It was found that water vapour had a considerable effect on the rate of diffusion, except in the case of carbonic acid, as is seen by comparing the numbers obtained with gases that had been passed through a tube of calcium chloride, with those found for moist gases that had been kept over water in a gasometer from which the water-vapour had not been removed.

	Coefficients of diffusion of ions in dry gases.		Coefficients of diffusion of ions in moist gases.	
	Positive ions.	Negative ions.	Positive ions.	Negative ions.
Air	·028	·043	·032	·035
Oxygen	·025	·0396	·0288	·0358
Carbonic acid	·023	·026	·0245	·0255
Hydrogen	·123	·190	·128	·142

The rates of diffusion of ions in air at various pressures, from 200 millimetres to 772 millimetres, were also determined, and it was found that the coefficient of diffusion was inversely proportional to the pressure, for both positive and negative ions.

This result is in accordance with the theory of diffusion when the size of the ions is not affected by the pressure, and the increase in the rate of diffusion is due to the increase of the lengths of the free paths. It may therefore be concluded that over the above range of pressure the dimensions of the ions remain constant.

The coefficients of diffusion of ions produced in air at atmospheric pressure by various methods * are given in the following table:

Coefficients of diffusion of ions into gases.

	Dry air.		Moist air.	
	Positive ions.	Negative ions.	Positive ions.	Negative ions.
Röntgen rays	·028	·043	·032	·035
Becquerel rays	·032	·043	·036	·041
Ultra-violet light	—	·043	—	·037
Point discharge	·0247	·037	·028	·039
	·0216	·032	·027	·037

* Phil. Trans. A, **195**, p. 259, 1900.

In the experiments with point discharges the ions were repelled into the stream of air from a point in a side tube projecting from an aperture in the large tube A, figure 28.

Two side tubes were used at different distances from the tubes T, so that the ions were in the gas for different lengths of time before measurements were made of their rates of diffusion. The numbers in the first line were found for the ions generated in the side tube nearest to the tubes T and are greater than those in the second line, which correspond to ions that had been in the gas for a longer time. The results show that both the positive and negative ions in this case tend to increase in size for some considerable time after they have been generated, which is probably due to the formation of oxides of nitrogen in small quantities by the point discharge. These oxides condense on the ions more readily than the molecules of air and reduce their rate of diffusion.

With regard to the other three methods of producing ions, it is most probable that the rates of diffusion of the negative ions would be exactly the same in each case if precautions were taken to have the air in the same state, as regards the amount of water vapour present, and free from dust.

A series of experiments has recently been made by Salles * in which tubing of different metals were used, and the same values of the coefficients of diffusion were obtained with tubes of German silver, brass, and steel. It was also found that the rates of diffusion were inversely proportional to the pressure, for pressures between 760 millimetres and 1,300 millimetres.

The following are the values of coefficients K_+ and K_- obtained for different gases at 760 millimetres pressure, the ionization being produced by a radio-active salt.

	$K+$	$K-$
Air	·032	·042
Nitrogen	·0295	·0414
Carbonic acid	·025	·026
Oxygen	·030	·041

* E. Salles, Thèses présentées à la Faculté des Sciences de Paris, 1913.

121. Comparison of the coefficients of interdiffusion of gases with the coefficients of diffusion of ions. An estimate of the sizes of the ions may be formed from the numbers obtained for the coefficients of diffusion. They may be compared with the coefficients of interdiffusion of gases, and it will be seen that the ions diffuse comparatively slowly. The following table gives the coefficients of inter-diffusion of some gases and vapours into air, carbonic acid, and hydrogen.

Coefficients of inter-diffusion of gases.

	Air.	Carbonic acid.	Hydrogen.	Observers.
Oxygen	—	·18	·721	Loschmidt
Carbonic acid	·142	—	·555	"
Ether	·077	·055	·29	Winkelmann
Alcohol	·101	·068	·378	"
Water	·198	·132	·687	"

These numbers are much larger than the rates of diffusion of ions into the same gases. Thus the mean rate of diffusion of the ions is about ·038 in air and ·146 in hydrogen. The experimental results show that if K is the coefficient of interdiffusion of two gases of densities ρ_1 and ρ_2, the product $K\sqrt{\rho_1 \rho_2}$ is approximately constant. It is to be noticed that the rates of diffusion of ions into gases are, to the same order of accuracy, inversely proportional to the densities of the gases. If the mass of a particle be inversely proportional to the square of its rate of diffusion, the mass of a negative ion in air is nine times the mass of a molecule of carbonic acid, and the mass of a positive ion about double that of a negative ion. This method of estimating the sizes of the ions is not accurate, for when a number of particles diffuse into a gas the molecules of which are small compared with the particles, the rate of diffusion depends principally on the linear dimensions and not on the mass of the larger particles.

If the effect of the electric forces between the ions and neutral molecules of the gas be neglected, the relative sizes of the ions and the molecules may be deduced from the coefficients of inter-diffusion and viscosity of gases. Various formulae for these coefficients are given in treatises on the Kinetic Theory of gases. The expressions for the coefficients of inter-diffusion are in many cases unsatisfactory, and some of the formulae give rates of diffusion which depend on the relative quantities of the two interdiffusing gases, and are not in agreement with experimental results. Also, another source of error arises from treating the effect of a collision as if all directions of motion of both particles are equally probable after the collision. When one of the molecules is of much larger mass than the other a large error may be introduced in omitting to consider that collisions which render all directions of motion of the smaller molecules equally probable may produce only a small deviation of the direction of motion of the larger molecules.

Langevin,* in his earlier discussion of this subject, gives an investigation which illustrates the principles on which the calculations of the relative masses of the ions and molecules may be made when the ordinary formulae for the coefficients of viscosity and interdiffusion are used. When the formulae are corrected for the persistence of the motion of particles after collisions these coefficients lead to a formula for the relative diameters of the molecules and the ions. A simple investigation of the dimensions of the ions by this method is given in Section 124.

122. Langevin's investigations of the mass associated with an ion. More recently Langevin † has made an important investigation of the coefficient of interdiffusion of two gases, with a view to finding an explanation of the small values of the coefficients of diffusion and mobilities of the ions. A complete solution of the problem of interdiffusion of two gases for any law of action between the molecules was obtained, and the method was applied to find the coefficient of diffusion when the forces between the molecules are the elastic impulses that arise from direct impact, and also to the more complicated case when the effect of the electric attraction of the ion on the neutral molecules is

* P. Langevin, Ann. de Chim. et de Phys. (7) **28**, p. 289, 1903.

† Id., (8) **5**, pp. 245-288, 1905.

taken into consideration. It is impossible to give an account of the investigations, as the mathematical work is necessarily very complicated and lengthy, but the conclusions to which they lead are very simple.

The electric attraction of the neutral molecules affects the motion of the ions only to a small extent; that is, if the mass of the ion were equal to that of a molecule of the gas, its rate of diffusion would be approximately the same as that of a molecule. In order to account for the small rate of diffusion of the ions it is necessary to suppose that a group of molecules is formed round the ion, which moves as a single charged particle, the molecules being held together by the charged centre. In dry air the ratio of the diameter of a positive ion to the diameter of a molecule is 2·8 : 1, and the diameter of the negative ion to that of a molecule 2 : 1.

In this investigation the forces arising from direct impact and the forces of attraction between the ions and the neutral molecules of the gas are taken into consideration. If the latter effect be omitted and the elastic forces that arise when ions come into contact with molecules be considered alone, then larger values of the diameters of the ions are obtained, which are 3 and 2·4 for positive and negative ions respectively in terms of the diameter of a molecule of air.

In order to show that a group of molecules may be formed round an ion, the attraction between an ion and a molecule was also investigated. The following method* of treating the problem gives only an approximate result, but it illustrates the principle on which the more complete investigation is based.

Let s be the specific inductive capacity of a gas. The moment of the electrostatic displacement of the electricity in the molecules is $\frac{(s-1)X}{4\pi}$ per cubic centimetre, when a force X is acting. The electric moment μ of a molecule polarized in an electric field of intensity X is therefore

* P. Langevin, Ann. de Chim. et de Phys. (7) **28**, p. 317, 1903.

$\frac{(s-1)}{4\pi} \cdot \frac{X}{N}$, N being the number of molecules per cubic centimetre. When the electric force arises from a charge e on the ion at a distance r from the molecule, the force f attracting the molecule is $f = \mu \frac{d}{dr}\left(\frac{e}{r^2}\right) = \frac{s-1}{2\pi N} \cdot \frac{e^2}{r^5}$. The work W done in diminishing the distance between the ion and the molecule from $r = \infty$ to $r = r_1$ is

$$W = \int_{r_1}^{\infty} f dr = \frac{s-1}{8\pi N} \cdot \frac{e^2}{r_1^4}.$$

Comparing this energy with the mean kinetic energy of a molecule, $w = \frac{1}{2} m V^2$, the ratio W/w becomes

$$\frac{W}{w} = \frac{s-1}{4\pi m N V^2} \cdot \frac{e^2}{r_1^4} = \frac{s-1}{12\pi P} \cdot \frac{e^2}{r_1^4},$$

P being the pressure of the gas.

For oxygen at atmospheric pressure, the value of s is 1.0006, and since $P = 10^6$, the ratio of the two quantities W and w is

$$\frac{W}{w} = \frac{10^{-10}}{2\pi} \cdot \frac{e^2}{r_1^4}.$$

If the ion be of the same dimensions as a molecule, the work done in bringing a neutral molecule into contact with the ion is found by taking r_1 as the diameter of a molecule.

Langevin substitutes in this expression for e and r the values that have been determined experimentally, and finds $W = 6w$ approximately.

A higher degree of accuracy is obtained by taking the value of e/r_1^2 deduced from the coefficient of viscosity. Jeans * has investigated the expression for the viscosity of a gas, taking into consideration the persistence of the velocity of the molecules after collision, and obtained an expression for the coefficient of viscosity which is approximately

$$\eta = \frac{\cdot 44 \rho V}{\sqrt{2}\pi N r_1^2},$$

* Jeans, Dynamical Theory of Gases, p. 250.

ρ being the density of the gas and V the mean velocity of agitation of the molecules.

Hence
$$\frac{e}{r_1^2} = \frac{\sqrt{2}\pi\eta Ne}{\cdot 44\rho V}.$$

At normal temperature and pressure $Ne = 1{\cdot}29 \times 10^{10}$, and the values of η, ρ, and V for oxygen are: $\eta = 1{\cdot}87 \times 10^{-4}$, $\rho = 1{\cdot}43 \times 10^{-3}$, and $V = 4{\cdot}2 \times 10^4$. When these numbers are substituted in the above formula the value of e/r_1^2* is found to be 4×10^5, and the corresponding value of the ratio W/w is 2·5.

Since W exceeds the energy of agitation, a group of a small number of molecules may be kept together by the electric force arising from the charged centre, and the group may be sufficiently stable to resist the disintegrating effect of a collision with a molecule having a kinetic energy less than one half of W.

It is thus possible to explain the rates of diffusion K and the velocities U under an electric force over a large range of pressures and forces when the values of K and U are small and inversely proportional to the pressure. The groups that are formed round the ions are not very stable, particularly those formed round the negative ions, as is seen from the results of experiments at high temperatures, or at ordinary temperatures when the pressure is reduced.

At ordinary temperatures the instability of the groups associated with the negative ions is shown by the rapid increase of the velocity of the ions under an electric force when the pressure is reduced. Thus in dry air at 25 millimetres pressure under a force of two volts per centimetre, the negative ions move with a velocity of 1,000 centimetres per second, and the positive ions with a velocity of 89·6 centimetres per second. These phenomena can only be explained on the hypothesis that the average size of the negative ion diminishes as the ratio X/p increases, and

* It is interesting to observe that $\frac{4e}{r_1^2}$ is the electric force at the surface of a charged molecule and is about 5×10^8 volts per centimetre for a molecule of oxygen.

that the mass of the positive ion is the same as at atmospheric pressure. What appears to be inconsistent with the simple theory that accounts for the formation of the groups is the difference in the stability of the groups formed round the positive and negative ions, and the fact that in the case of the negative ions the group should begin to disappear when the value of X/p is so small. The impulsive force tending to disintegrate the group when a collision occurs between an ion and a molecule depends on the velocities of the ions and the molecules. There is no increase in the velocity of agitation of the molecules when the pressure is reduced, and the velocity acquired by an ion under the electric force is small compared with the velocity of agitation of the ions. Thus if the group surrounding the negative ion is made up of nine molecules of air, the velocity of agitation of the group would be about $1 \cdot 5 \times 10^4$ centimetres per second, or one-third the velocity of agitation of a molecule of air. The velocity of this group in air at 25 millimetres pressure under a force of two volts per centimetre would be about 100 centimetres per second.

In this connection it is of interest to consider another property of the motion of negative ions that has been found experimentally, namely, that the velocity of agitation of the negative ions exceeds that of particles of equal mass in thermal equilibrium with the molecules, when the velocity under an electric force exceeds the value $k_2 X/P$, k_2 being the velocity under unit force at atmospheric pressure and P the pressure in atmospheres. Thus the velocity of agitation becomes abnormally great as X/p increases, and this may account for the fact that under similar conditions of force and pressure the tendency for groups to be formed round negative ions diminishes.

The most direct evidence of the formation of groups of molecules round the ions is probably to be found in the large effect that may be produced by small quantities of water vapour on the motion of the negative ions. When the negative ions move in dry air, or hydrogen, at low pressures under an electric force with a high velocity that may exceed the value $k_2 X/p$ by a factor of 100, the velocity is reduced to the normal value $k_2 X/p$ by admitting a small percentage of water vapour to the gas. In the dry gases the negative ions move as free electrons along some of their paths, but when water vapour or other molecules having a high specific inductive capacity are present, groups are formed of these molecules which are more stable than the groups that are formed of the molecules of air or hydrogen.

123. Mean free path of a particle moving in a gas. In order to calculate the dimensions of the ions from the formula given in Section 85 for the rate of interdiffusion of two gases, it is necessary to express the mean free path of a particle moving in a gas in terms of the radii of the particle and a molecule of the gas.

Let σ be the sum of the radii of a particle A and a molecule of a gas B. The volume traversed per second by a sphere of radius σ moving with a velocity u is $\pi\sigma^2 u$, so that if the molecules of B are at rest and the particle A moves with a velocity u, the number of collisions per second will be $\pi\sigma^2 uN$, N being the number of molecules of B per cubic centimetre. The number of collisions that actually take place depends on the relative motions of the particle A and the molecules of B, and they become more frequent when the molecules of the gas B are also in motion. The mean free path of the particle A then becomes shorter.

The determination of the length of the mean free path is more difficult under these conditions. The problem has been investigated by Maxwell,* who found that the mean free path l of a particle moving with a mean velocity u, in a gas containing N molecules per cubic centimetre, is

$$l = \frac{u}{N\pi\sigma^2\sqrt{u^2+u'^2}},$$

u' being the mean velocity of agitation of the molecules.

* J. C. Maxwell, Phil. Mag. (4) **19**, p. 28, 1860; Collected Papers, vol. i, p. 387.

The mean free path l' of a molecule of the gas, obtained by equating u and u', becomes

$$l' = \frac{1}{N\pi s^2 \sqrt{2}},$$

s being the diameter of a molecule.

Also it follows that the mean free path of a particle, moving at a very high speed through a gas, exceeds the mean free path of the same particle moving with a velocity equal to the velocity of agitation of a molecule of the gas by the factor $\sqrt{2}$.

The mean free path of a molecule is obtained from the viscosity η of the gas, by the formula

$$\eta = \frac{1{\cdot}25 m' N l' u'}{3},$$

m', l', u', being the mass, the mean free path, and the velocity of agitation of a molecule. The factor 1·25 is introduced in this expression to correct for the effect of persistence of velocity when molecules of equal mass collide.*

In what follows, the mean free path of a molecule will be taken as the value of l' obtained from this equation, which gives the mean free path of a molecule in air at 760 millimetres pressure and 15° C. as $7{\cdot}5 \times 10^{-6}$ centimetre.

124. Sizes of ions deduced from the coefficients of diffusion. When the effect of the electric force on the neutral molecules is neglected, and the collisions between ions and molecules are treated as ordinary collisions between uncharged molecules, the coefficient of diffusion of the ions is given by the formula

$$K = \frac{lu}{3} \cdot \frac{m}{m'},$$

l being the mean free path of an ion, u its velocity of agitation, m the mass of an ion consisting of a group of molecules held together by the charge, m' the mass of a molecule which is small compared with m. As has been

* Jeans, Dynamical Theory of Gases, p. 250.

explained in Section 85, the factor m/m' in this formula accounts for the effect of the persistence of the velocity of the larger mass after a collision.

Also since u is small compared with u' the mean free path of the ion is

$$l = \frac{u}{\pi\sigma^2 N u'},$$

where σ is the sum of the radii of an ion and a molecule.

Hence the coefficient of diffusion in terms of the radii becomes

$$K = \frac{mu^2}{m'u'} \cdot \frac{1}{3\pi\sigma^2 N} = \frac{u'}{3\pi\sigma^2 N},$$

since $mu^2 = m'u'^2$.

The coefficient of diffusion, therefore, only involves the linear dimensions of the larger mass.

Also, since the mean free path of a molecule of a gas is $1/\sqrt{2}\pi s^2 N$, the coefficient of viscosity is

$$\eta = \frac{1{\cdot}25}{3\sqrt{2}} \cdot \frac{\rho u'}{\pi s^2 N},$$

ρ being the density of the gas.

Hence

$$\frac{\sigma^2}{s^2} = \frac{\eta}{K\rho} \cdot \frac{\sqrt{2}}{1{\cdot}25}.$$

If the diameter of the ion be x times that of a molecule, then $2\sigma = (1+x)s$, and the equation for x becomes

$$\frac{1+x}{2} = \sqrt{\frac{\eta}{K\rho}} \times 1{\cdot}13.$$

The values of x thus obtained for ions in air are

$$x = 3{\cdot}6 \text{ for positive ions;}$$

$$x = 2{\cdot}7 \text{ for negative ions;}$$

the values taken for K being $\cdot 030$ for positive ions and $\cdot 045$ for negative ions, η the viscosity of air at 15° C. being taken as $1{\cdot}78 \times 10^{-4}$.

The viscosity η is independent of the pressure of the gas, and for a large range of pressures, when small electric forces are acting, the coefficient of diffusion is inversely proportional to the pressure, so that $K\rho$ is constant. Under these conditions the groups of molecules surrounding the atomic charges remain unaltered when the pressure varies.

As the pressure is reduced a point is reached, depending on the intensity of the electric force, at which the coefficient of diffusion increases more rapidly than the inverse of the pressure. The average size of the ion then diminishes, and at sufficiently low pressures the negative electrons are not attached to molecules of the gas.

Similar calculations for ions in hydrogen at the higher pressures give

$$x = 5{\cdot}2 \text{ for positive ions,}$$
$$x = 4{\cdot}0 \text{ for negative ions.}$$

Since the rate of diffusion of ions in any gas may be calculated from the mobilities if it be assumed that the charge on the ions is equal to the atomic charge, the ratio x of the diameter of an ion to the diameter of a molecule may be obtained for those cases in which the velocity due to an electric force has been determined.

125. Equality of atomic charges in liquids and gases. Values of the product Ne. The conclusions of most interest that may be deduced from the determinations of the rates of diffusion are in connection with the relation that the charge on an ion in a gas bears to the charge on a monovalent ion in a liquid electrolyte. When a quantity of electricity equal to one electromagnetic unit passes through a dilute acid, 1·22 cubic centimetres of hydrogen at 15° C. and 760 mm. pressure are evolved at the negative electrode. If E be the charge on an atom of hydrogen in the liquid, and N the number of molecules per cubic centimetre of a gas at atmospheric pressure and 15° C., the charge $2\,NE \times 1{\cdot}22$ is 3×10^{10} electrostatic units, or

$$NE = 1{\cdot}23 \times 10^{10}.$$

The values of Ne, where e is the charge on an ion in a gas, may be found by comparing the rates of diffusion K with the velocities U under an electric force X.

It has been shown, Section 84, that

$$\frac{U}{K} = \frac{Ne\,.\,X}{\Pi},$$

where Π is the pressure of a gas containing N molecules per cubic centimetre at the temperature 15° C., at which the quantities of K and U were determined. Taking $\Pi = 10^6$ as the pressure of the atmosphere in dynes per square centimetre, and U_1 the velocity due to a force of one volt per centimetre, the value of Ne becomes

$$Ne = \frac{3 \times 10^8}{K} \cdot U_1.$$

Substituting in this formula the values of the velocities obtained by Zeleny and the rates of diffusion given in Section 120, the following numbers are obtained for the product $Ne \times 10^{-10}$:

	Dry gas.		Moist gas.	
	Positive ions.	Negative ions.	Positive ions.	Negative ions.
Air	1·45	1·31	1·28	1·29
Oxygen	1·63	1·36	1·34	1·27
Carbonic acid	·99	·93	1·01	·87
Hydrogen	1·63	1·25	1·24	1·18

These results indicate that the charges on ions produced by Röntgen rays in gases are approximately equal to the atomic charge E. The numbers show that the values of Ne for positive ions are probably larger than those for negative ions, which may be explained on the hypothesis that some of the positive ions have twice the atomic charge.

The agreement with the theory, as shown by the above figures, is not so good in the case of carbonic acid, but this is probably due to errors introduced by using insufficiently pure gas in some of the experiments, as a more accurate determination of the product Ne by another method shows that in carbonic acid, as well as in other gases, the ions have atomic charges.

126. Lateral diffusion of a stream of ions moving under an electric force. In the above method of determining the quantity Ne the possible errors include the experimental inaccuracies in the determinations of U and K, and also errors arising from the fact that the same degree of dryness or purity of the gases may not be attained in each case. In order to avoid this difficulty, another investigation was made which provided a means of determining the quantity *Ne* from a single experiment. The method* practically consists in finding directly the ratio of the velocity of the ions under an electric force to the rate of diffusion when the two motions take place simultaneously. For this purpose ions are generated on one side of a thin metal sheet *A* with a circular aperture through which the ions move under an electric force, and are received by two electrodes at a suitable distance from *A*.

One of the electrodes was a disc of the same size as the aperture in *A* and immediately opposite to it, and the other a flat ring surrounding the disc, and separated from it by a narrow air-gap for the purpose of insulation. The ions move from the aperture in *A* under the electric force towards the electrodes, some being received on the disc and some on the ring. The spreading of the current is due to the diffusion of the ions, and since the time during which this process is acting is inversely proportional to the velocity of the ions in the field of force, the ratio of the charge received by the disc to that received by the ring depends on the value of the ratio $K/U_1 X$.

The apparatus which was used for the purpose of investigating the motion of ions generated by secondary Röntgen rays is shown in figure 29. The primary rays *P* enter the apparatus through an aluminium window in the brass cover and, passing through an annular opening in the thick plate *B*, fall on the plate *S*. The secondary rays ionize the gas in the space above *S*, and some of the ions moving under the electric force pass through the grating *G* and the circular aperture in the thin sheet of metal foil *A*. A stream of ions thus enters the lower field, the central portion falling on the disc *D*, and the outer portion on the wide ring *C*.

Outside the ring *C* and in the same plane was fixed another ring R_0 connected to earth, and above it, at inter-

* Proc. Roy. Soc. A, 80, p. 207, 1908.

vals of one centimetre, six similar flat rings were fixed. The rings were connected in series by equal resistances, the metal sheet A being connected to the upper ring and also to the plate S by two similar resistances. When the ring R_0 and the electrodes C and D are at zero potential and the plate S at a potential V, there is a uniform field in the space below the metal sheet A. In order to bring ions from the upper part of the apparatus to the grating, the plate B is charged to a potential V' above the plate S, so that the electric force is in the same direction on the two sides of the grating.

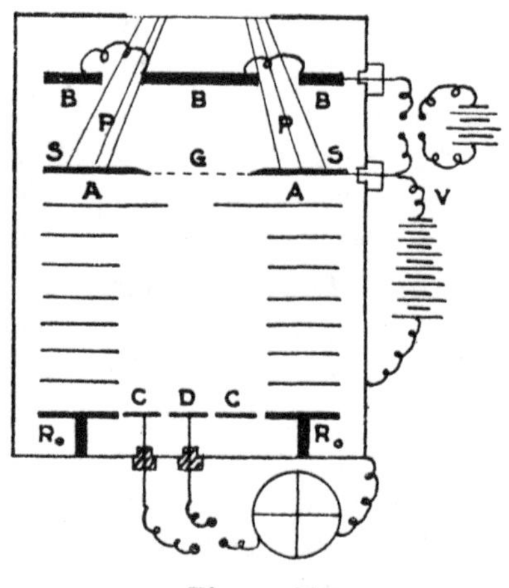

Figure 29.

The charges q_1 and q_2 received by D and C were measured accurately by a special form of induction balance. The quantities q_1 and q_2 include a small number of ions generated below the grating by scattered rays. In order to correct for this effect, the force in the upper field between S and B was reversed, and the charges q_1' and q_2' acquired by the electrodes due to ions generated in the space below the grating were determined. The charges $n_1 = q_1 - q_1'$ and $n_2 = q_2 - q_2'$ arriving on the disc D and the ring C, due to ions coming through the aperture in A, were thus found.

127. Mathematical investigation of the lateral diffusion of a stream. Direct determination of the product Ne. The mathematical investigation of the distribution of the ions below the sheet of metal A may be made on the assumption that the steady state is established, since the numbers that arrive at the electrodes when a uniform stream comes through the aperture in A is the same as when the ions come through the aperture intermittently according to any function of the time.

The differential equations for the partial pressure p of the ions in this case are

$$\frac{1}{K}(pu) = -\frac{dp}{dx},$$

$$\frac{1}{K}(pv) = -\frac{dp}{dy},$$

$$\frac{1}{K}(pw) = -\frac{dp}{dz} + nZe;$$

and the equation of continuity is

$$\frac{d}{dx}(pu) + \frac{d}{dy}(pv) + \frac{d}{dz}(pw) = 0.$$

On eliminating the velocities, the equation for n, the number of ions per cubic centimetre, becomes

$$\frac{d^2n}{dx^2} + \frac{d^2n}{dy^2} + \frac{d^2n}{dz^2} = \frac{NeZ}{\Pi} \cdot \frac{dn}{dz},$$

since $p = \frac{n\Pi}{N}$, N and Π referring to atmospheric pressure.

This shows that n is a function of NeZ which is determined by the solution of the differential equation that satisfies the surface conditions depending on the configuration of the apparatus. Hence the ratio of the charges n_1 and n_2 received by the disc D and the ring C is a function of NeZ which is independent of the pressure or the nature of the gas, and when the ratio of these charges is determined experimentally the value of Ne may be obtained.

For the form of apparatus that has been described, the

equations and surface conditions are as follows: Taking as origin of co-ordinates the centre of the aperture in A, and as the axis of z the line joining the origin to the centre of the disc D; then since the aperture is circular the distribution will be symmetrical with respect to z, and the equation for n becomes

$$\frac{1}{r}\frac{d}{dr}\left(r\frac{dn}{dr}\right)+\frac{d^2n}{dz^2}-\frac{NeZ}{\Pi}\cdot\frac{dn}{dz}=0, \quad . \quad . \quad . \quad (1)$$

r being the distance of any point from the axis of z.

At the plate A, n is a constant and equal to n_0 from $r = 0$ to $r = a$, a being the radius of the aperture, which in this case was 7·5 millimetres. From $r = a$ to $r = b$, n is zero, b being the inner radius of the rings R.

The six rings at intervals of one centimetre between the plate A and the ring R_0, were introduced in order to obtain a uniform electric field. They have the effect of discharging any ions that may diffuse out to a large distance from the axis. The theory shows that in general the number that would reach any of these rings is negligible, but that is not of importance as far as the solution of the problem is concerned. Since the rings discharge the ions that come into contact with them, the value of n is zero for all values of Z when $r = b$. The radius of b was 2·5 centimetres and the distance from the origin to the surface of the electrodes was 7 centimetres.

To solve equation (1) let $n = \phi\epsilon^{-\theta z}$, and the equation for ϕ becomes

$$\frac{d^2\phi}{dr^2}+\frac{1}{r}\frac{d\phi}{dr}=-\left(\theta^2+\theta\,.\frac{NeZ}{\Pi}\right)\phi=-\mu^2\phi \quad . \quad . \quad (2)$$

so that ϕ is the Bessel function $J_0(\mu r)$.

The general expression for n therefore becomes

$$n=\Sigma A_s J_0(\mu_s r)\,\epsilon^{-\theta_s z}. \quad . \quad . \quad . \quad . \quad . \quad . \quad (3)$$

The number n being zero for all values of z when $r = b$, the values of μ are given by the condition $J_0(\mu b) = 0$.

Hence $\mu_1 = x_1/b$; $\mu_2 = x_2/b$ etc. where $x_1, x_2 \ldots$ are the roots of the equation $\quad J_0(x) = 0$.

The corresponding values of θ are given by the equation

$$\theta_s^2 + \theta_s \cdot \frac{NeZ}{\Pi} = \frac{x_s^2}{b^2}.$$

The positive root of this equation

$$\theta_s = \sqrt{\frac{N^2e^2Z^2}{4\Pi^2} + \frac{x_s^2}{b^2}} - \frac{NeZ}{2\Pi},$$

must be used in the expansion (3), for a negative value of θ would make the exponential term infinite when z is infinite, which is impossible, since n diminishes as z increases.

The terms in the expansion of n, in equation (3), are thus fully determined except the coefficients A. These may be found from the surface conditions at the plane $z = 0$.

Equation (3), when $z = 0$, becomes

$$n = A_1 J_0(\mu_1 r) + A_2 J_0(\mu_2 r) + \ldots$$

Multiplying each side of this equation by $J_0(\mu_s r)\, rdr$ and integrating from $r = 0$ to $r = b$, the value of A_s is obtained from the equation

$$\int_0^b n J_0(\mu_s r)\, rdr = A_s \int_0^b J_0^2(\mu_s r)\, rdr \quad . \quad . \quad . \quad (4)$$

since the terms of the form $\int_0^b J_0(\mu_s r),\ J_0(\mu_{s'} r)\, rdr$ vanish when $\mu_s b$ and $\mu_{s'} b$ are roots of $J_0(x) = 0$.

Also, since $\int_0^b J_0^2(\mu_s r)\, rdr = \frac{b^2}{2} \cdot J_0'^2(\mu b)$; and $n = n_0$ from $r = 0$ to $r = a$, and $n = 0$ from $r = a$ to $r = b$, equation (4) becomes

$$n_0 \int_0^a J_0(\mu_s r)\, rdr = \frac{A_s b^2}{2} J_0'^2(\mu_s b) = \frac{A_s b^2}{2} \cdot J_1^2(\mu_s b).$$

$$\text{Hence } A_s = -\frac{2n_0 a}{\mu_s b^2} \cdot \frac{J_0'(\mu_s a)}{J_0'^2(\mu_s b)} = \frac{2n_0 a}{b^2} \cdot \frac{J_1(\mu_s a)}{\mu_s J_1^2(\mu_s b)}.$$

The value of n in terms of r and z thus becomes

$$n = \frac{2n_0 a}{b^2}\left[\frac{J_1(\mu_1 a) J_0(\mu_1 r)}{\mu_1 J_1^2(\mu_1 b)} \cdot \epsilon^{-\theta_1 z} + \frac{J_1(\mu_2 a) J_0(\mu_2 r)}{\mu_2 J_1^2(\mu_2 b)} \cdot \epsilon^{-\theta_2 z} + \ldots\right]$$

The ratio R of the charge received by the disc $\int_0^a 2\pi nrdr$ to the charge received by the disc and the ring $\int_0^b 2\pi nrdr$ is

$$R = \frac{a\left[\frac{J_1^2(\mu_1 a)\epsilon^{-\theta_1 z}}{x_1^2 J_1^2(\mu_1 b)} + \frac{J_1^2(\mu_2 a)\epsilon^{-\theta_2 z}}{x_2^2 J_1^2(\mu_2 b)} + \dots\right]}{b\left[\frac{J_1(\mu_1 a)\epsilon^{-\theta_1 z}}{x_1^2 J_1(\mu_1 b)} + \frac{J_1(\mu_2 a)\epsilon^{-\theta_2 z}}{x_2^2 J_1(\mu_2 b)} + \dots\right]},$$

where $z = 7$ centimetres, $a = \cdot 75$, $b = 2{\cdot}5$, and $\mu_1 b = x_1$, $\mu_2 b = x_2, \dots$

Hence $$R = \cdot 3 \cdot \frac{\frac{J_1^2(\cdot 3\, x_1)\epsilon^{-\theta_1 z}}{x_1^2 J_1^2(x_1)} + \frac{J_1^2(\cdot 3\, x_2)\epsilon^{-\theta_2 z}}{x_2^2 J_1^2(x_2)} + \dots}{\frac{J_1(\cdot 3\, x_1)\epsilon^{-\theta_1 z}}{x_1^2 J_1(x_1)} + \frac{J_1(\cdot 3\, x_2)\epsilon^{-\theta_2 z}}{x_2^2 J_1(x_2)} + \dots}$$

The quantities θ involve the product NeZ and the root x_s of the equation $J_0(x) = 0$, since

$$\theta_s = \sqrt{\frac{(NeZ)^2}{4\,\Pi^2} + \frac{x_s^2}{b^2}} - \frac{NeZ}{2\,\Pi},$$

so that when various values of $\frac{NeZ}{\Pi}$ are taken, the corresponding values of R may be obtained from the above equation.

For ordinary purposes values of R between $\cdot 4$ and $\cdot 6$ are required, but when dealing with the effects obtained in very dry gases the diffusion becomes abnormally great and it is necessary in many cases to find the values of R corresponding to small values of the variable $\frac{NeZ}{\Pi}$.

Haselfoot * has recently computed the ratio R for a large range of values, and the curve A, figure 30, Section 131, is drawn from the numbers that he has obtained. The ordinates of the curve are $y = R$ and $x = \frac{NeZ}{\Pi} \times \frac{1}{41}$. The

* C. E. Haselfoot, Proc. Roy. Soc. A, **87**, p. 350, 1912.

factor 1/41 is introduced so that the abscissae x may represent the force Z in volts per centimetre when

$$Ne = 1{\cdot}23 \times 10^{10},$$

and the rate of diffusion does not exceed the normal value.

For the purpose of determining the quantity Ne the ratio R in terms of the product NeZ is all that is required, but in other experiments, such as when the magnetic deflection of a stream of ions is being investigated, it is desirable to know to what extent the ions reach the boundary of the field.* In order that none of the ions may be deflected off the electrodes when the magnetic field is applied, it is necessary to use electrodes of large area to receive the ions, and it is useful to have a means of estimating the dimensions of the electrodes that would be required. Some information can be obtained on this point by finding what proportion of the ions come into contact with the boundary of the field in an apparatus of the dimensions of that used in the experiments for the determination of the product Ne. In this case the total number of ions that enter the field is $\int_0^a 2\pi rndr$ when $z = 0$, and the number received by the electrodes is $\int_0^b 2\pi rndr$ when $z = 7$ cm., the difference between these two quantities representing the number that diffuse outwards to a distance of 2·5 centimetres from the centre of the stream and are discharged by the boundary of the field. The ratio λ of the two integrals, or the fraction of the total number that reach the electrodes, has been calculated by Haselfoot for a large number of values of the quantity $\frac{NeZ}{\Pi}$. The following table gives the values of R and λ in terms of $\frac{NeZ}{\Pi}$:

$\frac{NeZ}{\Pi}$	2	4	8	16	24	32
R	·195	·197	·210	·265	·329	·384
λ	·1	·323	·661	·915	·975	·992

* See Section 104.

In the experiments for the determination of Ne the forces Z were chosen so that R should be greater than ·40, and in those cases only an inappreciable number of ions come near the outer edge of the ring C.

128. Effect of self-repulsion on the distribution of ions in a stream. The formulae given in the preceding section for the distribution of the ions are obtained on the hypothesis that the electric force is constant, and that the lateral expansion of the stream, arising from the mutual repulsion of the ions, shall be small compared with the effect of diffusion.

When the repulsive forces are small, the order of the correction to be applied to the charge n_1 acquired by the disc may easily be obtained, but it is difficult to estimate the correction accurately, so that it is necessary to ensure that it shall be small.

It has already been shown, Section 113, that the rate at which the density ρ varies at any point moving with the ions is given by the equation

$$\frac{\delta\rho}{\delta t} = 4\,\pi k\rho^2.$$

If s be the section of the aperture in the metal sheet A, kX the velocity of the ions, the electrical density in the central portion of the stream is $(n_1+n_2)/skX$ initially, and n_1/skX when the stream reaches the electrodes. The mean density being $\left(n_1+\frac{n_2}{2}\right)/skX$, the change in density $\delta\rho$ in the time $\delta t = z/kX$ is of the order

$$\delta\rho = 4\,\pi k\frac{\left(n_1+\frac{n_2}{2}\right)^2}{s^2k^2X^2}\cdot\frac{z}{kX},$$

z being the distance from the aperture in A to the electrodes. The amount to be added to n_1, the charge received per second by the disc, in order to compensate for the effect of the mutual repulsion of the ions is

$$skX\delta\rho = \frac{4\,\pi z\left(n_1+\frac{n_2}{2}\right)^2}{skX^2}.$$

This quantity must be small compared with n_1, and it is necessary to conduct the experiments at low pressures so that k may be large, otherwise only very small currents n_1 and n_2 could be used, and it would be difficult to measure them accurately.

129. Values of the product Ne obtained with negative ions. In order to test the theory, several experiments were made with ions generated by secondary Röntgen rays in air under various conditions. The results with negative ions are as follows:

Pressure of air in millimetres.	Force X in volts per centimetre.	Intensity of ionization in arbitrary units.	$Ne \times 10^{-10}$.
3	1·45	10	1·23
6	·98	8	1·23
6	1·47	16	1·25
6	1·96	13	1·15
6	2·96	17	1·27
12	1·00	8	1·20
12	·98	12	1·20
12	1·46	15	1·25
12	2·75	45	1·26
25	1·47	33	1·30
25	3·00	34	1·29

In the experiments at 25 millimetres the correction for mutual repulsion was about 10 per cent. of the observed value of n_1, so that the numbers deduced from them are the least accurate of the series. The mean of the other values of Ne is $1 \cdot 23 \times 10^{10}$, which is in good agreement with the value of NE, where E is the charge on a univalent ion in a liquid electrolyte.

The above experiments were made with secondary rays emitted by a tarnished surface at S. When a bright surface was used and the ions were produced by corpuscular rays, practically the same number, $1 \cdot 24 \times 10^{10}$, was obtained. With other gases * the values of $Ne \times 10^{-10}$ obtained for negative ions were 1·23, 1·24, and 1·23 for oxygen, hydrogen, and carbonic acid respectively.

* Proc. Roy. Soc. A, **80**, p. 207, 1908; **81**, p. 464, 1908; **85**, p. 25, 1911.

Similar experiments were made by Haselfoot * with ions generated by α and β rays emitted by a radio-active substance, and the mean values for the negative ions obtained in different sets of experiments were

$$1{\cdot}24 \times 10^{10} \text{ and } 1{\cdot}22 \times 10^{10}.$$

In all these experiments with negative ions a small quantity of water vapour was present, an amount probably of the order of one-tenth of a millimetre pressure, and it was found that results were then in accurate agreement with the theory. The ratio of the charges $n_1/(n_1+n_2)$ was independent of the pressure of the gas, and was the same for all gases when the force X was constant. When the gases are more completely dried by phosphorus pentoxide, contained in vessels connected through wide short tubes with the diffusion apparatus, different results are obtained. At the lower pressures the ratio $n_1/(n_1+n_2)$ becomes very much less than at the higher pressures when the same force is used. Under these conditions the motion of the negative ions does not follow the simple laws according to which the lateral diffusion of the stream of ions has been investigated, and it becomes impossible to estimate the value of the quantity Ne from the experimental results. The motion of the negative ions in dry gases at low pressures will be discussed more completely in another section.

130. Values of the product Ne obtained with positive ions. Experiments with streams of positive ions generated by secondary Röntgen rays and Becquerel rays were made under the same conditions as those with negative ions. They differ in two respects: the motion of the positive ions is not affected by traces of moisture, and in some cases the values of Ne for positive ions are much larger than the value corresponding to a single atomic charge.

The lateral diffusion of the stream of positive ions was the same for very dry gases as for gases containing traces of moisture over the range of pressures that were used, which in the case of air was from three to twelve milli-

* C. E. Haselfoot, Proc. Roy. Soc. A, **82**, p. 18, 1909; **87**, p. 350, 1912.

metres. When the ions are generated by Becquerel * rays the value of $Ne \times 10^{-10}$ found for positive ions was between 1·26 and 1·22. Similarly with positive ions generated by secondary Röntgen † rays emitted by a bright metallic surface S, the values of $Ne \times 10^{-10}$ were 1·26, 1·24, 1·26, 1·32, for air, oxygen, hydrogen, and carbonic acid respectively. These numbers, excepting the last, exceed the number corresponding to one atomic charge by an amount which may be due to experimental error, so that probably all the ions generated by Becquerel rays or by the electronic secondary Röntgen rays have monatomic charges. When, however, the ions are generated by secondary Röntgen rays emitted by a tarnished surface, the values of Ne are much larger than the above numbers. In a series of experiments in which the surface S, figure 29, was covered with a thin layer of vaseline, the numbers 2·03, 1·71, 1·84, 1·55, were obtained as the values of $Ne \times 10^{-10}$ for positive ions in air, oxygen, hydrogen, and carbonic acid respectively. Among the positive ions generated in these cases a large proportion of the ions appear to have double charges.

With regard to the charges on the positive ions generated by Röntgen rays at atmospheric pressure, it is difficult to reconcile all the results with those obtained at low pressures. The principal discrepancy is, that at atmospheric pressure such a large proportion of ions with double charges are not obtained, as at low pressures. Recently, Franck and Westphal ‡ made a series of determinations of mobilities and rates of diffusion of ions with a view to finding the percentage of positive ions with double charges at atmospheric pressure. With ions generated by Röntgen rays, they have found the coefficient of diffusion of the positive to be ·029, and with ions generated by α, β, and γ rays the coefficient of diffusion of the positive ions was much larger, being ·035, ·037, and ·035 for the three types of radiation. These

* Haselfoot, loc. cit.

† Proc. Roy. Soc., A, **81**, p. 464, 1908; **85**, p. 25, 1911.

‡ J. Franck and W. H. Westphal, Verh. d. D. phys. Ges., **11**, 146 and 276, 1909.

results confirm to a certain extent the earlier determinations in which it was found that the coefficients of diffusion of positive ions in air generated by Röntgen rays and by Becquerel rays were ·028 and ·032 respectively.

Franck and Westphal also found that the positive ions generated by Röntgen rays did not all diffuse at the same rate. When the process of diffusion takes place, the ions with the larger coefficient of diffusion are removed from the gas more rapidly than the others, and the mean rate of diffusion of those remaining in the gas is lower than ·029. In this way they found that the rate of diffusion of about 9 per cent. of the ions was as low as ·0175. These results may be explained on the hypothesis that the positive ions with double charges have larger groups of molecules associated with them, and therefore diffuse more slowly than those with single charges. No similar variation was found in the velocity under an electromotive force, which shows that positive ions with double or single charges move at the same rate in an electric field.

131. Effect of water vapour on the lateral diffusion of a stream. Abnormal diffusion of electrons in dry gases. It has already been mentioned that at low pressures small quantities of water vapour influence the lateral diffusion of a stream of negative ions in a marked degree. The effect is shown by the curves,* figure 30, which give the values of the ratio of the number of ions n_1 received by the disc C to the number $n_1 + n_2$ received by the disc and the ring D in the experiments made with the apparatus shown in figure 29. The abscissae represent the force Z in volts per centimetre in the space below the plate A through which the stream of ions is moving. The values of the ratio $n_1/(n_1 + n_2)$ are the ordinates of the curves. The three dotted curves represent the values of R for negative ions in dry air at the pressures 5·5, 11, and 16·5 millimetres. The curves show that the lateral diffusion increases as the pressure diminishes when the force is constant. The theo-

* Proc. Roy. Soc., A, 81, p. 464, 1908.

retical curve A represents the values of the ratio $n_1/(n_1+n_2)$ for different forces, and is the same at all pressures of any gas or any mixture of gases, when the motion of the ions follows the ordinary laws of diffusion and the ions have single atomic charges. When a small amount of water vapour having a partial pressure less than one-tenth of a millimetre is admitted to the diffusion apparatus the ratios

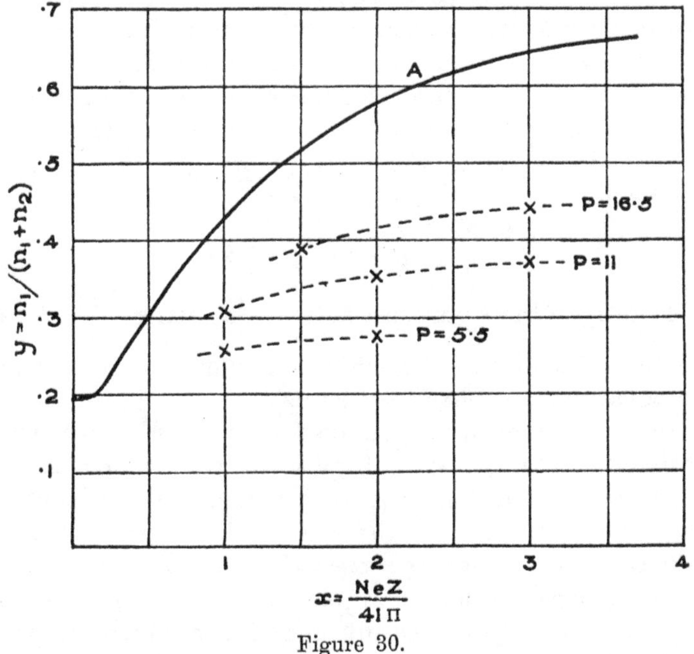

Figure 30.

$n_1/(n_1+n_2)$ increase and agree very closely with the theoretical values, given by the curve A.

With the same forces acting, these effects are obtained at much higher pressures in hydrogen than in oxygen or air, and in carbonic acid the effect was observed at lower pressures.

The increase in the lateral diffusion of the stream of negative ions shows that the rate of diffusion becomes abnormally large as compared with the velocity due to the electric force. This would happen if the charge on the ion diminished, since the electric force would then have a smaller effect on each particle, or it may be due to an increase in the velocity of agitation of the ions above the value that is attributed to it under ordinary circumstances. The first hypothesis may be dismissed as being too improbable, and moreover, it would be impossible to explain the large velocities obtained with electric forces in dry gases at low pressures if the charge e diminishes. It is necessary therefore to examine the equations of motion of the ions, to see at what point any modification is possible. These have been deduced, Section 85, from the expressions for the velocity due to an electric force and the rate of diffusion in terms of the velocity of agitation V of the particle, namely

$$U = \frac{Xe}{m} \cdot \frac{L}{V}, \text{ and } K = \tfrac{1}{3} LV.$$

The ratio of these quantities is taken as being $\frac{NeX}{\Pi}$, where Π is the value of $\frac{1}{3} NmV^2$, which is assumed to be constant, and equal to atmospheric pressure.

Since N denotes the number of molecules per cubic centimetre of a gas at that pressure, the average kinetic energy of an ion $\frac{1}{2} mV^2$ is thus assumed to be equal to that of a molecule of the surrounding gas. If it be assumed instead that the mean kinetic energy of agitation of a negative ion in an electric field is equal to that of a molecule multiplied by a factor k which increases as the force increases and the pressure diminishes, the equations of motion will then involve the quality $\frac{NeZ}{k\Pi}$ instead of $\frac{NeZ}{\Pi}$, and the ratio R obtained in Section 127 will be given by the curve A, figure 30, the abscissae being the values of $\frac{NeZ}{41k\Pi}$.

The quantity Ne may be assumed to have the normal value $1{\cdot}23 \times 10^{10}$, so that for any value of R determined experimentally with a force Z, the corresponding value of

Z/k is obtained from the theoretical curve. The values of k thus found for ions generated in air by secondary Röntgen rays are given in the following table, in which Z represents the electric force in volts per centimetre and p the pressure of the air in millimetres.

Z	1	1·5	1	2	3	1·5	3
p	11·0	16·5	5·5	11	16·5	5·5	11
k	1·85	1·83	2·86	2·90	2·85	3·75	4·00
Z/p	·091	·091	·182	·182	·182	·273	·273

The table shows that over this range of pressures the values of k depend on the ratio of the force to the pressure.

Further investigations on this subject have been made by Haselfoot*, with a similar apparatus, in which the air was reduced to lower pressures. In order that the number of ions in the stream should not fall off as the pressure is reduced, the ions were generated by the action of ultra-violet light on a metal plate.

The lateral diffusion continued to increase as the pressure was lowered, and for pressures lower than two millimetres the charge received by the disc D (figure 29) was less than a quarter of the charge received by the two electrodes when electric forces from one to four volts per centimetre were used. The values of $k-1$ were found to be approximately equal to $9X/p$, for a range of the variable X/p between the limits ·2 and 2·6.

132. Velocity of agitation of electrons. Values of the velocity of agitation in terms of the ratio X/p. When the ions are large compared with the molecules of the gas, the velocity W due to an electric force Z is given by the expression ZeT/m', where m' is the mass of a molecule and T the interval between collisions. The interval T depends on the linear dimensions of the larger particle and the velocity of agitation of the smaller, so that the velocity W involves the dimensions of the ion but is independent of its mass.†

When the negative ions are in the electronic state, the

* C. E. Haselfoot, Proc. Roy. Soc. 87, p. 350, 1912.

† See Section 122.

velocity due to an electric force depends on the mass of the electron, and is independent of its linear dimensions provided they are small compared with the dimensions of a molecule. The relation connecting the velocity with the mass and charge of the electron depends on the motion of agitation, and in order to investigate the mass of the electron by this method it is necessary to determine the value of k, when the conditions are the same as those under which the velocities were determined.

Since these motions are affected to a considerable extent by traces of water vapour, it is desirable to make both sets of experiments with the same gas. In order to satisfy these conditions, the values of k have been determined * with the apparatus illustrated in figure 23, which was used to find the velocity of electrons in dry air for a large range of forces and pressures. The value of k may be deduced from the ratio of the charge n_2 received by the electrode C_2 to the total charge $n_1+n_2+n_3$ when the magnetic force is zero.

The stream of ions, in this apparatus, came through a small narrow slit in the plate B, and was received by the electrodes C_1, C_2, and C_3, which formed portions of a disc 7 centimetres in diameter, the distance between B and the electrodes C being 4 centimetres. The central strip C_2 was 4·5 millimetres wide, and was separated from the two equal side-plates C_1 and C_3 by air-gaps ·5 millimetre wide. For practical purposes the central electrode may be taken as 5 millimetres wide and the side-plates may be considered to come within 2·5 millimetres of the central line.

The connection between the ratio $R = n_2/(n_1+n_2+n_3)$, and the quantity $\frac{NeZ}{\Pi k}$ may be determined as follows:

Let the origin of co-ordinates be taken at the centre of the slit, the axis of z being normal to the plate B, and the axis of y parallel to the length of the slit.

* J. S. Townsend and H. T. Tizard, Proc. Roy. Soc., A, **88**, p. 336, 1913.

The distribution in the electric field when the motion becomes steady is given by the equation

$$\nabla^2 n = \frac{Ne}{\Pi} \cdot \frac{Z}{k} \cdot \frac{dn}{dz}.$$

When the electric force Z is measured in volts per centimetre, and the value of the constants Ne and Π are inserted, the equation becomes

$$\nabla^2 n = \frac{41Z}{k} \cdot \frac{dn}{dz}.$$

The number of ions received on an area of the electrodes between the parallel lines x and $x+dx$ is proportional to $\int ndy$. Since the field is very wide, the ions do not extend to the boundary, so that n and dn/dy are zero at the limits of the integration.

The general equation for n, on integration with respect to y, becomes

$$\frac{d^2q}{dx^2} + \frac{d^2q}{dz^2} - \frac{41\,Z}{k}\frac{dq}{dz} = -\left[\frac{dn}{dy}\right] = 0,$$

where $q = \int ndy$, the integration being taken between two points on the boundary.

The series of sines that represents the solution of the differential equation for q converges very slowly, owing to the nature of the boundary conditions, and in order to obtain an accurate result it would be necessary to find a large number of terms of the series.

An approximate solution may be obtained in this case in a simple form, since the principal motion in the z direction is due to the electric force. The second term $\frac{d^2q}{dz^2}$ is small compared with the term $\frac{41\,Z}{k} \cdot \frac{dq}{dz}$ when Z is large, and with the values of Z/k that occur in the experiments the error introduced by neglecting the term $\frac{d^2q}{dz^2}$ is probably smaller than the experimental error. The equation for q then reduces to the form

$$\frac{d^2 q}{dx^2} = \frac{41\,Z}{k}\frac{dq}{dz}.$$

Considering the ions that pass at a uniform rate through a narrow strip in the centre of the slit, it appears from this equation that the distribution of the quantity q, as expressed in terms of x and z, is the same as the distribution of temperature ϕ, in terms of x and t, in an infinite solid initially at zero temperature throughout, except at the plane $x = 0$, where the temperature has a constant value ϕ_0 when $t = 0$.

The temperature is obtained from the equation

$$\frac{d^2\phi}{dx^2} = \frac{1}{K}\frac{d\phi}{dt},$$

and the solution given by Fourier is $\phi = At^{-\frac{1}{2}}\epsilon^{-x^2/4\,Kt}$.

In the problem of the distribution of ions in the space between the plates B and C, the surface conditions are $q = 0$ when $z = 0$, for all values of x except $x = 0$, so that q in terms of x and z is given by the equation

$$q = Az^{-\frac{1}{2}}\epsilon^{-\frac{41\,Z}{4\,kz}\cdot x^2}.$$

The distance z of the electrodes from the origin was 4 centimetres, and the central electrode was 5 millimetres wide, so that the ratio R of the charge n_2 received by the central electrode to the total charge is

$$R = \frac{n_2}{n_1 + n_2 + n_3} = \frac{\int_0^{\cdot 25}\epsilon^{-2\cdot 56\,Z_1 x^2}\,dx}{\int_0^{\infty}\epsilon^{-2\cdot 56\,Z_1 x^2}\,dx} = \frac{\int_0^{\cdot 4\sqrt{Z_1}}\epsilon^{-y^2}dy}{\int_0^{\infty}\epsilon^{-y^2}dy},$$

where $Z_1 = Z/k$.

When various values are substituted for Z_1 the corresponding ratios R may be obtained from the tables of definite integrals.

It is necessary to take into consideration the width of the slit; the proportion of the ions coming through a section of the slit near the edge that arrive on the central electrode is somewhat less than the proportion of those that come through at the centre. The exact proportion for any section of the slit is easily calculated, and the ratio R when the ions come through all sections of the slit, equally, may be found in terms of Z_1.

The curve representing R in terms of Z/k is very similar to the curve, figure 24, which gives the proportion of a cylindrical stream received by a central disc.

Thus when the ratio R is found experimentally for any force Z the value of $Z_1 = Z/k$ is obtained from the curve.

The values of k and W, corresponding to the same forces and pressures, were determined for a range of forces extending from 2 volts to 50 volts, and pressures from ·25 millimetre to 18·5 millimetres. It was found that both these quantities were functions of the ratio Z/p, as is seen by the following table, which contains examples of the results obtained when the ratio Z/p is approximately constant:

p	Z	Z/p	k	$W \times 10^{-6}$
18·5	40	2·16	24·0	1·75
12·0	30	2·5	26·0	1·97
1·8	4	2·2	24·0	1·9
3·7	40	10·8	46·0	5·5
1·8	20	11·1	46·5	5·6
·95	10	10·5	45·5	5·4

The usual expressions for the velocity W in terms of Z only apply when the velocity of agitation u of an electron is considerably greater than the velocity W, and it is necessary to find whether this condition is satisfied.

It will be seen from these investigations that when Z/p exceeds ·2, the ions in dry air are in the electronic state, so that the value of e/m may be taken as $5{\cdot}3 \times 10^{17}$. The value of e/m for a molecule of air being 10^{13}, the masses are in the ratio $5{\cdot}3 \times 10^4 : 1$, hence the velocity of agitation of an electron in thermal equilibrium with air is 230 times the velocity of agitation of a molecule of air, or 10^7 centimetres per second. The actual velocities of agitation when electric forces are acting exceed this value by the factor $\sqrt{k}$.

The following table gives the values of $u = \sqrt{k} \times 10^7$ and the velocities W in terms of Z/p:

Z/p	·2	·5	1	2	5	10	20	50	100	150	200
$W \times 10^{-6}$	·5	·9	1·25	1·75	3·0	5·2	9·0	17·3	27	35	44
$u \times 10^{-6}$	16	24	34	47	62	67	75	101	127	145	156
$Wu \times 10^{-12}$	8·0	21	42	82	186	350	675	1740	3430	5100	6900
$e/m \times 10^{-17}$	5·1	5·8	5·8	5·4	4·4	3·9	3·6	3·9	3·8	3·6	3·8

133. Values of e/m for electrons, obtained from the determinations of the velocity of agitation and the velocity under an electric force. When the mass m of the charged particle is small compared with that of a molecule of a gas the velocity W is given approximately by the formula $W = \frac{Ze}{m} \cdot \frac{l}{u}$, and is proportional to Z when u is constant. But the velocity of agitation of the electrons increases with Z, hence W does not increase in proportion to the force.

The results of the experiments are thus in accordance with the theory, since Wu is approximately proportional to Z when p is constant. The formula diminishes in accuracy as Z/p increases, since the velocity W approaches u.

The value of e/m for the charged particle may be deduced from the expression for W in terms of molecular quantities, but the above formula obtained under simplified conditions only gives approximate results. A more complete investigation of the rate of diffusion and the velocity under an electric force has been made by Langevin,* for the case in which the collisions between the charged particles and the molecules are of the type that occur between elastic spheres. The expression for the velocity W under these conditions is

$$W = \frac{3Ze}{8\sigma^2 N} \sqrt{\frac{h(m+m')}{\pi m m'}},$$

where m is the mass of the charged particle, m' that of a molecule of the gas, σ the sum of the radii of the particle and a molecule, N the number of molecules of the gas per

* P. Langevin, Ann. de Chim. et de Phys. (8) **5**, p. 245, 1905.

cubic centimetre, and h three-fourths of the reciprocal of the kinetic energy of agitation of the particle or a molecule of the gas.

Since the mass of the electron is small compared with that of a molecule, the ratio $(m+m')/mm'$ reduces to $1/m$, and $1/\pi\sigma^2 N$ is the mean free path l of the electron. The velocity W of an electron thus becomes

$$W = \frac{Ze}{m} \cdot \frac{l}{u} \times \cdot 815,$$

u being the velocity of agitation of the electron, and l its mean free path.

The value of e/m for the electrified particle may be obtained by eliminating the quantity u. Thus

$$W^2 = \frac{Z^2 e}{m} \cdot \frac{Ne}{mNu^2} \cdot l^2 \times \cdot 664,$$

where $Ne = 1{\cdot}23 \times 10^{10}$, and $mNu^2 = 3k \times 10^6$. At a millimetre pressure the mean free path of a molecule of air may be taken as $7.5 \times 10^{-6} \times 760$. The mean free path of the particle * exceeds that of a molecule by the factor 4, since the linear dimensions of the particle are small compared with those of a molecule, and by an additional factor $\sqrt{2}$, owing to the velocity of agitation being so great that the motion of the molecules may be neglected in estimating the free paths of the particle. Hence $lp = 3{\cdot}2 \times 10^{-2}$, and e/m is given by the formula

$$\frac{e}{m} = \left[\frac{Wp}{Z}\right]^2 \times \frac{k}{2{\cdot}8} \cdot$$

The values of e/m thus obtained from the determinations of W and k, corresponding to the different values of Z/p, are given in the table in the preceding section. The numbers do not differ very much from the value $5{\cdot}3 \times 10^{17}$ found by the more accurate methods under conditions in which the effects of collisions between electrons and molecules may be neglected.

The results show that the negative ions are in the

* See Section 123.

electronic state when moving in dry air in a field in which the value of Z/p exceeds ·2.

The differences between the values of e/m that were obtained are not entirely due to experimental errors, although the probable error in the final results is greater than that in the direct measurements, since the square of the velocity W is involved in the formula for e/m. The nature of the collisions between electrons and molecules may not resemble the collisions between elastic spheres to the same extent for different velocities of agitation u.

134. Theoretical investigation of the energy of agitation of charged particles in an electric field. The law of equipartition of energy applies in general when molecules react among one another, as in the kinetic theory of gases, but a change in the distribution of energy may be produced by external forces. In all cases where electric forces act on ions moving in a gas, their kinetic energy exceeds that of the molecules. For small values of Z/p this excess is small, but as the preceding experiments show the negative ions may acquire an abnormal kinetic energy of agitation when Z/p increases.

A theoretical investigation of this phenomenon has recently been made by Pidduck,* after the method used by Maxwell † in his later papers on diffusion.

The problem considered was that of the final steady motion of a stream of ions moving under a constant force.

If W is the velocity under the electric force and Ω the velocity of agitation of the molecules of the gas, k is given by the equation

$$k-1=\frac{W^2}{\Omega^2}.$$

This result is obtained on the hypothesis that on collision the law of force is that of the inverse fifth power of the distance between the centres of the ion and molecule, no assumption being made as to their relative masses.

The velocity W is much larger for electrons than for

* F. B. Pidduck, Proc. Roy. Soc. A, **88**, p. 296, 1913.

† J. C. Maxwell, Phil. Trans. **157**; Collected Papers, vol. ii, p. 26.

positive ions, and it follows from this equation that, with a given value of Z/p, the energy of agitation of the electrons may be abnormally great when that of the positive ions is the same as the energy of agitation of the surrounding molecules.

The theoretical values of k obtained by substituting the values of W and Ω in the formula are much greater than those obtained experimentally, but a more satisfactory result is obtained by assuming that there is a loss of energy on collision. When the collisions are of the type occurring between imperfectly elastic spheres, and the mass m of the ion is small compared with the mass M of a molecule of the gas, the following expression for k is obtained,

$$k-f+\frac{kM}{m}(1-f)=\frac{W^2}{\Omega^2}\cdot\frac{4f-3}{2-f},$$

where $f=\frac{1}{2}(1+e)$, e being the coefficient of restitution.

If the collisions are perfectly elastic ($f=1$), this equation gives

$$k-1=\frac{W^2}{\Omega^2}$$

as before.

The values of $k-1$ obtained experimentally, for four different values of Z/p, are given in the second line of the following table; the values obtained from the formula when $f=1$ are given in the third line.

A very slight deviation of f from unity will give much smaller values of k. The values of f in the last line are those which give the values of k obtained experimentally.

Z/p	·2	2	20	150
$k-1$ (observed)	1·8	20	55	210
$k-1=W^2/\Omega^2$	100	12×10^2	32×10^3	49×10^3
f	·9993	·9988	·99	·96

This investigation shows that the loss of energy on collision increases with the velocity of the electrons.

135. Magnetic deflection when the mass of the particles undergoes changes. The preceding investigations apply to cases in which the value of Z/p is comparatively large, and the electrons move freely between the molecules of the gas. Under these conditions the magnetic deflection of a stream, moving under a constant electric force, increases continuously with the magnetic force. With smaller values of Z/p the magnetic force has a different effect, as is shown by the experiments described in Section 105. When the pressure of the air exceeds 10 millimetres and a constant electric force of 1 volt per centimetre is acting, the magnetic deflection increases at first, and after attaining a maximum value, the deflection diminishes as the magnetic force increases. In these cases the ions are in a transition stage, and the electrons may be detached from the molecules with which they are usually associated, so that for certain intervals the electrons move freely between the collisions with molecules. In order to explain the effect of the magnetic force on the stream, it is necessary to consider the average velocity of a particle when the mass undergoes large variations from time to time.

Let a * be the velocity of a free electron in the electric field when the magnetic force is zero, and b the velocity when the electron is associated with a molecule or a group of molecules, the velocity b being small compared with a. Let $x = \frac{HeT}{m}$, where H is the magnetic force, e the charge and m the mass of an electron, and T the mean interval between two consecutive collisions of an electron with molecules of the gas. Also, let the sum of the intervals during which the electron moves freely be t_1, and t_2 the sum of the remaining intervals when the electron is associated with a comparatively large mass. When the magnetic force is acting, the velocity in the direction of the electric force is $\frac{a}{1+x^2}$ during the time t_1, and during the time t_2 the velocity is b, since the magnetic force has practically no effect on the motion of the larger mass. The distance S travelled in the direction of the electric force in the time $t_1 + t_2$ is therefore

$$S = \frac{at_1}{1+x^2} + bt_2.$$

* See Sections 92 and 93.

The velocity in a direction perpendicular to the electric force is $\frac{ax}{1+x^2}$ for the time t_1, and during the time t_2 the magnetic deflection is inappreciable. Hence the distance h that the charge travels in this direction in the time t_1+t_2 is

$$h = \frac{axt_1}{1+x^2}.$$

The angle of deflection of the stream is therefore given by the equation

$$\tan\theta = \frac{h}{S} = \frac{axt_1}{at_1+bt_2(1+x^2)}.$$

If it be assumed for simplicity that the ratio t_1/t_2 is not affected by the magnetic force, then x is the only quantity in the expression for $\tan\theta$ that involves H, and the angle θ has a maximum value when $d\theta/dx$ vanishes. Hence the value of H for which θ is a maximum is given by the equation

$$at_1+bt_2-x^2bt_2 = 0,$$

or

$$x^2-1 = \frac{at_1}{bt_2},$$

and θ_m the maximum value of θ is given by the equation

$$\cot^2\theta_m = 4\frac{bt_2}{at_1}\left(1+\frac{bt_2}{at_1}\right).$$

As the pressure is increased, t_2/t_1 increases and the maximum deflection diminishes, which agrees with the observations. Also the maximum value of θ may be expressed in terms of x as follows:

$$\tan\theta_m = \frac{x^2-1}{2x}, \text{ or } x = \tan\theta_m + \frac{1}{\cos\theta_m}.$$

Thus when the maximum value of θ is small the magnetic force required to produce the maximum deflection is given approximately by the equation

$$x = 1, \text{ or } \frac{He}{m}.T = 1.$$

At 10 millimetres pressure the mean free path of the electron is of the order $3{\cdot}2\times10^{-3}$ centimetres, and the velocity of agitation is $1{\cdot}4 \times 10^7$ centimetres per second, assuming e/m to be $5{\cdot}3 \times 10^{17}$ and $k = 1{\cdot}8$. The value of T may therefore be estimated, and the maximum deflection should, according to this formula, be obtained with a force of the order of 250 electromagnetic units. The force that was observed to give the maximum deflection was about 100 electromagnetic units, which is of the right order, but a closer agreement with the theory could not be expected, owing to the uncertainty of the effect of the magnetic field on the ratio t_1/t_2.

136. Mean velocity as obtained by different methods. In connection with this theory it will be noticed that the velocity U of an ion due to an electric force, as obtained by a direct method, is not in general the same as the velocity deduced from the magnetic deflection $U_h = \dfrac{X \tan\theta}{H}$. They represent average velocities taken in different ways, and are accurately the same only when the mass associated with the charge does not undergo variations during its motion through the gas. If the variations in the mass were small the two velocities would be approximately equal, but when the mass may change from that of an electron to that of a molecule of the gas they may be widely different.

The distance S traversed by the charge in the time $t_1 + t_2$, when $H = 0$, is

$$S = at_1 + bt_2,$$

and the velocity of the ion as obtained by a direct method is

$$U = \frac{at_1 + bt_2}{t_1 + t_2}.$$

The velocity as obtained by the magnetic deflection is

$$U_h = \frac{X}{H}\tan\theta = \frac{X}{H} \cdot \frac{axt_1}{at_1 + bt_2(1 + x^2)},$$

where $x = \dfrac{He}{m} . T$; and a the velocity of the electron due to the electric force is $\dfrac{Xe}{m} . T$.

Hence
$$U_h = \frac{a^2 t_1}{a t_1 + b t_2 (1 + x^2)}.$$

If the value of H be small, so that the ratio t_1/t_2 will be the same in the two cases, then, neglecting x^2, the value of U_h becomes
$$U_h = \frac{a^2 t_1}{a t_1 + b t_2}.$$

When $a t_1$ and $b t_2$ are of the same order, U_h is much greater than U, but they become equal when t_2 is zero, that is, when the ions are in the electronic state throughout the motion.

CHAPTER VI

RECOMBINATION

137. Methods of determining the rate of recombination. When ions of both signs are generated in a gas, the number that remain free to move under an electric force diminishes, owing to the recombination of the positive and negative ions. There are many experiments in which the currents between electrodes in a gas are regulated to a great extent by this process. The currents obtained with different electric forces when ions are generated at a uniform rate in a gas at rest between parallel-plate electrodes afford a simple illustration of the effect of recombination. When the ions are generated at the same rate at each point of the gas, it is possible to find expressions for the currents obtained with various forces, and to determine the coefficient of recombination by comparing the experimental determinations with the theoretical formulae. The principal difficulty in obtaining accurate results in this case is that the ions are seldom produced at the same rate at each point between the electrodes, for a beam of rays that ionizes a gas is always divergent, and when the rays fall on an electrode secondary radiation is given off which increases the rate at which ions are generated near the electrodes. In order to avoid these difficulties, a method has been devised by Langevin in which a large number of ions are generated, almost instantaneously, between parallel plates by a single discharge through a Röntgen-ray tube, and the coefficient of recombination is easily deduced from the charges acquired by the electrodes when different forces are acting. The advantage gained by this method is that the results are independent of the initial distribution, provided that the charge received per unit area of the electrodes is constant over the electrodes. The formula for the number of ions that arrive at the electrodes when various forces are acting applies accurately only when the forces are so large that the number that recombine is a small proportion of the total number generated. The variation of the charge with the force is therefore small, and in order to obtain accurate results it is necessary to measure these small variations to a high degree of accuracy. From an experimental point of view the method is therefore somewhat difficult, but when special arrangements are made for measuring small differences of two charges, very consistent results are obtained. Probably the simplest method, but one which only gives approximate results, is that used by McClelland in his investigation of the conductivity of gases rising from flames, and also by Rutherford, who showed that the rate of recombination of ions is proportional to the square of the conductivity, as the theory indicated. The principle of this method is to ionize a stream of gas moving along a wide tube, and to measure the conductivity between electrodes at various distances from the point at which the gas is ionized. The conductivity in these cases depends on various factors; but it is easy to arrange experiments in which the loss of conductivity during the time the ions are in the gas is due principally to the process of recombination, and an approximate value of the coefficient of recombination may be obtained.

138. Approximate determination of the coefficient of recombination. A form of apparatus that may be used to determine the coefficient of recombination, from measurements of the conductivity of a stream of air, is shown in figure 31.

A stream of gas is passed through the wide tube A and is ionized by a beam of Röntgen rays, that enter the apparatus through the aluminium window W in the large chamber C. The secondary radiation from the surfaces on which the primary rays fall is confined to the space C, and no additional ions are generated in the gas as it passes down the tube A. The electrode E is carried in a tube B, which has a gauze of fine wire fixed to the end, the tube being made to fit exactly into A, so that it may be placed with the gauze at various distances from the chamber C. The tubing is raised to a high potential V, and the electrode E is connected to the insulated quadrants of an electrometer, and its potential does not differ much from zero during an experiment.

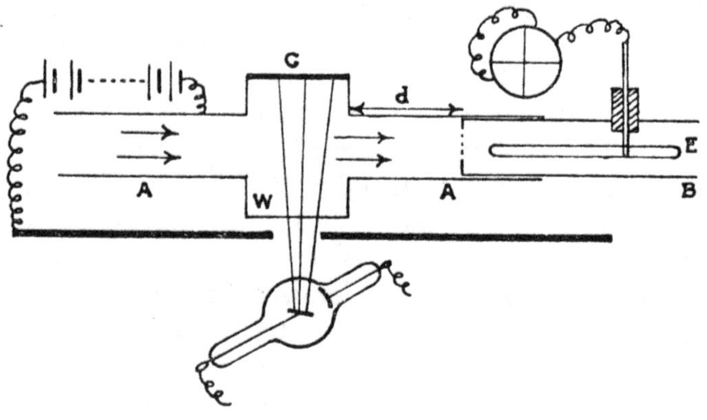

Figure 31.

The field of force is limited by the gauze to the space between the electrode and the tube B, so that the positive and negative ions are not separated by the electric force until they come through the gauze. The ions recombine while the gas passes from C to the gauze in front of the electrode, and when the gauze is reached practically all those with a charge of the same sign as the potential V pass through the gauze and are collected on the electrode E.

The charges Q_1 and Q_2 acquired by the electrode while the rays are acting for a certain time θ are measured, with the gauze at distances d_1 and d_2 centimetres from the chamber C.

If n is the number of positive or negative ions per cubic

centimetre at a point moving with the gas, the rate of change of the charge $q = ne$ is given by the equation

$$\frac{dq}{dt} = -\frac{\alpha}{e}q^2,$$

where α is the coefficient of recombination.

If q changes from q_1 to q_2 in the time T, then

$$\frac{1}{q_2} - \frac{1}{q_1} = \frac{\alpha}{e}.T.$$

The charges q_1 and q_2 per cubic centimetre of the gas in the tube may be deduced from the charges Q_1 and Q_2. If u be the mean velocity of the gas in the tube the time T is equal to $(d_1-d_2)/u$, and the quantity α/e is given by the formula

$$\frac{\alpha}{e} = \left(\frac{1}{q_2} - \frac{1}{q_1}\right)\frac{u}{d_1-d_2} = \frac{su^2\theta}{d_1-d_2} \cdot \frac{Q_1-Q_2}{Q_1 \cdot Q_2},$$

s being the section of the tube, and $qsu\theta = Q$.

It is necessary to consider the effect of diffusion, since some of the ions are lost to the surface of the tube A in the distance d_1-d_2. The quantity to be added to the smaller charge Q_2 to compensate for this effect diminishes in importance as the radius of the tube and the density of ionization are increased, so that it is possible to reduce the correction to a very small quantity. The principal inaccuracy arises from the variation of velocity of the gas in the tube, since the equations that have been used only apply accurately when the density of ionization and the velocity are constant over each section of the tube.

The coefficients α/e as found by this method * are 3420, 3380, 3500, and 3020 for ions produced in air, oxygen, carbonic acid, and hydrogen respectively, and are in good agreement with the numbers found by other methods.

139. Theory of Langevin's method. In Langevin's method the problem to be investigated theoretically is to find the quantity Q of electricity arriving at an electrode when ions are generated instantaneously in a field of force of intensity

* Phil. Trans. A, **193**, p. 157, 1899.

X between two parallel plates, the initial distribution being any function of the distance from one of the plates. The solution of this problem may be found by considering the following simple example of two streams of ions moving past each other.

A large number of positive ions N_1, distributed throughout a cylinder of unit section and moving with a velocity u along the axis of the cylinder, meet a small number of negative ions n_0 travelling in the opposite direction with a velocity v; to find the number of ions that recombine.

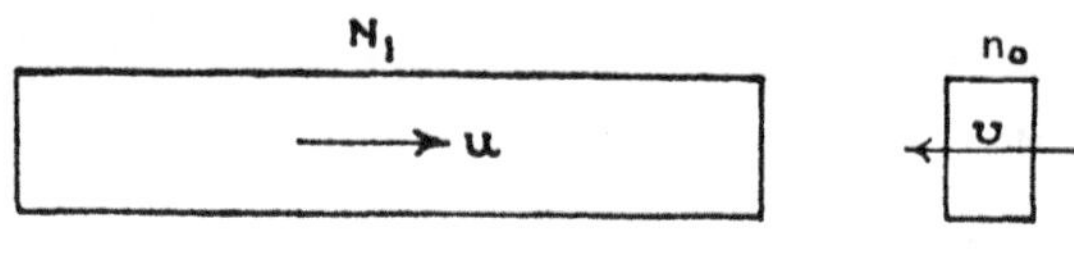

Figure 32.

Let ν_1 be the number of positive ions per unit volume at a distance x from the end of the cylinder, n the number of negative ions that enter the section at x at the time t. Then the number that recombine in the time dt is

$$dn = -\alpha n \nu_1 dt.$$

But $dx = (u+v)dt$.

Hence
$$\frac{dn}{n} = -\frac{\alpha \nu_1 dx}{u+v},$$

which on integration gives

$$\log \frac{n}{n_0} = -\frac{\alpha}{u+v}\int_0^x \nu_1 dx.$$

If l be the whole length of the cylinder, the integral $\int_0^l \nu_1 dx$ is equal to N_1, and the number n_0' of negative ions that come through the stream of positive ions is

$$n_0' = n_0 \epsilon^{-\frac{\alpha N_1}{u+v}}.$$

Hence the number of positive or negative ions that recombine is

$$n_0 (1 - \epsilon^{-\frac{\alpha N_1}{u+v}}).$$

This number is independent of the distribution of the positive ions along the axis of the cylinder and is a function of the total number N_1, and the relative velocity of the ions.

In order to simplify the calculation in the practical case where a number n_0 of positive and negative ions are generated initially per unit volume in the space between the plates, it will be supposed that the negative ions are at rest and that the positive ions move with a velocity $u+v$. Let Q/e be the number of positive ions that pass through unit area of the plane parallel to the electrodes at a distance x from the positive electrode. These move with a velocity $u+v$ through the $n_0 dx$ negative ions contained between the two planes x and $x+dx$, and the number that recombine in that space is

$$n_0 dx\left(1-\epsilon^{-\frac{\alpha Q}{e(u+v)}}\right).$$

Thus the total number $(Q+dQ)/e$ of positive ions that reach the plane $x+dx$ is

$$\frac{Q}{e}+\frac{1}{e}\frac{dQ}{dx}\cdot dx=\frac{Q}{e}-n_0 dx\,(1-e^{\frac{\alpha Q}{e(u+v)}})+n_0 dx,$$

the latter term being the number of positive ions generated initially in the layer of thickness dx.

Hence $$\frac{dQ}{dx}=n_0 e\epsilon^{-\frac{\alpha Q}{e(u+v)}}.$$

This equation on integration gives

$$\frac{u+v}{\alpha}\left(\epsilon^{\frac{\alpha}{e}\frac{Q}{u+v}}-1\right)=\int_0^x n_0 dx.$$

When $x=l$, Q is the charge that reaches unit area of the negative electrode, and the integral $\int_0^l n_0 dx$ is Q_0/e the total number of positive or negative ions, per unit area of the electrodes, generated initially in the gas.

Hence $$\epsilon^{\frac{\alpha}{e}\cdot\frac{Q}{u+v}}=1+\frac{\alpha}{e}\frac{Q_0}{u+v}.$$

If the intensity of the electric field X be written $X = 4\pi\sigma$, the above equation becomes

$$\epsilon^{\frac{bQ}{\sigma}} = 1 + \frac{bQ_0}{\sigma},$$

where $b = \dfrac{\alpha}{4\pi(k_1+k_2)e}$, since $u = k_1 X$ and $v = k_2 X$,

or
$$\frac{bQ}{\sigma} = \log\left(1 + \frac{bQ_0}{\sigma}\right).$$

Langevin * gives the equation for Q in this form.

It has been assumed in this investigation that the velocities u and v are constant; and therefore the separation of the positive and negative charges was not supposed to disturb the uniformity of the electric field. In order to avoid any serious error that might arise from that cause it is necessary to estimate the effect approximately, and to arrange the experiments so as to eliminate as far as possible the inaccuracy that is thus introduced. The disturbance of the field will be greatest when the two streams have just passed each other. If the velocities of the positive and negative ions are equal, and if the initial distribution is uniform, at that instant positive ions will be distributed in one half of the field and negative ions in the other half, the total charge on either side of the central plane being $Q/2$ approximately. These charges give rise to no force at the surface of either electrode, but they produce a force $2\pi Q$ in the centre of the field. It is necessary therefore that Q should be less than σ, and Langevin estimates that the coefficient b is determined to an accuracy of 1 per cent. when the ratio Q/σ is less than ·25.

In order to satisfy this condition, the electric force $4\pi\sigma$ must be large, and Q cannot differ from the maximum charge Q_0 by more than a few per cent. In air at atmospheric pressure the quantity b is of the order ·27, so that

* P. Langevin, Ann. de Chim. et de Phys. (7) **28**, p. 433, 1903.

$\frac{bQ}{\sigma}$ will be of the order ·07 and the equation

$$1 + \frac{bQ_0}{\sigma} = \epsilon^{\frac{bQ}{\sigma}},$$

on expanding the exponential term, gives the following expression for $Q_0 - Q$:

$$\frac{Q_0 - Q}{Q} = \frac{bQ}{2\sigma} + \frac{b^2 Q^2}{6\sigma^2}, \text{ approximately.}$$

The equation shows that $Q_0 - Q$ is a small charge of the order of 4 per cent. of the maximum charge.

140. Arrangement of apparatus in Langevin's experiments. For the purpose of determining the small difference between two charges obtained with different forces, an arrangement similar to that illustrated in figure 20 was used. The apparatus consists of two pairs of parallel plates: the insulated plates AB and $A'B'$ were connected together and to an electrometer and the opposite plates CD and $C'D'$ were maintained at equal and opposite potentials $+V$ and $-V$ sufficiently great to give values of Q/σ of the required order. The apparatus was adjusted so that the electrometer gave no deflection when the air between the two pairs of parallel plates was ionized by rays emitted by a Röntgen-ray tube. The force on one side of the apparatus was then increased to a large value, and a quantity approximately equal to $Q_0 - Q$ was obtained on the insulated system at each discharge of the tube. The charge $Q_0 - Q$ is thus measured directly, and variations in the intensity of the rays affect the value thus found in the same proportion as the measurements of the larger charge Q are affected. Having found $Q_0 - Q$ and Q by this means to the same order of accuracy, the value of b may be obtained from the above formula.

If Q_0, the charge obtained with the large force $4\pi\sigma_0$, be not sufficiently near the actual maximum M, the amount by which they differ may be found from the equation

$$M - Q_0 = \frac{Q_0{}^2 b}{2\sigma_0},$$

in which the approximate value of b is used. A more accurate value of b may then be found by taking M as the maximum instead of Q_0.

The coefficient of recombination α/e may be found from the coefficient b when the velocities k_1 and k_2 of the positive and negative ions, under a force of one electrostatic unit, are known $[\alpha/e = 4\pi b(k_1+k_2)]$.

The following values of b for air and carbonic acid at various pressures p are given by Langevin*:

Air.

p in atmospheres.	b	b/p^2
·20	·01	·25
·49	·06	·25
1·00	·27	·27
2·04	·62	·15
3·05	·80	·09
5·00	·90	·036

Carbonic acid.

p in atmospheres.	b	b/p^2
·178	·01	·32
·465	·13	·60
·72	·27	·515
1·00	·51	·50
2·05	·95	·24
3·13	·97	·10

The third column in the table for air shows that when p is less than an atmosphere, b is approximately proportional to p^2, and since the velocities k_1 and k_2 are inversely proportional to p, the coefficient $\alpha/e = 4\pi(k_1+k_2)b$ is proportional to the pressure.

The corresponding numbers in the experiments with carbonic acid do not follow this law so accurately, but in another set of experiments † Langevin gives the following values of b/p^2: ·52, ·50, and ·51 for carbonic acid at pressures of ·5, ·74 and one atmosphere respectively.

The coefficients of recombination α/e at atmospheric pressure obtained from the above values of b are—

* P. Langevin, Ann. de Chim. et de Phys. (7) **28**, p. 433, 1903.
† P. Langevin, Comptes rendus, **137**, p. 177, 1903.

	b	k_1+k_2	α/e
Air	·27	930	3200
Carbonic acid	·51	530	3400

141. McClung's determinations of the rate of recombination. Values of the coefficient of recombination in different gases. McClung* has also determined the coefficient of recombination by another method.

When ions are generated by rays at a constant rate in a gas between two electrodes at the same potential, the number of ions per cubic centimetre increases until the rate at which ions are lost by recombination is equal to the rate at which they are generated by the rays. A certain number of ions are also lost by diffusion to the electrodes, but if this effect be neglected, the increase in the number of ions per cubic centimetre n/e is given by the equation

$$\frac{1}{e}\frac{dn}{dt}=\frac{Q}{e}-\frac{\alpha n^2}{e^2},$$

Q/e being the number generated per cubic centimetre per second by the rays. When the steady state is reached, $\frac{dn}{dt}$ is zero, and the coefficient of recombination α/e, obtained from the maximum value of n, is

$$\frac{\alpha}{e}=\frac{Q}{n^2}.$$

In order to determine the maximum value of n, ions are generated by the rays for a certain time when the electrodes are at the same potential, and immediately after the rays cease to act, a high electromotive force V is applied to one of the electrodes, the other electrode being insulated and connected to a suitable capacity, so that its potential should not rise to a value of the same order as V. All the ions of the same sign as the applied potential are collected on the insulated electrode and the charge n may be measured by an electrometer. The experiment should be repeated with the rays acting for various times in order to find the maximum value of n.

The quantity Q is found from the current passing through

* R. K. McClung, Phil. Mag. (6) **3**, p. 283, 1902.

the gas while the rays are acting and a large potential difference is maintained between the electrodes.

The value of α/e found by this method for air at atmospheric pressure was 3384, and the coefficient was found to be the same at all pressures from three atmospheres to ·125 atmosphere. There is thus a great discrepancy between the results obtained by the two methods at the lower pressures, Langevin having found the coefficient of recombination to be approximately 800 at a pressure of ·125 atmosphere. In the method used by McClung the quantity n is estimated on the hypothesis that recombination is the only process taking place that reduces the number of ions in the gas before the electric force is established between the electrodes. The loss of ions due to diffusion to the electrodes is not considered, and since the rate of diffusion is inversely proportional to the pressure the effect rises in importance as the pressure is lowered. Also the relative effects of recombination and diffusion depend on the density of ionization, since the loss of ions due to recombination is proportional to the square of the density of ionization, and the loss due to diffusion is proportional to the density. A considerable error may therefore arise at low pressures, when the loss by diffusion is neglected, and the effect on the final result would be accentuated, since α involves the factor n^2.

Langevin's experiments are not subject to an error of this kind, since the positive and negative ions move under the electric force away from the electrodes charged with electricity of the same sign as soon as they are generated, and the velocity with which they move increases like the rate of diffusion as the pressure is reduced. His results at the lower pressures are therefore the more reliable

At the pressure of one atmosphere the values of the coefficient α/e, as obtained by the three methods, are as follows:

	1	2	3
Air	3420	3200	3384
Oxygen	3380	—	—
Carbonic acid	3500	3400	3492
Hydrogen	3020	—	2938

142. Rate of recombination of ions generated by α rays. Initial recombination. The results of the various experiments are in good agreement as far as the determinations at atmospheric pressure are concerned, but some experiments on currents which are controlled to a great extent by the process of recombination indicate that the rate of recombination depends on the method by which the ions are generated. Thus the currents obtained between parallel-plate electrodes with moderate electric forces are much greater when the ions are generated by Röntgen rays than when they are generated by the α rays from a radio-active substance, although the total number of ions generated initially by the rays is the same in each case. The number that recombine in the latter case is larger than what might be expected if the ions were generated uniformly between the plates and recombination took place according to the ordinary laws.

This phenomenon has been investigated by Rutherford, also by Bragg and Kleeman, and they attributed the effect to a special recombination that takes place between the two ions that are generated from the same molecule. On this hypothesis an electron emitted by a molecule with a comparatively small velocity recedes only to a small distance from the remaining positively-charged ion, and the two ions are attracted to each other and recombine when the external electric force acting on the gas is not very large. The rate of recombination would therefore depend on the method by which the ions are generated, as the electrons would not be emitted with the same velocity from the molecules, under the action of various types of radiation. As the external force tends to separate the two ions generated from the same molecule, the number of recombinations of this kind that take place diminishes as the external force increases, so that the current between the electrodes also increases.

Thus in addition to the general recombination that takes place between all the ions that are present, which is proportional to the square of the number per cubic centimetre, there is another special kind of recombination, proportional to the number of ions that are generated in the gas, which may become the principal effect in some cases.

In order to explain the variations that have been observed in the currents between parallel plates it is necessary to suppose that forces of the order of 100 volts per centimetre would have considerable effect in stopping the recombinations of the latter kind.

Langevin points out, as an objection to this theory, that if such recombinations do occur, they would not be affected to an appreciable extent by the external forces, and consequently there would be no noticeable increase in the current with the electric force. He suggests that the large effect produced by recombination when the ions are generated by α rays is due to the fact that the gas is not ionized uniformly, as there is a large concentration of ions along the path of each α particle. The number of ions per cubic centimetre is thus very large in the narrow columns of gas ionized by the α particles, so that general recombination takes place very rapidly at first between positive and negative ions in the same column. As the ions diffuse away from the track of the α particle recombination then proceeds less rapidly.

An effect of this kind would certainly take place when the initial distribution of ions is concentrated in narrow columns in the gas, and probably the large electric forces that are required to obtain saturation currents are due to that cause. The question as to whether there is a special process of recombination of the kind that was first suggested will be discussed later.

143. Recombination between ions generated by the same α particle. The evidence in favour of Langevin's explanation is very conclusive, and experiments show that there is very rapid general recombination between the ions generated by the same α particle. According to this theory, the ions are generated in narrow columns along the path of each α particle, where the number per cubic centimetre is very large, so that initially recombination takes place more rapidly than would be the case if the ions were distributed uniformly throughout the gas. Subsequently, when the ions diffuse through the gas, general recombination between ions generated by the different α particles takes place.

When the total ionization is small, recombination between ions in the same column is the predominating effect, and the number that disappear from any column will be independent of the number of these columns.

It is easy to see that under ordinary conditions the intensity of the currents between parallel-plate electrodes may be adjusted so that an inappreciable number of ions will be lost by general recombination of ions produced by the different α particles. For example, let the electrodes be 10 square centimetres in area and 3 centimetres apart, one of the electrodes being of thin aluminium through which the α particles may pass. With forces exceeding 20 volts per centimetre the ions will remain in the space between the plates for less than one-tenth of a second, and if the number of columns of ions does not exceed three or four per square centimetre of the electrodes at any instant, ions from one column will not diffuse to another and the effect produced by recombination between ions from different columns will be inappreciable. This condition will therefore be satisfied if the total number of particles traversing the gas per second does not exceed 300. According to Rutherford's* measurements, each α particle generates $1 \cdot 7 \times 10^5$ ions along its trajectory through air, so that the number generated per centimetre of its path is of the order 2×10^4.

Hence if 300 particles enter the space between the electrodes per second, the charge on all the ions generated in 10 seconds would be ·08 electrostatic unit, a quantity which may be easily measured with an ordinary electrometer.

144. Moulin's experiments on the currents between parallel plates. Theoretical investigation of the effect of recombina-

* E. Rutherford, Phil. Mag. (6) **10**, p. 206, 1905.

tion when the ions are distributed in columns. The results obtained by Moulin * in the experimental study he has recently made of the subject are in complete agreement with this theory.

When the electric force between two parallel plates is constant, the ratio of the corresponding current to the maximum current is independent of the distance between the electrodes, and of the maximum current, when that current is small.

This result is in agreement with the theory, but the experiments which are most conclusive are those in which it was found that the currents depended on the inclination of the paths of the α particles to the direction of the electric force. The number of ions that recombine, according to this theory, ought to be greater when the direction of the electric force is parallel to the trajectories of the α particles than when it is perpendicular to them. In the first case the columnar distribution is not disturbed by the motion of the ions under the electric force, but in the second case the positive and negative ions move in opposite directions and are completely separated from each other after the ions have travelled the width of the column.

The process of diffusion tends to produce a uniform distribution of the ions, and for this reason the difference between the numbers that recombine in the two cases is not so marked as it otherwise would be. The experiments, however, show that the currents between parallel-plate electrodes are decidedly greater when the paths of the α particles are at right angles to the direction of the force than when they are parallel to the force, or inclined at an angle of 45° to the force.

The curves given by Moulin,† figures 33 and 34, represent the results of the experiments obtained with air and carbonic acid. The curves give the ratio i/I of the current corresponding to a field X to the maximum current, as

* M. Moulin, Thèses présentées à la Faculté des Sciences de Paris, 1910, Gauthier-Villars.

† Moulin, loc. cit.

a function of the intensity of the field X which is measured in volts per centimetre. The maximum current I was in each case 10^{-3} electrostatic unit. For each gas two sets of curves are given, corresponding to different scales of the electric force. The curves I correspond to the case where

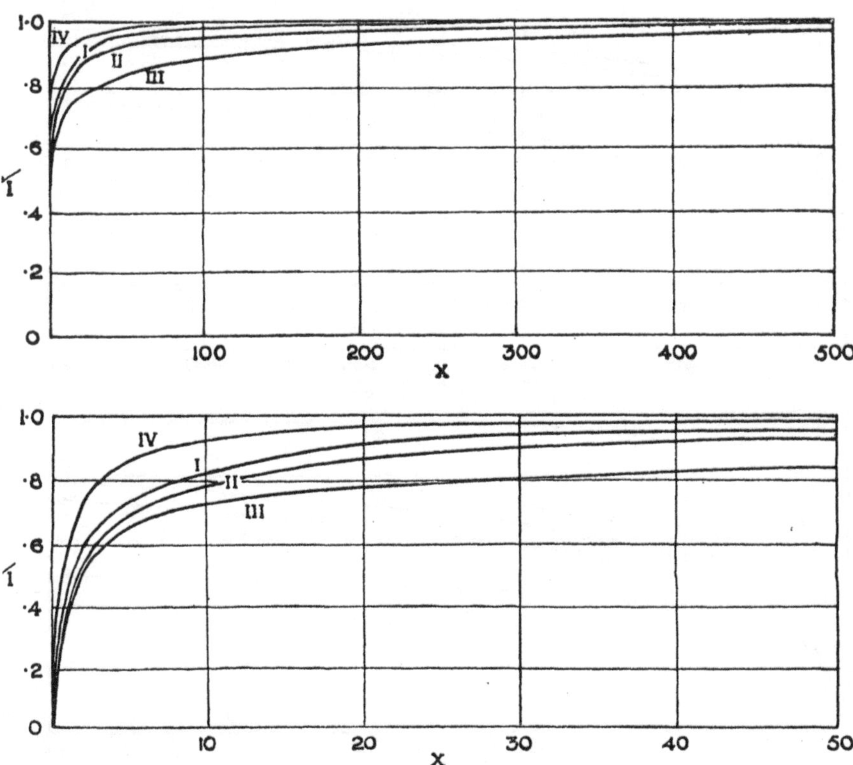

Figure 33. Current-electric-force curves in air at 750 mms. pressure and 16° C. X in volts per centimetre.
I. α rays perpendicular to the force. III. α rays parallel to the force.
II. α rays at an angle of 45°. IV. Penetrating rays.

the rays are perpendicular to the direction of the electric force, the curves II to the case where they are at an angle of 45°, and the curves III to the case where they are parallel to the field. With hydrogen the effect of the initial recombination is very small, and only very slight differences were observed in the currents when the direction of the trajectories of the α rays was altered.

In addition, the curves IV, representing the currents obtained with penetrating γ rays, producing the same maximum current as the α rays, are given in each of the figures.

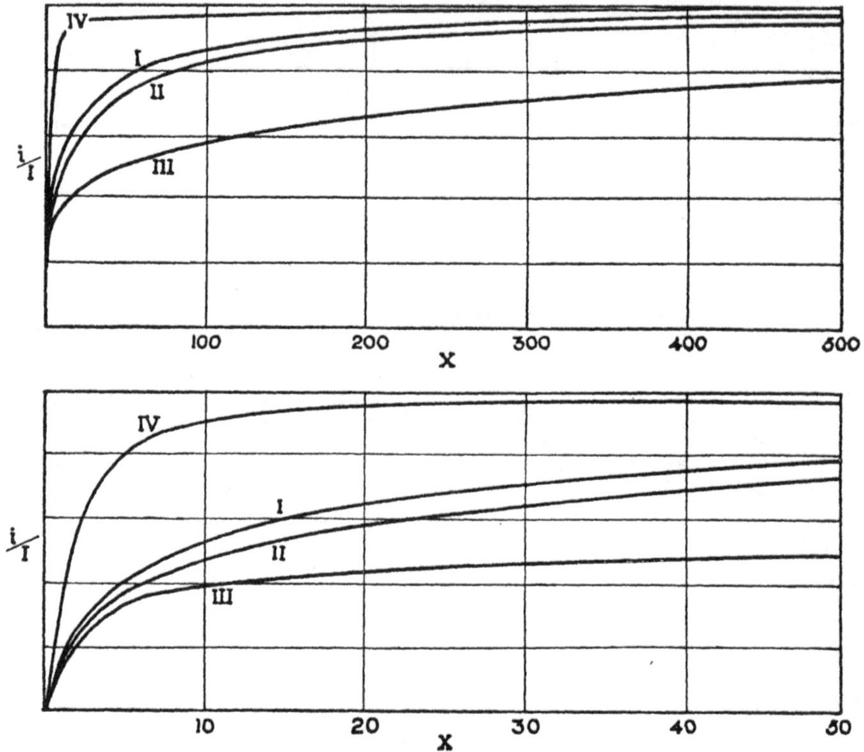

Figure 34. Current-electric-force curves in carbonic acid at 750 mms. pressure and 16° C. X in volts per centimetre.
I. α rays perpendicular to the force. III. α rays parallel to the force.
II. α rays at an angle of 45°. IV. Penetrating rays.

The ions in this case are produced by β particles passing through the gas. Each of these particles produces a comparatively small number of ions per centimetre of its path, so that recombination among the ions generated by the same particle is inappreciable.

According to Geiger and Kovarik * the average number of ions produced by a β particle, of high speed, per centimetre of its path in air at atmospheric pressure is 67. This is smaller than the value first given for that number by Durack,† which is 127. The number, in either case, is very small compared with the number produced by an α particle, and in order to obtain the same number of ions between the plates as when α rays are used it is necessary to have a large number of β particles per square centimetre of the gas. In this case, therefore, general recombination takes place between all ions present, as the initial distribution is practically uniform.

The relative position of the four curves obtained for each gas is in complete agreement with the theory. Also the large effect observed in carbonic acid, and the small effect in hydrogen as compared with air, may be explained by considering how the initial recombination must vary in the three cases. In carbonic acid, the number of ions generated per centimetre by each α ray is greater than in air, and the number generated per centimetre in air is greater than in hydrogen. Also the rate of diffusion of the ions is different in the three gases, being greatest in hydrogen and least in carbonic acid. On this account the ions generated in any column diffuse into the surrounding gas and reduce the number of ions per cubic centimetre in such a way as to diminish the effect of initial recombination in hydrogen more than in air, and in air more than in carbonic acid.

The currents obtained with a given force when the initial ionization is distributed in columns have been investigated mathematically by Jaffé.‡

The differential equation for the change of distribution at any point is given by the equation

$$\frac{dn}{dt} = K\left(\frac{d^2n}{dr^2} + \frac{1}{r}\frac{dn}{dr}\right) - \alpha n^2,$$

* H. Geiger and A. F. Kovarik, Phil. Mag. (6) **22**, p. 604, 1911.
† J. J. E. Durack, Phil. Mag. (6) **5**, p. 550, 1903.
‡ G. Jaffé, Ann. der Phys. (4) **42**, p. 303, 1913.

when the force X is zero; the term in brackets indicates the change due to diffusion, and αn^2 the loss by recombination.

Hitherto an exact solution of this equation has not been obtained, but Jaffé shows that, in this case, the effect of recombination will be almost the same if, instead of considering α to be constant, the quantity

$$\tfrac{1}{2}\left(\alpha\, \epsilon^{\frac{r^2}{4Kt+b^2}}\right)$$

is substituted for α. A complete solution of the new differential equation is then found.

When the electric force is parallel to the axis of the column, the time required to separate the positive and negative ions is

$$T = \frac{d}{2uX},$$

d being the length of the column, and the total loss by recombination is easily obtained.

The case in which the force is inclined at any angle ϕ to the axis of the column is also investigated, and an expression is obtained from which it is possible to calculate the ratio of the current corresponding to any force X to the saturation current.

Jaffé represents his results in the form of curves which agree very closely with the experimental curves obtained by Moulin for the cases in which ϕ is zero, $\pi/4$ and $\pi/2$.

The theory is also found to be in good agreement with the results of experiments * on the conductivity produced by α rays in insulating liquids such as pure hexane, and also with some recent experiments made by Mies † on the conductivity produced by α and β rays in gases at high pressures.

145. Investigation of the process of recombination. Effect of attraction between positive and negative ions. In order

* G. Jaffé, Ann. d. Phys. (4) **25**, p. 282, 1908; Le Radium, **10**, p. 126, 1913.

† Mies, Le Radium, **10**, p. 136, 1913.

to apply the principles of the kinetic theory of gases to obtain a theory of the process of recombination when the distribution is uniform, it is necessary to take into consideration the motion of agitation of the ions and also the motion due to their mutual attraction when a positive and negative ion are near together. The problem of finding the number of collisions per second between positive and negative ions is therefore more complicated than the corresponding problem for uncharged particles.

The latter case has been investigated by Maxwell, and an expression was found for the number of collisions that occur between two sets of particles, or the number of times per second a particle of one set comes within a distance σ of a particle of the other set. This expression may easily be found from the formulae given in section 123. If the particles are of the same masses as the ions in air, then their velocities of agitation u_1 and u_2 may be taken as being approximately 9×10^3 and $1 \cdot 5 \times 10^4$ centimetres per second; for if the diameter of the positive ion be three times the diameter of a molecule, and the diameter of a negative ion twice that of a molecule, their masses will be respectively twenty-seven times and eight times the mass of a molecule, and the corresponding velocities of agitation will be approximately 1/5th and 1/3rd of the velocity of agitation of a molecule of air. Hence if the electric forces between positive and negative ions be neglected, the number of occasions during the time δt, in which the distance between a positive ion and a negative ion will be less than σ, is

$$\pi \sigma^2 \sqrt{u_1^2 + u_2^2}\, n_1 n_2 \,\delta t,$$

n_1 and n_2 being the number of positive and negative ions per cubic centimetre.

If the radius of a molecule of air be taken as $1 \cdot 4 \times 10^{-8}$ centimetre,* a collision between a positive and a negative ion will occur when σ is five times the radius of a molecule. Substituting 7×10^{-8} for σ in the above formula, and for u_1

* Jeans, Dynamical Theory of Gases, p. 251.

and u_2 the values obtained above, the number of collisions that occur in the time δt becomes

$$2{\cdot}7 \times 10^{-10}\, n_1 n_2 \delta t.$$

The number of ions that recombine in the time δt is $\alpha n_1 n_2 \delta t$, where $\alpha = 3400 \times e$, so that if e, the charge on an ion, be taken as $4{\cdot}5 \times 10^{-10}$ electrostatic unit, the number that recombine is

$$1{\cdot}5 \times 10^{-6} \times n_1 n_2 \delta t.$$

It thus appears that the number of collisions that actually occur between positive and negative ions is a quantity of a much larger order than the number of collisions between uncharged particles of the same size and mass.

It is therefore necessary to consider the effect of the electric attraction between positive and negative ions as well as the motion of agitation, and the most satisfactory method of proceeding is to apply the general equations of motion to investigate the movement and the distribution of ions of one sign at various distances from an ion of opposite sign. Since the equations do not apply for large electric forces such as are obtained at points very near an ion, it will be supposed that a small sphere surrounds the ion and the number of ions of opposite sign that come within various distances of it may be determined.

146. Investigation of currents converging on a spherical electrode. The process of recombination may be investigated by finding an expression for a current to a spherical electrode of radius a, at a potential E/a; since the current is proportional to the number of ions, of opposite sign to the charge E, that collide with the sphere.

Let the charge E be positive and let the number of negative ions (n per cubic centimetre) be small, so that the forces X, Y, and Z acting on the ions will be

$$\frac{d}{dx}\left(\frac{E}{r}\right),\quad \frac{d}{dy}\left(\frac{E}{r}\right),\ \text{and}\ \frac{d}{dz}\left(\frac{E}{r}\right)$$

respectively, r being the distance of any point from the centre of the sphere.

The general equations of motion given in Section 85 may be written

$$nu = -K\frac{dn}{dx} + nkX,$$

$$nv = -K\frac{dn}{dy} + nkY,$$

$$nw = -K\frac{dn}{dz} + nkZ,$$

K being the coefficient of diffusion, and k the velocity due to unit electric force.

It will be supposed that no ions are generated in the field under consideration, the supply being maintained from an external source by a stream converging on the electrode. In that case, when the steady state is reached, the equation of continuity becomes

$$\frac{d}{dx}(nu) + \frac{d}{dy}(nv) + \frac{d}{dz}(nw) = 0\,;$$

and on substituting for nu, nv, and nw their values given by the equations of motion, the equation for n becomes

$$-K\nabla^2 n + k\left[\frac{dn}{dx}\cdot\frac{d}{dx}\left(\frac{E}{r}\right) + \frac{dn}{dy}\cdot\frac{d}{dy}\left(\frac{E}{r}\right) + \frac{dn}{dz}\cdot\frac{d}{dz}\left(\frac{E}{r}\right)\right] = 0,$$

or, since n is a function of the distance r,

$$\frac{K}{r^2}\frac{d}{dr}\left(r^2\frac{dn}{dr}\right) + \frac{kE\,(x^2+y^2+z^2)}{r^4}\frac{dn}{dr} = 0.$$

Hence $$K\frac{d}{dr}\left(r^2\frac{dn}{dr}\right) + kE\frac{dn}{dr} = 0.$$

The integral of this equation is

$$n = c_1 + c_2\,\epsilon^{\frac{c}{r}},$$

c_1 and c_2 being arbitrary constants, and

$$c = \frac{kE}{K} = 1{\cdot}23 \times 10^4 \times E.$$

Since the sphere discharges the ions, the number per cubic centimetre in the neighbourhood of the surface is very small compared with the number in other parts of the field, hence $n = 0$ when $r = a$, and by means of this condition one of the constants may be eliminated and the value of n becomes

$$n = c_1 \left[1 - \epsilon^{c\left(\frac{1}{r} - \frac{1}{a}\right)}\right].$$

Let n_0 be the value of n when r is large compared with a, then $c_1 = \dfrac{n_0}{1 - \epsilon^{-\frac{c}{a}}}$.

The current i flowing across unit area of the surface of the sphere of radius r is given by the equation

$$\frac{i}{e} = K\frac{dn}{dr} + \frac{knE}{r^2} = \frac{kEc_1}{r^2};$$

and the total number of ions I/e that reach the electrode per second is

$$4\pi kEc_1 = \frac{4\pi kEn_0}{1 - \epsilon^{-\frac{c}{a}}}.$$

Since c represents the quality $1{\cdot}23 \times 10^4 \times E$, the value of c/a is $41\,W_a$ where W_a is the potential of the electrode in volts.

If, therefore, the potential of the electrode exceeds that of the surrounding space by more than an eighth of a volt the quantity $\epsilon^{-\frac{c}{a}}$ may be neglected in comparison with unity, and c_1 may be taken as being equal to n_0. The current to the electrode I then becomes $4\pi kEn_0 e$, which is proportional to the charge E and is independent of the radius of the sphere.

Also, since the quantity $\left(\frac{c}{a} - \frac{c}{r}\right)$ is proportional to the fall of potential $W_a - W_r$ from the electrode to a point at a distance $r - a$ from the surface, the value of n is

$$n = n_0\left(1 - \epsilon^{-41(W_a - W_r)}\right).$$

Hence the value of n is practically constant and equal to n_0 at points where the potential differs from that of the surface by more than ·125 volt. At points nearer the surface the density n diminishes, and when the potential fall from the electrode is ·025 volt the density is $n_0(1-\epsilon^{-1})$. This result is independent of the radius of the sphere and applies to a stream of ions approaching a plane electrode.

For small potentials the quantity $\epsilon^{-\frac{c}{a}}$ cannot be neglected and the general expression for the current must be used, namely,

$$\frac{I}{e} = \frac{4\pi k E n_0}{1-\epsilon^{-\frac{k}{K}\cdot\frac{E}{a}}}.$$

In this case when E is constant the current increases with the radius of the electrode.

In the limit when E is very small the denominator of the above fraction becomes $\frac{k}{K}\frac{E}{a}$, and the current is then

$$I = 4\pi K n_0 ae.$$

This is the diffusion current representing the number of ions that come into contact with an uncharged electrode when the distribution is given by the formula

$$n = n_0\left(1-\frac{a}{r}\right).$$

147. Diffusion currents converging on small spherical conductors. The above results have been obtained on the assumption that the number of ions in the immediate neighbourhood of a sphere is so small compared with n_0 that n may be taken as zero when $r=a$. When the radius of the sphere is small compared with the mean free path of the ions it has very little effect on the distribution in its neighbourhood, since the number $\pi a^2 n_0 u$ that strike it per second diminishes as the square of the radius; u being the velocity of agitation of the ions. The investigations of the preceding section, in which the number that strike a large sphere was found to be $4\pi K n_0 a$, do not apply to small spheres, and it is necessary to make a more general investigation so as to include cases in which a is small compared with the mean free path L.

In general, let ν be the number of ions per cubic centimetre at the surface of the sphere, and n_0 the constant distribution at a long distance from the surface. Considering first the case in which the sphere is uncharged, the equation for n becomes $\frac{d}{dr}\left(r^2 \frac{dn}{dr}\right) = 0$, which on integration gives $n = n_0 - \frac{(n_0 - \nu)\,a}{r}$. The diffusion current per unit area of the sphere is $i = \frac{K\,(n_0 - \nu)\,ae}{r^2}$ and the total number of ions that collide with the sphere per second is $4\pi Ka\,(n_0 - \nu)$.

Another expression for the number of ions that collide with the sphere may be obtained from Maxwell's formula, which shows that the number of uncharged particles that collide with a sphere per second is $\pi a^2 n'u$, n' being the number of particles per cubic centimetre at the surface, and u the velocity of agitation. The number n' includes those particles which are moving away from the sphere after colliding with it, so that only one-half of the uncharged particles near the surface are moving towards the sphere at any instant, with a mean velocity $u/2$ in the direction of the normal.

In the case of the diffusion of ions, the distribution is not uniform, and at the surface there are no ions in the gas moving away from the sphere, since they become discharged by colliding with the sphere. In order, therefore, to apply the formula $\pi a^2 n'u$ to the case of the diffusion of ions the quantity n' must be taken as the number per cubic centimetre at a short distance h from the surface where the ions are moving in all directions The length h may be taken as $2L/3$ or $2K/u$, which is the mean value of the projections on the normal of the free paths that are terminated by the sphere.

When the ions are diffusing towards the spherical conductor, the distribution n is given by the expression $n = n_0 - \frac{(n_0 - \nu)\,a}{r}$, so that at the distance h from the surface the value of n is $(n_0 h+\nu a)/(a+h)$, and the number that collide with the surface is

$$\frac{\pi a^2 u (n_0 h+\nu a)}{a+h}.$$

Equating this number to the number $4\pi K a(n_0-\nu)$, obtained by the previous calculation, the following relation between ν and n_0 is obtained:

$$\nu = n_0 \frac{4K(a+h)-auh}{4K(a+h)+a^2 u}.$$

When a is small compared with the path L, a may be neglected in comparison with h and the equation for $n_0-\nu$ becomes

$$n_0-\nu = \frac{auhn_0}{4Kh} = \frac{aun_0}{4K},$$

which is independent of h, and the number that collide with the sphere is

$$4\pi K a(n_0-\nu) = \pi a^2 n_0 u.$$

When a is large compared with h the value of ν becomes $\nu = \dfrac{4K-hu}{au} n_0$. The distance h may be assumed to be $2L/3$, so that $\nu = \dfrac{2K}{au} n_0$, and since for negative ions in air $u = 1{\cdot}5\times 10^4$ and $K = {\cdot}045$, the value of ν is ${\cdot}03\, n_0$ when $a = 10^{-4}$ centimetre. For spheres of larger radii ν may be neglected in comparison with n_0 and the number of ions that collide with the sphere becomes $4\pi K a n_0$, as above.

The interpretation of these results as applied to a sphere surrounded by a uniform distribution of uncharged particles that rebound from the surface is as follows: When the steady state is reached, the number of different particles that collide with the sphere per second is $4\pi K(n_0-\nu)a$, the number $\pi n_0 a^2 u$ is the total number of collisions per second with the sphere, and their ratio $aun_0/4K(n_0-\nu)$ is the average number of collisions made by each particle that collides with the sphere before it diffuses out into space or comes into contact with the outer boundary of the gas.

148. Current converging on a small charged sphere. When a current I flows towards a sphere with a charge E, the number of ions per cubic centimetre in the gas being ν at the surface and n_0 at points remote from the sphere, the current is given by the expression

$$\frac{I}{e} = \frac{4\pi kE\left(n_0 - \nu\epsilon^{-\frac{c}{a}}\right)}{1 - \epsilon^{-\frac{c}{a}}}.$$

Since the investigation is of interest in connection with the problem of recombination, the charge may be taken as being equal to the atomic charge e, and in that case the constant c becomes $5{\cdot}5 \times 10^{-6}$. When a is small of the order 10^{-6} centimetre, the exponential terms in the above expression may be neglected and the value of I is independent of the quantity ν.

In order to determine the values of I for larger values of a, it will be assumed that the ions that collide with the sphere approach the surface with a mean normal velocity $\frac{u}{2} + \frac{kE}{a^2}$.

The number I/e that collide with the sphere per second is then $4\pi a^2 \nu\left(\frac{u}{2} + \frac{kE}{a^2}\right)$, and by equating this expression to that given above, the following equation to determine ν is obtained:

$$4\pi kE\left(n_0 - \nu\epsilon^{-\frac{c}{a}}\right) = 4\pi a^2\nu\left(\frac{u}{2} + \frac{kE}{a^2}\right)\left(1 - \epsilon^{-\frac{c}{a}}\right),$$

the values of the constants being

$$e = 4{\cdot}5 \times 10^{-10},\ u = 1{\cdot}5 \times 10^{4},\ \text{and}\ k = 550$$

for negative ions in air at atmospheric pressure, and

$$c = 5{\cdot}5 \times 10^{-6}.$$

The values of I/e may thus be found for large spheres of radii exceeding 10^{-4} centimetre, and an approximate value of I/e may be found when $a = 10^{-5}$ centimetre.

The number of negative ions that come within a given

distance of a positive ion may be determined from the above formula. If the positive ion is at rest, the number of negative ions that come within the distance a is the same as the number I/e that would collide with a sphere of radius a, the charge on the sphere being equal to the atomic charge. When the volume of the sphere is small compared with $1/n_0$ it is improbable that two negative ions will be in the neighbourhood of the sphere at the same time, so that the force acting on any negative ion near the sphere may be taken as e^2/r^2. This condition will therefore be satisfied if a is less than 10^{-4} centimetre, since the number of ions per cubic centimetre does not exceed 10^8 in most cases.

The values of $\frac{I}{e} \cdot \frac{1}{4\pi k n_0 e}$ are given in the second column of the following table. The number corresponding to any distance a is proportional to the number of negative ions that come within that distance of a fixed positive ion.

In the third column the quantity D represents the number that come within the distance a of an uncharged fixed point, by the process of diffusion, namely,

$$D = 4\pi K a (n_0 - \nu) = 4\pi k e (n_0 - \nu) \frac{a}{c} \cdot$$

The numbers corresponding to the radii 10^{-4} and 10^{-5} were calculated by use of the formula $\nu = 2K/(2K + au)$, which is obtained on the same hypothesis as that on which the values of ν were determined in the expression for I. The number given for the radius 10^{-6} was found from the expression given in the preceding section for the quantity $(n_0 - \nu)$ when au is small compared with $4K$, that is,

$$n_0 - \nu = \frac{n_0 au}{4K} \cdot$$

a in centimetres.	$\frac{I}{e} \cdot \frac{1}{4\pi k n_0 e}$	$\frac{D}{4\pi k n_0 e}$
10^{-4}	17	17
10^{-5}	1·8	1·1
10^{-6}	1·0	·015

A comparison between the numbers in the two columns

shows that the number of ions that come within the distance 10^{-4} centimetre of a positive ion is practically the same as the number that arrive within that distance of an uncharged centre by the process of diffusion. The electric force has therefore very little effect on the motion of the ions outside the sphere of radius 10^{-4} centimetre. At the distance 10^{-5} centimetre the electric force becomes distinctly appreciable, and the number that come within the distance 10^{-6} centimetre is about seventy times as great as the number that would come within that distance by the process of diffusion.

There is a limit to the distances for which the formula for the number I/e may be used, since the electric force increases rapidly as the radius a diminishes. The mean free path of an ion in air at atmospheric pressure is of the order 10^{-6} centimetre, so that in traversing the distance from $a = 2 \times 10^{-6}$ to $a = 10^{-6}$ centimetre it would acquire under the electric force a velocity of the same order as the velocity of agitation. The ordinary equations of motion are not accurate under these conditions, and it would be necessary to consider the accelerations of the ions in order to estimate the probability that a negative ion would continue to approach the charged centre when the distance 10^{-6} is attained.

149. Rate at which ions of opposite sign approach each other. When the positive ion is free to move, the numbers in the above table are all increased approximately in the same proportion by the factor $(k_1 + k_2)/k$. Hence if there be n_1 positive and n_2 negative ions per cubic centimetre in air at atmospheric pressure the number of ions of one sign that come within the distance 10^{-4} centimetre of an ion of opposite sign, or the number of encounters that take place at that distance, in the time δt is

$$4\pi (k_1 + k_2) e n_1 n_2 \delta t \times 17.$$

Substituting for $k_1 + k_2$ the value 930 (the sum of the velocities of positive and negative ions under unit electrostatic force in air at atmospheric pressure) the above expression becomes

$$1{\cdot}2 \times 10^4 \times e n_1 n_2 \delta t \times 17.$$

The number that recombine in the time δt as found experimentally is

$$3{\cdot}4 \times 10^3 \times e n_1 n_2 \delta t.$$

Hence, before an ion recombines, about sixty ions of opposite sign may come within the distance 10^{-4} centimetre, and on an average six of these approach within the distance 10^{-5} centimetre.

Langevin has investigated the problem of recombination by a simpler method, in which the effect of diffusion is neglected, and found that the quantity $4\pi(k_1+k_2)e n_1 n_2 \delta t$ represents the number of collisions between ions, and is independent of the distance. He concludes that this number is the upper limit of the number of possible recombinations, and the ratio $\alpha/4\pi(k_1+k_2)e$ has been defined as the ratio of the number that recombine to the number that come into collision.

A very simple result is thus obtained, but it is impossible on this theory to explain why all encounters do not result in recombination. For if the number of encounters at the distance a is equal to the number for a smaller distance a' then every pair of ions that are within the distance a must approach each other until they reach the distance a'. They could not subsequently move apart, if they move towards each other with the velocity $(k_1+k_2)\,e/r^2$ in the space between the spheres of radii a and a', and eventually they would recombine.

It may be seen from the above investigation that as the distance a diminishes, the effect of the electric force becomes large compared with the effect of diffusion, and if the ordinary equations of motion apply to very small values of a, the quantity I/e assumes the constant value $4\pi k e n_0$. But the effect of diffusion may be neglected only when $\epsilon^{-\frac{c}{a}}$ is small compared with unity, and since

$$c = \frac{ke}{K} = 5 \cdot 5 \times 10^{-6},$$

the distances at which I/e can become constant are of the order 10^{-6} centimetre.

In general, the equations of motion that have been used for the above calculations do not apply to distances of this order, since the velocities of the ions in gases at atmospheric pressure, due to the force e/r^2, is as great as the velocity of agitation when $r = 10^{-6}$ centimetre. Hence it is only when the pressure is very high that the number of encounters can be taken as being independent of the distance, and in that case α becomes equal to

$$4\pi(k_1 + k_2)e.$$

150. Recombination in an electric field. The question as to whether the coefficient of recombination is the same when the ions are in an electric field, as when no external forces are acting, may be considered by estimating the effect of an increase in the relative motion of the positive and negative ions on the number of encounters that take place between them. The external force X produces an average relative velocity $u_1 + u_2 = (k_1 + k_2)X$, and if this velocity is small compared with the velocity of agitation V, the effect of the electric force will be the same as if the velocities of agitation were increased in the proportion $1 + u^2/3V^2$, u standing for either u_1 or u_2. If the force X is less than 300 volts per centimetre, the ratio $u^2/3V^2$ will be less than 1/700, so that forces of that order would not alter the coefficient of recombination to an appreciable extent.

The same result follows from the experiments at atmospheric pressure, since the coefficient of recombination in an electric field, as found by Langevin's method, is practically the same as the coefficients found by the other methods where no electric force is applied while recombination takes place.

Also it appears from these considerations that an external

field of the order of 100 volts per centimetre can have practically no effect in altering the probability of recombination between a negative ion and the positive ion from which it was originally separated.

When ionization takes place by the emission of an electron from a molecule, the former may travel through a small distance without colliding with another molecule, and return to the original positive ion after having reached a maximum distance r.

The initial kinetic energy corresponding to the distance r is $e^2\left(\frac{1}{r_1}-\frac{1}{r}\right)$, r_1 being the radius of a molecule; hence if the electron comes to rest under the force of attraction at a distance exceeding 10^{-5} centimetre, the initial kinetic energy lies between the values e^2/r_1 and $e^2\left(\frac{1}{r_1}-10^5\right)$. Taking the radius of a molecule of air to be $1{\cdot}4\times10^{-8}$ centimetre, the limiting values of the initial kinetic energy are

$$\frac{e^2}{r_1} \text{ and } \frac{e^2}{r_1}(1-1{\cdot}4\times10^{-3}).$$

Since these limits are so narrow it is very improbable that the initial energy of any considerable number of the electrons will lie between them, and for the large majority of cases it may be assumed either that the trajectories are very short, or that the electron collides with molecules and a negative ion is formed.

In the first case the electric field, due to the force e/r^2 towards the positive ion, is of a high intensity, so that an additional force of the order of 100 volts per centimetre in a fixed direction will have no appreciable effect on the chance of recombination taking place.

In the second case, when a negative ion is formed, moving with a velocity of agitation in thermal equilibrium with the surrounding molecules, it may at first be only a short distance from the particular positive ion from which it was separated originally, but the probability of its recombining is not affected by the external force. In air at atmospheric pressure the probability that recombination will take place between two ions is about 1 : 60 when they are at a distance apart equal to 10^{-4} centimetre, so that even if all the negative ions were formed at such short distances from the particular positive ions from which they were originally separated, there would be no remarkable increase in the coefficient of recombination.

CHAPTER VII

THE FORMATION OF CLOUDS AND THE DETERMINATION OF THE ATOMIC CHARGE

151. Conditions under which clouds are formed. The formation of clouds in gases by the condensation of vapour takes place under various circumstances, but one condition which facilitates the process of condensation in a remarkable manner is the presence of small particles such as dust or the clusters of molecules that form round ions. These particles may be so small as to be invisible, but they act as nuclei round which vapour condenses, and as they grow in size they become sufficiently large to produce a visible cloud in the gas. Generally it is necessary to have the vapour supersaturated before condensation takes place on the small particles, and this state may be obtained either by a steam jet or by cooling a saturated gas by an adiabatic expansion. In some cases cooling by expansion is not required and clouds are produced in the presence of unsaturated vapour. For example, when moist phosphorus oxidizes in air, a cloud is formed which may be removed by bubbling the gas through sulphuric acid, but the cloud appears again if the air is passed into a vessel containing water. The same effect is obtained from newly prepared gases given off from acid solutions. In many cases these gases become cloudy in appearance when they are passed into a vessel containing water, but the cloud disappears almost completely when the gas is dried. The particles on which the moisture condenses in these cases probably contain a large percentage of acid, or other substance soluble in water, which absorbs the moisture from the surrounding gas, and the drop continues to grow until the solution of which it is composed becomes so dilute that the vapour pressure outside the drop is in equilibrium with the vapour pressure in the gas. A stable cloud which does not evaporate at ordinary temperatures is thus obtained.

When the particles in the gas are ordinary dust particles, or contain only a very small quantity of a substance soluble in water, the gas must be supersaturated in order to obtain condensation, the degree of supersaturation required depending on the size and nature of the particles.

152. Effect of small particles observed by Coulier and Aitken. If ordinary air is compressed in a vessel containing water, a cloud is formed when the air becomes saturated and is allowed to expand rapidly. This effect was observed by Coulier,* and he found that it was due to the dust that was present in the air of the room, for when the vessel was filled with air that had been filtered through cotton-wool no cloud was formed when the expansion took place. Coulier also found that after the dust was removed clouds were again obtained by expansion if a small hydrogen flame burns in the air for a short time, or if a platinum wire in the air is raised to a high temperature. The effect produced by the flame or the hot wire was also removed by filtering the air through a tube containing cotton-wool.

Similar results were obtained by Aitken† in a series of experiments in which two jets of vapour were directed into two vessels, one containing ordinary air and the other air that had been filtered through cotton-wool. It was found that the first becomes filled with a cloud when the vapour enters the vessel, but in the second the air remains clear and transparent. If the cloud is allowed to fall, some of the dust particles are removed, but many remain in the gas, for when the steam jet is admitted the second time another cloud is formed. Several such experiments may be made with the same gas, the cloud becoming less dense each time, and eventually only a few drops are formed, which may be seen to

* M. Coulier, Journal de Pharmacie et de Chimie, **22**, pp. 165 and 254, 1875.

† John Aitken, Trans. Roy. Soc. of Edin. **30**, p. 337, 1880–1.

fall like fine rain-drops. The experiments may also be made with a vessel connected to an air-pump and provided with a tube with a stop-cock through which filtered air may be admitted to the vessel. A small quantity of water is kept in the vessel to saturate the air, and when the stop-cock is closed supersaturation is obtained by reducing the pressure of the air with the pump.

When ordinary air was used a cloud was formed, but when the vessel was filled with air that had been introduced through a tube containing cotton-wool no cloud was produced on expansion.

A cloud may also be produced by either of these methods with air coming from a gas flame, when both the gas and the air have been freed from dust before combustion takes place. The particles on which the vapour condenses are therefore produced in the flame.

153. Effect of large ions obtained from flames. It has already been shown that the ions generated by a flame or hot wire increase in size as the gas cools, and thus act as nuclei on which the moisture tends to condense.

The velocity of these ions under a force of one volt per centimetre is of the order ·01 centimetre per second, so that they are much larger than the positive ions generated by Röntgen rays, which move with a velocity of 1·46 centimetre per second.

It follows from the expression, obtained in Section 85, for the velocity U of a charged particle moving through a gas under a force X, that the velocity is inversely proportional to the square of the diameter of the particle when it is large compared with a molecule of the gas. The expression for the velocity is $U = \frac{Xe}{m}.T$, T being the interval between collisions with molecules of the gas, and since the mass of the particle is large compared with a molecule, m in this case is the mass of a molecule of the gas. Also $T = 1/\pi a^2 nu$, u being the velocity of agitation of a molecule, n the number of molecules per cubic centimetre, and a the sum of the radii of the particle and a molecule. Hence the velocity is inversely proportional to a^2, and since the radius of a positive ion generated in air by Röntgen rays is about $3{\cdot}5\,r$, where r is the radius of a molecule, the radius of an ion in air obtained from a flame or a hot wire is of the order $50r$. The large ions are very efficient as nuclei for the condensation of water, and the degree of supersaturation required in order to obtain a cloud in air containing some of these ions is easily obtained.

154. Condensation on ions from a point discharge. The simultaneous appearance of ions and nuclei on which moisture tends to condense occurs in many experiments, and the phenomenon is well illustrated by the observations made by R von Helmholtz,* Richarz,† and other experimenters on the effect of an electric discharge from a metal point at a high potential in air on the appearance of a steam jet.

When water vapour is expelled in the form of a jet from a narrow orifice into the air it presents the well-known appearance of a greyish cloud. The exact form of the jet depends on the velocity of the vapour and various other circumstances, but in general appearance the colour does not alter under normal conditions. If a metal point connected with an electrostatic machine be brought near the orifice the appearance of the jet alters in a remarkable manner when an electric current passes through the vapour. R. von Helmholtz observed that the jet appears to be blue when a large quantity of electricity escapes from the point, which indicates the formation of a large number of small drops of uniform size.

If the current from the electrified point diminishes, the blue becomes tinged with white, which shows that larger drops are produced; then purple, red, yellow, and green tints appear, and finally, when the current is very small, pale blue of a higher order (of diffraction) reappears.

Richarz‡ afterwards found that condensation is produced

* R. von Helmholtz, Wied. Ann. **32**, p. 1, 1887.
† R. von Helmholtz and F. Richarz, Wied. Ann. **40**, p. 161, 1890.
‡ F. Richarz, Wied. Ann. **59**, p. 592, 1896.

in a steam jet by the action of Röntgen rays. The ions generated by the rays also act as nuclei on which condensation takes place when a saturated gas expands suddenly. This effect has been obtained by Wilson with an apparatus in which the expansion takes place very rapidly, and a high degree of supersaturation is produced. The rate at which expansion takes place in the apparatus used by Coulier and Aitken is too slow to produce any condensation on the ions generated by Röntgen rays or Becquerel rays, as these ions are much smaller than those generated by a flame or hot wire.

155. Apparatus used by C. T. R. Wilson to produce rapid expansion. The principle of the method devised by C. T. R. Wilson* to investigate the properties of the various kinds of nuclei that may be generated in a gas is illustrated in figure 35. The upper vessel A in which the condensation is observed may be of any shape to suit the conditions of the problem under investigation, but the rest of the apparatus containing the parts necessary to produce the expansion was constructed on the same principle in each case.

The upper vessel A is connected through a wide tube to the glass cylinder B (internal diameter about 2·7 cm.). The lower end of this cylinder is closed with a rubber stopper, through the centre of which a glass tube C passes (about 1 cm. diameter), the upper end of the tube being provided with an enlargement to serve as a guide to the light glass plunger P, which slides freely over it. The plunger is made from a thin-walled test-tube, the open end of which has been cut perpendicular to the sides and ground smooth. Its lower edge is always immersed in the mercury which fills the lower part of B, and thus the gas in A and the upper part of B is completely cut off from the air inside P. The external diameter of the plunger is 2 mm. less than the internal diameter of the outer tube; there is thus a space of 1 millimetre all round the tube. When the

* C. T. R. Wilson, Phil. Trans. A, **189**, p. 265, 1897; A, **192**, p. 403, 1899.

tap T_1 is open and there is thus free communication between the space inside P and the atmosphere, the plunger rises till the pressure in A only differs from the atmospheric pressure by an almost negligible amount, depending on the difference between the weight of the plunger and of the mercury displaced by the immersed part of its walls. If now communication with the atmosphere be cut off (by closing the tap

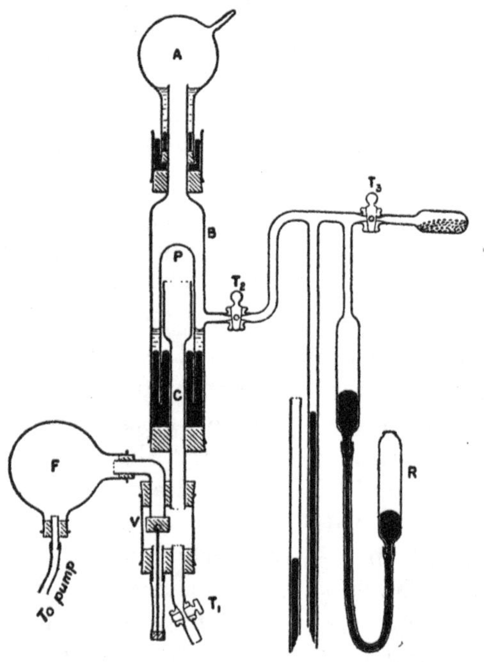

Figure 35.

T_1) and the space below the plunger be suddenly connected with the vacuum in F by means of the valve V, the plunger is driven through the mercury till it strikes the india-rubber, against which it remains tightly held by the pressure of the air above it. The mercury remains practically stationary while the thin edge of the plunger cuts its way through it. If T_1 be again opened, readmitting air into the space below the plunger, the latter rises to its original position and an expansion of the same amount may be repeated as often as may be required. To arrange for an expansion of any given amount the tap T_2 must be opened while the plunger is in contact with the india-rubber, that is, in the position it occupies immediately after an expansion. The mercury reservoir R is then fixed at such a level that the pressure in A, indicated by the gauge, is the desired amount below that of the atmosphere; the tap T_2 is then closed and the plunger made to rise by opening the tap T_1.

If B be the barometric pressure, then the pressure of the gas before expansion is

$$P_1 = B + m - \Pi,$$

where Π is the vapour pressure at the temperature of experiment, and m is the pressure (amounting to 1 or 2 mm. of mercury) required to keep the walls of the plunger immersed in the mercury (m is measured by finding the pressure which has to be applied to the air in A to keep the piston immersed to the same depth when the space below it is in communication with the atmosphere).

The pressure of the gas after expansion is

$$P_2 = B - p - \Pi,$$

where p is the difference of pressure indicated by the open mercury gauge when put in connection with A before the previous contraction.

Then the ratio of the final to the initial volume of the gas is (if Boyle's law holds)

$$\frac{V_2}{V_1} = \frac{P_1}{P_2} = \frac{B + m - \Pi}{B - p - \Pi}.$$

P_2, it will be noticed, is the pressure not at the moment when the expansion is completed but after the temperature has risen to its original value.

As the initial pressure P_1 in these experiments is always approximately equal to the atmospheric pressure, it is sufficient for many purposes to take $P_1 - P_2$ as a measure of the expansion without further reduction.

156. Experiments with dust-free air. The results obtained with the apparatus constructed on this principle are as follows, the ratio v_2/v_1 of the final to the initial volume of the gas being taken as a measure of the expansion.

When dust-free air saturated with water vapour expands adiabatically, condensation takes place throughout the gas if v_2/v_1 exceeds the value 1·25. Very few drops are formed, less than 100 per cubic centimetre, provided a second limit $v_2/v_1 = 1{\cdot}38$ is not exceeded. This rain-like condensation takes place in air, oxygen, nitrogen, and carbonic acid when the expansion lies between the above limits, and the phenomenon indicates the presence of nuclei other than the molecules of the gas or vapour. The number of drops that are formed depends on the nature of the gas, and none were observed in hydrogen. No direct experimental evidence was obtained to show that the nuclei are charged particles, although from theoretical considerations it might be supposed that the nuclei are the ions, which are always present in small numbers.

When the expansion exceeds the limit 1·38 a very dense cloud is produced whatever gas is present, and there is a rapid increase of the density with increasing expansion, the number of particles being probably many millions per cubic centimetre. In this case condensation takes place independently of any nuclei other than the molecules of water vapour or those of the gas with which it is mixed.

In conducting gases, or in gases traversed by rays, the phenomena of cloud formation are in some cases very complicated. The simplest and most definite effects are obtained when a gas is traversed by Röntgen rays or by rays from a radio-active substance. When the gas in the chamber A is ionized by these rays a comparatively dense cloud is obtained on expansion when the ratio v_2/v_1 exceeds 1·25, even when the radiation is very weak, but no condensation takes place with smaller expansions.

157. Condensation on ions generated by Röntgen rays. In order to show that the nuclei on which the moisture condenses are the ions that are generated by the rays, the expansion was produced in an apparatus of the form shown in figure 36, connected to the cylinder *B* which contained a plunger of the type illustrated in figure 35.

The upper part of the vessel *A* was covered with an aluminium plate fixed by means of sealing-wax to make the vessel air-tight. The water in the lower part of *A* and the aluminium plate were connected to the terminals of a battery of small accumulators, and a potential difference was thus established between the plate and the surface of the water; in the apparatus that was used the distance between the aluminium and the water was 1·6 centimetres.

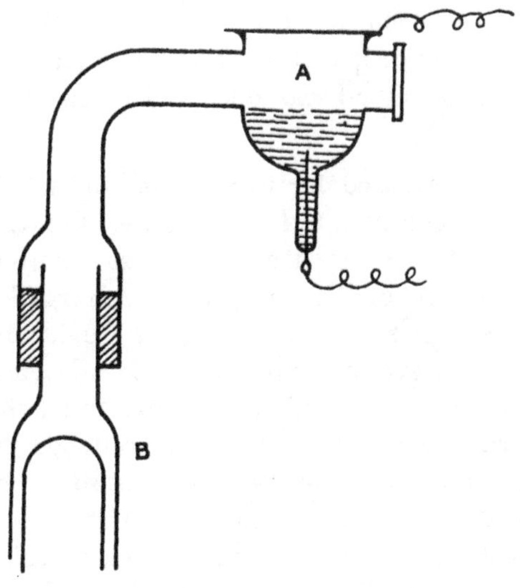

Figure 36.

On exposing the air to Röntgen rays, the expansion being between 1·25 and 1·38, a dense cloud was obtained in the absence of electromotive force, but when a difference of potential of 240 volts was established between the water and the aluminium plate only a very slight cloud appeared on expansion. An expansion of the same amount made 3 seconds after the rays were cut off gave a cloud in the absence of any difference of potential, whereas when the potential difference was 240 volts no effect due to the rays could be detected when the expansion was brought about 2 seconds after cutting off the rays.

Hence, since the nuclei on which the water condenses are removed from the gas by an electric force, they must be identical with the ions which give rise to the conductivity.

Similar results were obtained when the nuclei were produced by the action of uranium rays. In hydrogen the minimum expansion required to cause condensation on the ions was the same as in air. In carbonic acid the expansion found to be necessary to produce rain-like condensation in the absence of Röntgen rays was $v_2/v_1 = 1{\cdot}36$. Dense condensation began at the limit $v_2/v_1 = 1{\cdot}53$. When the rays were acting a few drops were observed with an expansion 1·339, and a dense cloud was obtained when the expansion was 1·45.

158. Nuclei generated by ultra-violet light. The phenomena are more complicated when gases are traversed by ultra-violet light. In moist air or moist oxygen, the light has a direct action on the gas, and uncharged nuclei are produced of various sizes depending on the intensity of the light and the length of exposure. Clouds may be thus obtained with very small expansions, and under the influence of strong ultra-violet light clouds are produced without any expansion even in unsaturated air. In some experiments of Lenard and Wolf* light rich in ultra-violet rays was admitted through a quartz window into a vessel containing moist dust-free air. They found that if the air was allowed to expand after being exposed for some minutes to the light a cloud was produced, showing that nuclei of some kind had been produced by the action of the ultra-violet rays. Similar results were obtained in steam-jet experiments. At first these experiments were regarded as indicating that under the action of ultra-violet rays small particles are given off from the surface of the vessel which act as nuclei for condensation.

159. Action of ultra-violet light on gases. Wilson's

* P. Lenard and M. Wolf, Wied. Ann. **37**, p. 443, 1889.

experiments, while confirming the experimental results obtained by Lenard and Wolf, show that the effect is produced by the action of the light on the moist air, and cannot be attributed to particles given off from the surface of the vessel containing the gas. In support of his explanation of this phenomenon, Wilson points out that the nuclei grow in size under continued exposure to the light, and also, when a long tube is traversed by a narrow beam of ultra-violet light, the nuclei are produced throughout the volume of the moist air, and are not confined to the neighbourhood of the quartz window or of the glass walls of the tube. In hydrogen, nuclei were produced under the influence of the ultra-violet rays, but never in very large numbers, and always requiring great supersaturation to make water condense upon them however long the exposure. Wilson suggests that the growth of the drops in oxygen is due to the formation of hydrogen peroxide, and in the case of air the formation of nitric acid may contribute to the effect.

An effect was also produced by sunlight in moist air contained in a vessel provided with a quartz window directly facing the sun, the ultra-violet rays being cut off when desired by means of a glass plate in front of the quartz window. No nuclei were produced on which moisture condensed with small expansions, even when the ultra-violet rays were concentrated by means of a quartz lens. But when the expansions exceeded $v_2/v_1 = 1{\cdot}25$ clouds were obtained in which the drops were plainly more numerous when no glass was interposed than when the glass screen was used to cut off the ultra-violet rays.

The glass did not, however, eliminate the whole effect produced by sunlight, for when the glass screen was interposed and expansions made, with and without an additional screen of black paper, a very marked difference was observed between the results of the expansions. Thus sunlight differs from the light from an arc lamp, or from a spark gap in a condenser discharge, since the light from the artificial sources does not facilitate the process of cloud formation after passing through a glass screen.

Further researches on the nuclei produced by ultra-violet light in air have recently been made by Lenard and Ramsauer.* They concluded that nuclei capable of acting on a steam jet were only produced in air containing traces of certain vapours. On freezing out the impurities no nuclei could be found, either with the long-waved light that passes through quartz or with the highly absorbable rays of the Schumann region. The nuclei reappeared if the purified air was allowed to come into contact with contaminating substances such as cotton-wool or india-rubber tubing.

160. Condensation on ions set free by ultra-violet light from a metal surface. When ultra-violet light falls on a negatively electrified metal plate, the ions which are set free also act as nuclei on which moisture tends to condense. Lenard and Wolf (loc. cit.) found that dense condensation takes place in a steam jet when a negatively electrified metal plate near the jet is exposed to the action of the light.

The effect of the ions may be also shown by means of expansions similar to those required to produce condensation on the ions generated by Röntgen rays. The phenomenon is more easily observed in hydrogen than in air, since the effect of the light throughout the volume of the gas is very slight. When light falls on a metal plate and an expansion just exceeding the critical value $v_2/v_1 = 1{\cdot}25$ takes place in the surrounding gas the density of the cloud that is produced depends on the potential of the plate. When the plate is positively charged no effect is observed due to the incidence of the light, but when the plate is negatively charged the density of the cloud increases as the negative charge is increased. For a certain value of the potential the density attains a maximum value, and for high potentials the effect of the light disappears. These observations show that the ions set free by the light act as nuclei for the condensation, since the number of ions in the gas at any instant varies with the electric force in the same way as the density of the

* P. Lenard and C. Ramsauer, Sitzungsberichte der Heidelberger Akademie, 5 parts, Aug. 1910–Aug. 1911. See also Sections 49, 50.

cloud. The same degree of supersaturation is required to obtain condensation on the ions set free from the metal surface by the light, as is required with ions generated by Röntgen rays. The ions produced by the two methods are therefore of the same order of magnitude.

161. Variations in the size of ions produced by point discharges. Wilson also investigated the ions repelled from a point discharge, which, as R. von Helmholtz had observed, give rise to condensation in a steam jet. For this purpose a glass vessel with a metal point projecting into the centre,

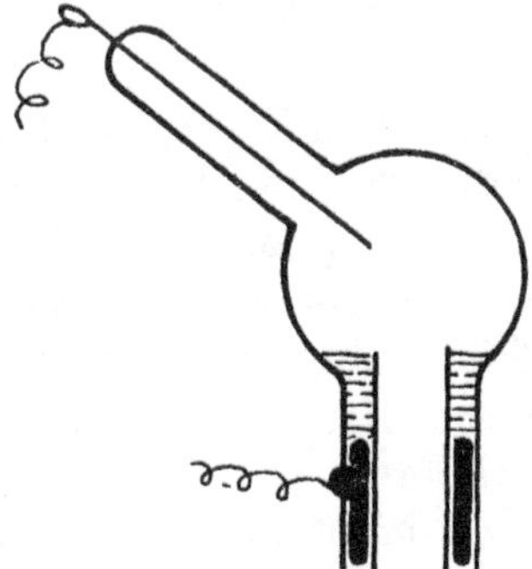

Figure 37.

as illustrated in figure 37, was connected to the expansion apparatus described above. The lower part of the vessel contained water, and connection was made through it to the moist surface of the glass surrounding the point. The current through the gas was maintained by a Wimshurst machine, one terminal being connected to the pointed wire and the other to the water in the bottom of the vessel.

The phenomena are simplest when the expansion takes place while the current is passing through the gas. In air and in hydrogen no drops are formed so long as the expansion is below the limit $v_2/v_1 = 1{\cdot}25$, but with expansions slightly above this limit very dense clouds are obtained. When the potential of the wire is negative clouds are always obtained with slightly lower expansions than when the point is positive. The cloud only lasts for a short time, since the charged particles are removed quickly from the gas by the high electric force required to maintain the discharge.

When the discharge is stopped before expansion takes place, by short-circuiting the wires leading to the electrical machine, the exact degree of supersaturation required to produce condensation in air is not so well marked. The value of the ratio v_2/v_1, at which condensation begins, diminishes as the time that elapses between the cessation of the current and expansion of the gas increases. During that interval the nuclei appear to grow in size, and after about thirty seconds a few drops are obtained when the expansion $v_2/v_1 = 1{\cdot}06$ takes place.

In pure hydrogen no corresponding increase was observed in the size of the nuclei that remain in the gas after the discharge is stopped. The effect in air is probably due to the formation of small quantities of nitric acid or of hydrogen peroxide which condense on the nuclei.

The tendency of ions to become larger while they are in air through which the discharge has passed is also shown by the change in the rate at which they diffuse. When ions are repelled by a point discharge into a stream of air in a tube, the coefficient of diffusion of those that have been carried a long distance by the stream is less than the coefficient of diffusion of ions that have travelled a short distance.

162. Relative efficiency of positive and negative ions, as nuclei for condensation of moisture. In his experiments with point discharges, Wilson observed a difference between the expansions required to produce condensation on ions coming from points that were positively and negatively charged, and subsequently he found that the positive and negative ions generated by Röntgen rays differed in a similar manner.* The apparatus used for this investigation consisted of a glass cylinder (figure 38) provided with a pair of horizontal plates, *A* and *B*, that acted as electrodes, one of the plates being used to close the upper end of the

* C. T. R. Wilson, Phil. Trans. A. **193**, p. 289, 1899.

cylinder. Connection was made with the expansion apparatus by a tube passing through the stopper in the lower end of the cylinder. The upper part of the cylinder was screened by a sheet of lead from the action of the rays, and ionization was produced only in a layer of air immediately above the lower electrode. When the electric force is established between the plates the ions of one sign move through the gas and are discharged by the upper electrode, and those of opposite sign are brought into contact with the lower electrode after passing through a very short distance in the gas. The gas therefore contains principally ions of one sign, those moving towards the upper electrode, at any instant while the current is passing.

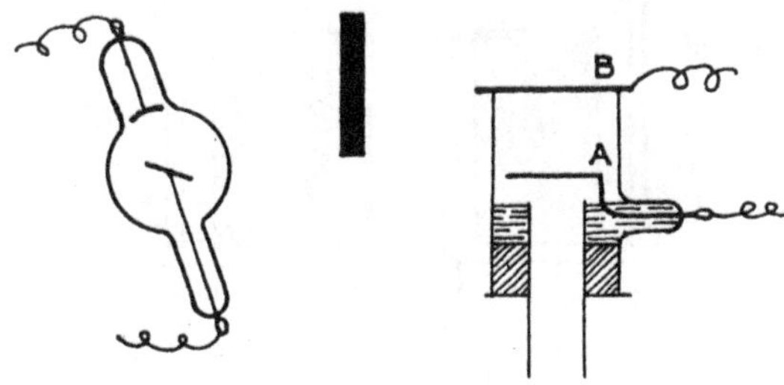

Figure 38.

Careful observations were made of the exact effect of expansions in which the ratio v_2/v_1 was increased gradually from the value 1·25 to 1·38.

When the upper plate was positive and an excess of negative ions was distributed throughout the gas, comparatively dense clouds were obtained when the expansion slightly exceeded the value $v_2/v_1 = 1{\cdot}25$, the density of the cloud being the same for all expansions between the limits 1·28 and 1·38.

When the electric force was reversed and positive ions were distributed throughout the gas, very few drops were obtained until the ratio v_2/v_1 attained the value 1·31; comparatively dense clouds were obtained for the expansion $v_2/v_1 = 1{\cdot}33$, and when the value 1·35 was attained the clouds were as dense as those obtained with negative ions.

163. Wilson's method of making visible the paths of ionizing particles in a gas. An interesting series of investigations has recently been made by C. T. R. Wilson * with an improved form of apparatus by which it is possible to render visible the tracks of ionizing rays through a gas by means of a cloud formed on the ions.

A shallow cylindrical cloud-chamber (figure 39) was used, 16·5 centimetres in diameter and 3·4 centimetres high.

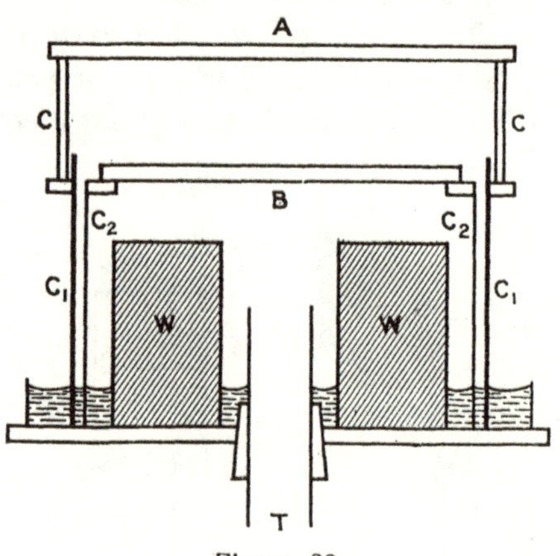

Figure 39.

The glass plate A formed the cover of the glass cylinder CC, which was fixed to the longer brass cylinder C_1C_1. The floor B, also of glass, was fixed to the brass cylinder C_2C_2, of the same length as the cylinder C_1C_1, but somewhat smaller in diameter, so that the floor and smaller cylinder acted as a movable plunger. When the inner cylinder rests on the base of the instrument, the plates are at their maximum distance apart d. The distance between the two plates A and B may be adjusted to any value $d-x$ by regulating the pressure of the air below the plunger. A sudden expansion of the amount $d/(d-x)$ is then obtained by connecting the space below the plunger through the wide tubing T with an exhausted vessel. The space below the

* C. T. R. Wilson, Proc. Roy. Soc. A. **85**, p. 285, 1911; **87**, p. 277, 1912.

Figure 40. α rays from a small quantity of radium.

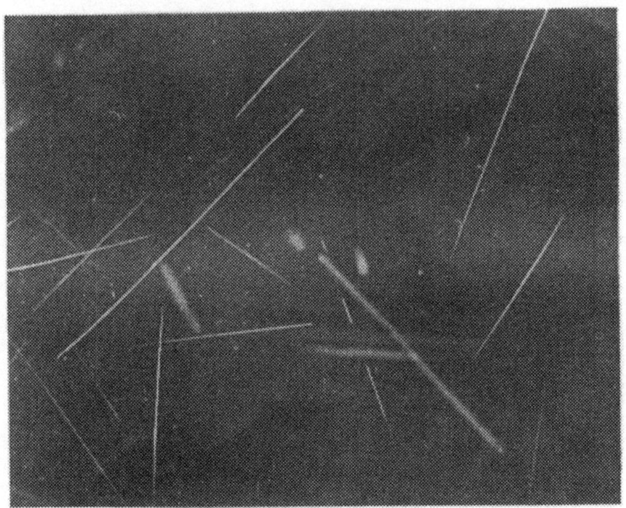

Figure 41. α rays from radium emanation.

Figure 42. β rays from radium.

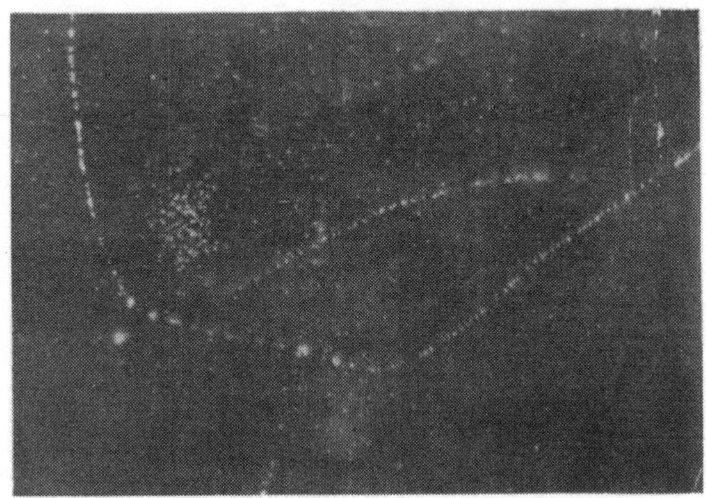

Figure 43. β rays produced by γ rays.

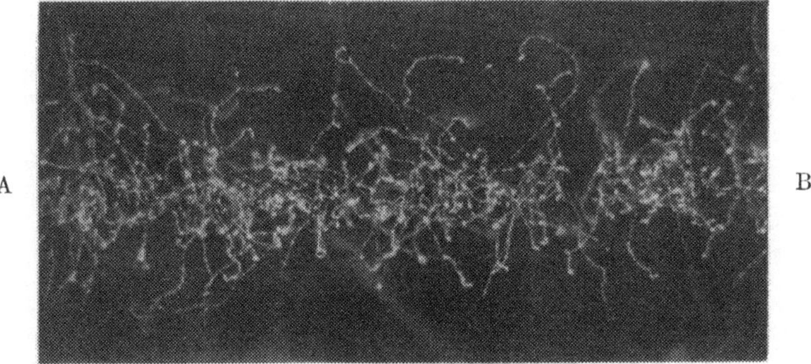

Figure 44. Effect of narrow beam of Röntgen rays along the line *AB*.

plunger is partially filled with a wooden cylinder W to reduce the volume of air passing through the tube T.

When expansion takes place, the motion at each point of the gas between the plates is in a vertical direction, and there are no irregular disturbances that would appreciably alter the positions of the ions after they are formed.

In order to remove the ions generated before the expansion takes place a strong electric field is maintained between the plates. The ions generated just after expansion in the supersaturated gas acquire a large mass very quickly, and do not move appreciably under the electric force. Using a strong light to illuminate the cloud chamber, Wilson has succeeded in photographing the clouds that are produced by various kinds of radiation. In some cases the pictures are obtained in sufficient detail to show the single drops formed by individual ions. Some examples of the photographs of the clouds formed along the tracks of ionizing rays are given in the accompanying illustrations.

The effect obtained with α rays emitted from a small quantity of radium on the end of a wire is shown in figure 40. The paths of the α particles are very distinct, since a large number of ions are generated by each ray, and a dense cloud is obtained along the path of each particle. The drops are so numerous that the cloud when illuminated appears to be a continuous bright line.

The α rays are generally straight over the greater part of their length, but they are nearly all bent, often abruptly, in the last two millimetres of their course. The length of the path as shown by this method is in good agreement with that previously found by experiments on the conductivity at various distances from the source.

Figure 41 illustrates the effect of the ionization produced by radium emanation. The bright lines originating at various points in the gas represent the tracks of the α particles emitted by molecules of the radio-active gas.

The tracks produced by the β rays (figure 42) are not so distinct, as a small number of ions are produced per centimetre of the path of each β particle. At irregular intervals along the path, groups of twenty or thirty ions appear to have been generated within a small area. The occurrence of a group may be interpreted as indicating that in certain cases when an electron is separated from a molecule by a β ray, the electron may have sufficient energy to ionize other molecules in its neighbourhood. The path of a β particle is nearly straight when the velocity is high, but as the velocity diminishes the number of ions generated per centimetre of its path increases, and at the same time the ray is more readily deviated by colliding with a molecule. At a later stage of slower velocity, the path of the particle becomes curled up and resembles the paths of the secondary corpuscular rays generated by Röntgen rays.

Figure 43 shows the effect of passing a narrow beam of γ rays through the cloud chamber. The tracks in this case are doubtless those of β particles starting from the walls of the vessel.

The effect obtained by a narrow beam of Röntgen rays traversing the cloud chamber after the expansion of the gas is shown in figure 44. The magnification was 2·45 and the width of the beam about 2 millimetres. The ionization is produced along a number of bent or curled-up lines, originating in the path of the primary beam. These are the tracks of electrons emitted with a high velocity from molecules of the gas that are ionized by the direct action of the rays. The illustration shows that the larger number of ions produced by Röntgen rays are due to secondary corpuscular rays. The penetrating power of the secondary rays emitted from the molecules of the gas is of the same order as that of the corpuscular secondary rays emitted by a metal surface when the rays fall upon it.

These experiments show the accuracy of the previous conclusions deduced by electrical methods, and in addition afford a method of deciding many uncertain points in connection with the processes of ionization which were not very clearly explained.

164. Principle of the method used to determine the charge on an ion. Determination of the number of drops

in a cloud. The atomic charge may be found by a direct method when condensation takes place on the ions, and the charged drops are made visible. The theory of the method, as it was first applied, is as follows. When a cloud is formed in a gas, the size of the drops may be deduced from the velocity of the drops as they fall through the gas under the action of gravity. The formula given by Stokes for the velocity v of a sphere of radius a moving uniformly through an incompressible fluid of viscosity μ, under the action of a constant force F, is $6\pi\mu av = F$. This formula is obtained on the hypothesis that the velocity of the fluid at the surface of the sphere is the same as that of the sphere, or in other words, that there is no slipping at the surface of the sphere. When the sphere consists principally of water, as in the case of the drops forming the clouds that may be obtained in newly-prepared gases, or when an expansion takes place in a moist gas, the density may be taken as unity and the force acting on the sphere is $\frac{4}{3}\pi a^3 g$. An approximate value of the radius of the drop may therefore be obtained from a determination of the velocity by means of the formula $2a^2g = 9\mu v$.

The weight w of each drop may therefore be calculated, and the number of drops n per cubic centimetre of the gas may be found from a determination of the weight of the cloud contained in a given volume. Thus $nw = W$, W being the weight of the cloud per cubic centimetre of the gas.

The charge on the ions in the gas may be found in electrostatic units with a sensitive electrometer, and if E be the charge per cubic centimetre the charge on one ion is E/n.

165. Electrical properties of gases evolved by electrolysis. The above method of finding the charge e was first used to determine the charge on a drop in the clouds formed in oxygen given off by electrolysis.* The electrical properties of the gases evolved by a current passing through a dilute

* Proc. Camb. Phil. Soc. 9, p. 244, 1897.

solution of sulphuric acid or caustic potash are similar to those of newly prepared gases obtained by other methods, which in many cases contain a large number of ions associated with particles that are comparatively large. The rate of diffusion of the charged particles is therefore very slow, and the gas retains a large proportion of the ions after passing through tubing or after bubbling through a liquid.

In the presence of water vapour the particles increase in size and form a visible cloud, but when the gas is dried the water evaporates from the drops and the gas becomes clear. The particles that remain after drying contain a substance that is soluble in water and reduces the vapour pressure, since a cloud is again formed in the presence of water vapour. Generally it is not possible to find the chemical composition of the nuclei in the dry gas, as very small quantities are involved, and traces of acid vapour, or other substances that would tend to condense on ions generated initially in the gas, may be formed during the chemical reaction.

When a cloud is formed in the lower part of a vessel, the gas in the upper part remains quite clear and the surface of the cloud sinks slowly, at the rate of about a centimetre in three minutes, which shows that the drops are very small, their radii being of the order 7×10^{-5} centimetre.

In the gases given off by electrolysis, as frequently happens when gases are evolved from solutions, there is a large excess of ions of one sign, which gives rise to a distribution of electricity throughout the volume of the gas, and the charge may be measured by introducing the gas into a large insulated vessel. Different intensities of electrification may be obtained, as the charge carried in the gas increases as the temperature of the electrolyte is raised. In some of the experiments made with oxygen evolved from a solution of caustic potash there was no appearance of a cloud, and no charge could be detected in the gas when the temperature of the cell was 10° centigrade, but when the temperature was raised above 20° a faint cloud appeared, and when the gas entered an insulated vessel a small electrometer deflection was produced. With higher temperatures the clouds become much denser, the weight of the water forming the drops being much greater than the weight of the vapour required to saturate the gas. It is possible, therefore, to find the amount of water forming the drops in a given volume of the gas by weighing, and the total number of drops may be compared with the charge in the same volume.

166. First determinations of the elementary charge. The method adopted was to pass the gas through a vessel containing water and then through an insulating tube into a series of small bulbs containing sulphuric acid, which removed the water vapour and also the cloud that had been formed as the gas passed through the water. The vessel containing the water was maintained at a constant temperature, so that the weight W_1 of the vapour required to saturate the gas was known. The increase of the weight of the acid W_2 varied with the density of the cloud, and in some cases W_2 was five times as great as W_1, the difference $W_2 - W_1$ being the weight of the water forming the cloud. The weight w of each drop was calculated from observations of the rate of fall of the cloud, and the number of particles $(W_2 - W_1)/w$ was determined. The number of particles thus found was very large, being of the order 10^7 per cubic centimetre of the gas.

In order to find the charge carried by the gas as it enters the drying apparatus, the bulbs containing the sulphuric acid were insulated, and the dry gas was passed into a large insulated receiver. The charges q_1 and q_2 acquired by the drying apparatus and the receiver were determined, the charge on the cloud carried into the drying apparatus being $q_1 + q_2$. The density of electrification E was thus found to be proportional to the number of particles n per cubic centimetre as deduced from the weight of the cloud.

Assuming that all the particles have charges of the same sign, the charge on each is E/n. The value of the atomic charge e as first obtained on this hypothesis * was 3×10^{-10} electrostatic units, approximately.

Subsequently, in 1898, the author found that there was an error in this number, as particles having charges of both signs were present in the gas. The charge measured was proportional to $e(n_1 - n_2)$ whereas the total number of particles was $(n_1 + n_2)$. The ratio $n_1 : n_2$ was determined by experiments on the conductivity of the gas, and n_1 was found to be four times n_2; hence $(n_1 + n_2)/(n_1 - n_2) = 5/3$. When this factor was taken into consideration the charge e was found to be 5×10^{-10} electrostatic unit.†

167. J. J. Thomson's determinations of the elementary charge. Charges on ions generated by Röntgen rays and ultra-violet light. The above principle was subsequently applied by Thomson to determine the charge on one of the drops in a cloud formed by expansion when the gas is ionized by Röntgen rays, and also when negative ions are set free by ultra-violet light from a metal surface. In the experiments with Röntgen rays the expansion was produced in a vessel similar to that shown in figure 36. The radius of the drop was determined from the rate at which the cloud fell through the gas. The weight W of the cloud in unit volume of the gas was estimated by calculating the amount of water vapour in the gas before expansion and the amount required to saturate the gas when its lowest temperature was attained. The number of drops N per cubic centimetre was thus deduced, namely $N = W/w$, w the weight of a drop being found by observing the rate at which the cloud fell. The total charge on the ions in unit volume of the gas was determined from the current between the upper electrode and the surface of the water in the lower part of the vessel. If N be the total number of ions, positive as well as negative, per cubic centimetre of the gas, U_0 the mean of the velocities of the ions under a potential gradient of one volt per centimetre, the current

* Proc. Camb. Phil. Soc. **9**, p. 244, 1897.
† Phil. Mag. (5) **45**, p. 125, Feb. 1898.

per square centimetre of the electrode is NeU_0X. The values * of e thus obtained are $e = 6{\cdot}5 \times 10^{-10}$ for ions generated in air and $e = 6{\cdot}7 \times 10^{-10}$ for ions generated in hydrogen. The negative ions set free from a metal surface by ultra-violet light,† were similarly found to have a charge $6{\cdot}8 \times 10^{-10}$ electrostatic unit.

As it was uncertain that the expansions used in the earlier experiments were sufficient to produce condensation on all the positive ions generated in the gas, another investigation was made, in which the expansion was carefully measured, and the fact ‡ that condensation takes place only on negative ions when the expansion v_2/v_1 is between the limits 1·25 and 1·31 was taken into consideration. In these experiments the gas was ionized by rays from a radioactive substance, and the value $3{\cdot}4 \times 10^{-10}$ electrostatic unit was found for the atomic charge.§ This value was considered to be more accurate than that previously obtained when the ions were generated by Röntgen rays.

Considerable difficulties are met with in these experiments, so that the final results are not very accurate, but they show that the charge on an ion generated by ultraviolet light, $6{\cdot}8 \times 10^{-10}$, is a quantity of the same order as that on ions generated in gases by Becquerel rays, $3{\cdot}4 \times 10^{-10}$. The experiments on the diffusion of ions show that these charges are equal.

168. H. A. Wilson's improvement in the method of estimating the charge e. An improvement in the above methods was introduced by H. A. Wilson, which eliminates the errors that occur in the determinations of the weight of the cloud and the charge on the ions in unit volume of the gas. The principle consists in applying an electric force to increase or diminish the velocity of the charged cloud. If the velocity v_1 of the upper surface of a cloud falling under the action of gravity be determined, the radius or the mass m of a drop may be found by Stokes's formula, the force acting

* J. J. Thomson, Phil. Mag. (5) **46**, p. 528, Dec. 1898.

† Ibid. (5) **48**, p. 547, Dec. 1899.

‡ See Section 161.

§ J. J. Thomson, Phil. Mag. (6) **5**, p. 346, 1903.

on the drop being $\frac{4}{3}\pi a^3 g$. In an electric field of intensity X the force on the drop is $Xe+mg$, and if v_2 be the corresponding velocity, then the charge e may be determined from the equation

$$\frac{v_1}{v_2}=\frac{mg}{mg+Xe},$$

since the velocities are proportional to the forces. In the experiments made on this principle the ions were generated by Röntgen rays in the space between two horizontal plane electrodes, contained in a glass vessel connected with an expansion apparatus similar to that illustrated in figure 35. The expansion was adjusted to produce condensation principally on the negative ions, so that the electric force should act in the same direction on all charged drops forming the cloud. When a potential difference of the order of 2000 volts was established between the electrodes at one centimetre or half a centimetre apart, the velocity of the particles, as determined by the motion of the upper surface of the cloud, was considerably altered. Thus in an experiment in which the velocity was ·0193 centimetres per second while the drops fell under the action of gravity, the velocity obtained with an electric force of 16 electrostatic units was ·031 per second. The mean value of the charge e deduced from these experiments was $3{\cdot}1 \times 10^{-10}$ electrostatic unit.*

169. Millikan's recent determination of the charge e. Recently, a higher degree of accuracy has been obtained in the determination of the value of e, and the most reliable results do not differ by 10 per cent. from the number $4{\cdot}5 \times 10^{-10}$.

An interesting series of investigations has been made by Millikan,† who, in collaboration with Begeman, made experiments at first on the clouds formed by expansion, and determined the charge by the method devised by H. A. Wilson, obtaining the value $4{\cdot}06 \times 10^{-10}$ for the atomic charge.

* H. A. Wilson, Phil. Mag. (6) **5**, p. 429, 1903.
† R. A. Millikan and L. Begeman, Phys. Rev. **26**, p. 198, 1908.

Subsequently,* new modifications were introduced, and instead of observing the surface of a cloud, a single drop was kept under observation and its motion in an electric field was determined with a short-focus telescope. Individual drops in the clouds formed by expansion were thus found to have different charges, but in each case the charge was an exact multiple of the smallest charge that was observed, which was $4 \cdot 65 \times 10^{-10}$. For the purpose of these calculations a special determination was made of the coefficient of viscosity of saturated air. It was found to be higher than that of dry air by about 2 per cent., and the low value of e found in the previous determination was attributed to the error in the value of μ used in Stokes's formula for the velocity in terms of the radius when the drop is falling under the action of gravity.

Further experiments were made with oil drops made by a sprayer, instead of the drops formed by the expansion of a moist gas. The effect of evaporation was thus eliminated and it was possible to observe a drop for several hours. In order to charge the drop, the air in which it was suspended was ionized by radium or Röntgen rays, and when a potential difference was established between the electrodes, the addition of one or more ions to the charge on the drop was indicated by a change in the velocity. Experiments were made with different charges on the same drop, and, as before, each charge was found to be an exact multiple of the same atomic charge.

The accuracy of Stokes's formula for the velocity of a sphere through a viscous liquid was also tested, since in its simple form it applies only to the case in which the velocity of the fluid at the surface of the sphere is equal at each point to the velocity of the surface. It is easy to see that when the radius of the sphere is small compared with the mean free path of a molecule of the gas, the velocity that it acquires under a constant force is inversely proportional to the square of the radius, so that Stokes's formula cannot apply to very small spheres.

* R. A. Millikan, Phil. Mag. (6) **19**, p. 209, 1910.

In the case of the drops used in Millikan's experiments and also in the earlier experiments with clouds, the radii of the drops were large compared with the mean free path, so that the velocity is given approximately by Stokes's formula, but in order to reach a high degree of accuracy it is necessary to introduce a correction depending on the ratio of the mean free path of a molecule of the gas to the diameter of the sphere.

170. Correction to Stokes's formula. This problem has been investigated mathematically by Cunningham,* who found expressions for the velocity of a sphere in a gas when the collisions of the molecules with the sphere are of the nature of impacts between two smooth elastic bodies, and also for the case of collisions where the velocities of the molecules on rebounding from the surface are unconnected with the velocities with which they collide. If a certain fraction f of the collisions are of the first type and the remainder of the second type, the force F required to maintain a velocity V in a gas of viscosity μ is given by the formula

$$F\left[1+\frac{1 \cdot 63\, l}{(2-f)\, a}\right]=6\pi\mu a V,$$

l being the mean free path of a molecule and a the radius of the sphere. Millikan found from his experiments that the charges obtained with different spheres were the same, when V is given by an expression of the above form and the additional term is $\left(1+\cdot 817\frac{l}{a}\right)$. In other words, the formula that is indicated by the experiments is practically of the same form as that given by Cunningham when the quantity f is zero.

When these corrections are made, Millikan † gives the number $4 \cdot 891 \times 10^{-10}$ as the value of e. This value has subsequently been corrected to $4 \cdot 774 \times 10^{-10}$.

171. Millikan's experiments on positive and negative charges. Millikan also made an investigation to determine

* E. Cunningham, Proc. Roy. Soc. A. **83**, p. 357, 1910.
† R. A. Millikan, Phys. Rev. **32**, p. 349, 1911.

whether any of the ions generated in the gas had double charges. In his experiments on the clouds formed by expansion, many of the drops were observed to have more than one atomic charge, a result that was also obtained by H. A. Wilson. This may be due to different causes, such as the coalescence of drops with single charges, and from his later experiments Millikan concludes that if any ions with double charges are generated in the gas they are very few under any circumstances; much less than 9 per cent. of the total, as observed by Franck and Westphal for positive ions generated by Röntgen rays in their experiments on diffusion in air at atmospheric pressure. The method finally adopted was to observe the variations of the motion of an oil drop when the air was ionized by rays of low intensity, but differing in penetrating power, and it was found that the changes in the velocity correspond to an increase in the charge by the amount e in the large majority of cases, and only on very rare occasions was the charge increased by the amount $2e$.

It is difficult to reconcile the results of the different investigations of this problem, and at present it is not clear under what conditions particles with double charges may be produced. The experiments on diffusion show that all ions generated by the different kinds of rays have single atomic charges, except the positive ions generated by Röntgen rays of a certain type.

172. Rutherford and Geiger's determination of the charge on an α particle emitted by a radio-active substance. A special method has been devised by Rutherford and Geiger to find the positive charge on the α particle emitted by a radio-active substance. The principle of the method is to measure the positive charge lost by a specimen of the radio-active material, and also the total number of the α particles that are emitted. The latter quantity was determined by two methods which gave the same result. The α particles were admitted through a small aperture in a vessel containing a pair of electrodes and also a fluorescent screen opposite the aperture. (The aperture was covered with very thin foil through which the α particles penetrated, and after traversing the gas between the electrodes impinged on the screen.) The advent of each particle was indicated by the appearance of a bright spark on the screen, such as are seen in the spinthariscope designed by Crookes to show the emission of α particles from radium. The number of particles that passed through the aperture in a given time were counted by the number of bright spots that occur on the screen. It was also found that when the gas was reduced in pressure and a high potential was established between the electrodes, a comparatively large instantaneous current passed between the electrodes when an α particle traversed the gas. The total charge on the ions produced by each particle along its path between the electrodes was not sufficient to produce an appreciable effect, but when high potentials were used a large additional number of ions were generated by the collisions with the molecules, so that the number that arrived at the electrode was increased by a large factor. The electrometer deflections that occur when each α particle passes through the gas are not necessarily the same, as the particles may traverse different paths, but that is unimportant, as in this case it is only necessary to observe the number of electrometer disturbances that occur in a given time.

The detector was placed in different positions in the field traversed by the rays, and the number of α particles emitted per second in various directions was determined, and an estimate was thus made of the total number emitted by the radium in a given time. The total positive charge lost by the radium in the same time having been determined, the charge on each particle was found. The value of this charge* is $9{\cdot}3 \times 10^{-10}$, which is probably twice the atomic charge, so that these experiments lead to the value $4{\cdot}65 \times 10^{-10}$ for e.

* E. Rutherford and H. Geiger, Proc. Roy. Soc. A, **81**, p. 141, 1908.

173. Regener's determination of the charge on an α particle. A determination of the charge on the α particle was also made by Regener,* who found the value $4{\cdot}79 \times 10^{-10}$ as half the charge on an α particle emitted by polonium. As in the above method, the number of α particles were estimated by the number of scintillations produced on a fluorescent substance, and the charge on each was obtained from the total charge on all the particles that were emitted.

174. Perrin's determination of N, the number of molecules per cubic centimetre of a gas. Application of the kinetic theory to particles suspended in a solution. The charge e may also be found from the number N of molecules per cubic centimetre of a gas, since the product Ne has been determined accurately. A large number of approximate estimates have been made of N, depending on determinations of various atomic quantities, but there are very large differences in the values found by the different methods. The most reliable method is probably that recently devised by Perrin, which is founded on a study of the Brownian movement of small particles suspended in a liquid. These experiments are of interest in several respects, as they exhibit the nature of the motions of systems of particles and afford direct proof of many of the results indicated by the kinetic theory. The particles being large compared with molecules, their motion is comparatively slow, and by the aid of a microscope they may be observed moving about in a clear liquid. In the case of the Brownian movement the particles are uncharged and the forces that act are gravitational. Also, since the density of the surrounding fluid is of the same order as that of the particles, the vertical component of the hydrostatic pressure must be taken into consideration, so that the force acting on each particle is $(m-m')\,g$, where m is the mass of a particle and m' that of the displaced fluid. Under these conditions Maxwell's equation of motion in the vertical direction z becomes

$$\frac{pw}{K} = -\frac{dp}{dz} + n\,(m-m')\,g,$$

* E. Regener, Sitz.-Ber. der K. Preuss. Akad. der Wiss. **38**, p. 948, 1909.

p being the partial pressure and n the number of particles per cubic centimetre.

When the steady state is reached, the layers of equal density are horizontal, and the velocity w is zero. The corresponding distribution is obtained by integrating the above equation, and if z be the distance between two planes at which the distributions are n_1 and n_2, the ratio n_1/n_2 is given by the formula

$$\frac{1}{z}\log\left(\frac{n_1}{n_2}\right) = \frac{3(m-m')g}{mV^2}.$$

When the mass m of a particle and the mass m' of an equal volume of water are known, the kinetic energy of translation $mV^2/2$ may be found by measuring the ratio n_1/n_2.

175. Number of granules in an emulsion. An emulsion suitable for these researches may be obtained by dissolving some gamboge in distilled water. This solution contains particles of various sizes, and it is necessary first to prepare an emulsion containing particles all of approximately the same size. By a method of fractional centrifuging the smaller particles of diameter less than a certain value a_1 were removed, and from the remaining sediment a uniform emulsion was obtained consisting of particles of diameter approximately equal to a_1.

The mass m of a particle and the density were determined by different methods which gave the same result. One of the methods was as follows: let μ_1 be the mass of distilled water that fills a specific gravity flask, μ_2 the mass of the same volume of the emulsion, and μ_3 the mass of the granules remaining in a vessel after the water has been removed by evaporation. The mass of the intergranular water is $\mu_2-\mu_3$ and the mass of water that occupies the same volume as the granules is $\mu_1-\mu_2+\mu_3$. From these measurements the mass m of each particle and the mass m' of an equal volume of water may be found, when the total number of particles in a given volume of the solution is obtained. The number of particles per cubic centimetre of the emulsion may be found by counting the particles in a small volume of the preparation contained in a shallow vessel. It would be impossible to count all the particles while they are moving about in the liquid, but in some of the experiments Perrin observed that in a slightly acid solution the granules adhere to the glass vessel that contains the emulsion. A specimen of the emulsion under observation was therefore mixed with a certain volume of acidulated water and a small quantity of the mixture was placed in a shallow vessel such as is used for microscopic observation. The granules disappear from the solution and become attached to the bottom of the vessel and the surface of the cover-glass. The glass surfaces were marked in small squares of known area, so that when the particles contained in some of the squares were counted and the average number obtained, it is possible to calculate the number ν per cubic centimetre of the original emulsion.

176. Distribution of granules in a shallow vessel. Observation of Brownian movement. When the granules are suspended in pure water they do not adhere to the glass but move about in the liquid, and in a shallow vessel the distribution may be studied with the aid of a microscope. The particles collect near the bottom of the vessel, but Perrin observed that in layers at small distances from the bottom some of the particles move about in the liquid.

In order to find the distribution at different heights, the preparation was placed in a shallow cell (·1 millimetre in depth) and mounted on the stage of a microscope. Only those granules can be seen clearly which are present in the very thin horizontal layer that is in focus. By raising or lowering the microscope the granules in another layer may be seen. The vertical distance z between the layers which enters into the equation of distribution may be obtained by multiplying the displacement z' of the microscope by the relative refractive index of the two media which the cover-glass separates.

For the larger granules (exceeding in diameter 5×10^{-5} centimetre), the numbers in the different layers were obtained by instantaneous photographs, and the ratio n_2/n_1

was found by counting the images of the particles. For the smaller granules this method was not practicable and it was necessary to make direct observations. This was done by limiting the field of view so that only five or six particles at the most should be visible at the same time. It was then possible to count the number in the small area at any instant, and the mean of a large number of these observations was taken as being proportional to the number n_1 in the layer that was in focus. Similar observations were then made at a different level and a number proportional to n_2 was found. The value of the quantity mV^2 was thus determined from the equation

$$\frac{1}{z}\log\left(\frac{n_1}{n_2}\right) = \frac{3(m-m')g}{mV^2}.$$

Assuming that the kinetic energy of the granule is equal to the mean kinetic energy of translation of a molecule in thermal equilibrium with the fluid, the number of molecules per cubic centimetre N of a gas at atmospheric pressure was found from the formula for the pressure $\frac{1}{3}mNV^2 = 10^6$. The corresponding value of e obtained by Perrin * from his first experiments was $4{\cdot}1 \times 10^{-10}$ electrostatic unit, and on repeating the investigation † the value $4{\cdot}25 \times 10^{-10}$ was obtained.

177. Einstein's investigation of the displacement of a particle. Another experimental determination was made on a different principle, suggested by Einstein's theoretical investigation of the displacement of a particle that moves in a solution and exerts a partial pressure equal to that of a molecule in thermal equilibrium with the solution.

In this investigation it was shown that if a particle start from a point P_0, and arrive at a point P in a given time t, the mean square of the distance between the two points $(\overline{x^2}+\overline{y^2}+\overline{z^2})$ may be expressed in terms of gas constants. The mean square $\overline{x^2}$ of the perpendicular distance of the

* J. Perrin, Comptes rendus, **145**, p. 967, 1908.
† Ibid., **152**, p. 1380, 1911.

point P from a plane through P_0 is given by the formulae *

$$\overline{x^2} = 2Kt = t\left(\frac{RT}{N}\right)\frac{1}{3\pi\mu a},$$

$\left(\frac{RT}{N}\right)$ being $\frac{2E}{3}$, where E is the kinetic energy of a molecule in a gas at temperature T, μ the coefficient of viscosity of the liquid, and a the radius of the granule. The second equation was obtained on the hypothesis that the velocity of a small particle due to an applied force is given by Stokes's law.

178. Smoluchowski's investigation of the displacement. A more direct investigation is that given by Smoluchowski,† which is founded on the theory of probability. The collisions between a particle and the molecules of a surrounding gas occur at various distances from the initial position of the particle, and an expression may be found for the probability that the nth collision should take place within any given area. The probability $p_n(x)\,dx$ that in the time t the displacement in the direction of the axis of x should lie between the values x and $x+dx$ was thus found to be

$$p_n(x)\,dx = \frac{\beta}{\sqrt{\pi t}}\,\epsilon^{-\frac{\beta^2 x^2}{t}}\,dx,$$

from which the mean distance $\bar{x}$ of the particle from the plane in the time t was found to be

$$\bar{x} = \frac{1}{\beta}\sqrt{\frac{t}{\pi}},$$

where

$$\beta = \sqrt{\frac{3}{4\lambda c}}.$$

In this notation the coefficient of diffusion K is $\frac{\lambda c}{3}$ so that $\beta = \frac{1}{\sqrt{4K}}$.

This result is the same as that found by Einstein, since

* A. Einstein, Ann. der Phys. **17**, p. 549, 1905, and **19**, p. 378, 1906.

† M. Smoluchowski, Bull. Int. de l'Acad. des Sc. de Cracovie, p. 202, 1906.

the mean square $\overline{x^2}$ of the displacement of the particle in the time t, as obtained from the value of $p_n(x)$, is

$$\overline{x^2} = \frac{t}{2\beta^2} = 2Kt.$$

The case in which a particle is large compared with a molecule was also considered, and the following formula was obtained for the mean displacement Δ per second of a sphere of radius R, when the velocity due to an applied force is given by Stokes's law,

$$\Delta = \frac{4\sqrt{2}}{9\sqrt{\pi}} \frac{c\sqrt{m}}{\sqrt{\mu R}},$$

$\frac{mc^2}{2}$ being the kinetic energy of translation of a molecule of the gas and μ the coefficient of viscosity. This formula is only approximate, as it involves the assumption that the radius of the sphere is small compared with the mean free path, and Stokes's law does not apply accurately in that case.

Smoluchowski * also explains the fallacy in an objection, often considered decisive, to the application of the kinetic theory to particles which are large compared with molecules of the surrounding gas or liquid. It has been pointed out that the velocity communicated to a spherical particle of diameter ·003 millimetre by a collision with a molecule of hydrogen is only 2×10^{-6} centimetre per second, which would not be visible under a microscope, and it has been stated that the impacts acting on all sides would cancel and no perceptible motion would ensue. This resembles the error in supposing that the total amount lost or gained by a player in a game of chance does not exceed the amount staked on each event. It is well known that in general the chances do not balance exactly, and the total amount lost or gained may increase with the number of times the game is played.

Smoluchowski illustrates this remark by a simple calculation based on the supposition that the gains and losses are

* M. Smoluchowski, loc. cit., p. 577.

equally probable, the total number being n. In considering all possible combinations, the probability of the occurrence of m gains and $n-m$ losses, or an excess of $2m-n$ gains over losses, is

$$\frac{n!}{2^n\, m!\, (n-m)!},$$

and when n is large the mean total gain or loss ν is

$$\nu = \sqrt{\frac{2n}{\pi}}.$$

When a particle of large mass M is surrounded by molecules of mass m, the velocity transmitted to the mass M by one collision with a molecule is of the order mc/M, c being the velocity of the molecule. When M is the mass of a particle of an emulsion the velocity mc/M is of the order 10^{-7} centimetre per second, but in a gas more than 10^{16} collisions occur per second, and in a liquid more than 10^{20}, some accelerating and some retarding the motion of the mass M. The mean excess of the number of collisions of either kind is of the order 10^8 or 10^{10} in one second; hence the velocity may rise to 10 or 10^3 centimetre per second. This shows that the objection to the theory is not justified, but the final result is not exact, since the change in the velocity C of the large mass is not the same at each collision, it depends on the absolute value of C, also the collisions that retard the motion occur more frequently than collisions that accelerate when C is large. These two factors oppose the unlimited increase of the velocity C. The final result obtained on the principle of the equipartition of energy is that the mean kinetic energy of translation of the sphere $\frac{MC^2}{2}$ is equal to the mean kinetic energy of translation of a molecule $\frac{mc^2}{2}$.

179. Value of N deduced from determinations of the displacements of a particle. An experimental investigation of the displacement of a particle in suspension has led to a confirmation of the theoretical conclusions. When the emulsion consisting of particles of known diameter and mass is sealed in a capillary tube, which is placed in a vertical position in a thermostat to prevent convection currents, the emulsion leaves the upper part of the liquid and descends the same amount each day. Perrin thus found that the vertical velocity w_1 due to gravity was in agreement with Stokes's law, $6\pi\mu a w_1 = (m-m')g$.

Also, in collaboration with Chaudesaigues and Dabrowski, Perrin determined the mean square of the horizontal displacement per second of a particle of the emulsion by observing the position of the particle in a camera lucida at definite intervals. The mean value $\overline{x^2}/t$ for particles of known diameter was thus found, and by means of Einstein's formula

$$\frac{\overline{x^2}}{t} = \left(\frac{RT}{N}\right) \cdot \frac{1}{3\pi\mu a},$$

the kinetic energy $E = \frac{3}{2}\left(\frac{RT}{N}\right)$ was determined. The result * of this investigation gives $e = 4{\cdot}2 \times 10^{-10}$, which is in good agreement with that obtained by the first method from determinations of the distribution in a shallow vessel when equilibrium is established.

180. Theory of Brownian movement. Determination of K for particles in suspension. With regard to these experiments it may be mentioned that the equation given by Einstein for the displacement of a particle is a particular example of a more general formula which applies to any particle and does not involve Stokes's law which applies only to spheres. The general expression for the kinetic energy of agitation of a particle may be obtained from elementary considerations, in terms of the velocity w_1 due to gravity and the quantity $\overline{x^2}/t$.

The ordinary definition of the coefficient of diffusion K may be adopted, which is that $K\dfrac{dn}{dx}$ represents the excess of the number of particles that move across unit area of

* J. Perrin, Comptes rendus, **152**, p. 1380, 1911.

a plane perpendicular to the axis of x per second in one direction, over the number that cross in the opposite direction when no external forces are acting. It follows immediately from this definition that the rate of change of the mean square R^2 of the distances of the particles in any distribution from any point is $6K$, provided that the particles do not come into contact with a boundary*. Since $R^2 = \overline{x^2} + \overline{y^2} + \overline{z^2}$, and the motion is symmetrical with respect to the three rectangular co-ordinates, the above result may be expressed by the equations $\frac{dR^2}{dt} = 3\frac{d\overline{x^2}}{dt} = 6K$.

The latter equation gives on integration $\overline{x^2} - \overline{x_0^2} = 2Kt$, or, if the particles start from the plane $x = 0$,

$$\overline{x^2} = 2Kt. \qquad (1)$$

Hence, when one particle is considered, the mean of the squares of the distances traversed in the time t, in a given horizontal direction x, is $2Kt$. The rate of diffusion K is thus found directly from the observations of the positions of a single particle at various intervals of time t.

181. Remarks on Perrin's determinations. Values of e recently obtained by different methods. The connection between the rate of diffusion and the kinetic energy of translation of the particle may be deduced from the general proposition established in Section 84, which applies to particles moving either in gases or in liquids. It may be stated as follows: If a particle of mass m move irregularly through a medium with a velocity of agitation V, the mean value of the rate of change of the square of its distance from any point dR^2/dt is connected with the velocity U acquired under a force F by the equations

$$\frac{dR^2}{dt} = 2(l+\lambda)V$$

and

$$U = \frac{F}{m} \cdot \frac{(l+\lambda)}{V}.$$

* See Section 80.

Eliminating the unknown factor $(l+\lambda)$, and substituting for $\frac{dR^2}{dt}$ its value $\overline{3x^2}/t$, the following relation is obtained:

$$\frac{\overline{x^2}/2t}{U} = \frac{mV^2}{3F}. \qquad (2)$$

When U is the vertical velocity w_1 due to the force $(m-m')g$ acting on a particle in a solution, equation (2) becomes

$$\frac{\overline{x^2}/2t}{w_1} = \frac{K}{w_1} = \frac{mV^2}{3(m-m')g}. \qquad (3)$$

In his first researches Perrin determined the equilibrium distribution of an emulsion in a shallow vessel. In that case the number of particles crossing a horizontal plane in an upward direction due to diffusion is balanced by the downward motion due to gravity. Hence n is given by the equation

$$K\frac{dn}{dz} = nw_1,$$

or

$$\frac{1}{z}\log\left(\frac{n_1}{n_z}\right) = \frac{w_1}{K} = \frac{3(m-m')g}{mV^2}.$$

Thus Perrin's results do not involve the hypothesis that the particles are spherical or that the velocity w_1 due to gravity is accordance with Stokes's law. It is unnecessary to determine the radius of a particle in the emulsion, or the mass m, in order to find the kinetic energy $\frac{mV^2}{2}$. When the difference between the mass of a particle and the mass of an equal volume of water $(m-m')$ is found, the energy $\frac{mV^2}{2}$ may be determined either from observations of the distribution in the steady state, which give the value of $\frac{1}{z}\log\frac{n_1}{n_2}$ as in Perrin's first experiments, or from the determinations of the quantities $\overline{x^2}/2t$ and w_1 that were made in the subsequent experiments.

The most accurate determinations of the atomic charge,

therefore, appear to be the following: $4 \cdot 2 \times 10^{-10}$ and $4 \cdot 25 \times 10^{-10}$, deduced by Perrin from determinations of the number of molecules per cubic centimetre of a gas, $4 \cdot 65 \times 10^{-10}$ obtained by Rutherford and Geiger, and $4 \cdot 79 \times 10^{-10}$ obtained by Regener as half the charge on an α particle, and the value $4 \cdot 77 \times 10^{-10}$ obtained by Millikan from measurements of the charge on an ion in a gas.

CHAPTER VIII

IONIZATION BY COLLISIONS OF NEGATIVE IONS WITH MOLECULES OF A GAS

182. Conductivity of ionized gases at low pressures. Ionization by collision. In the preceding chapters a number of experiments are described in which currents are obtained between electrodes in a gas when the ions are generated by rays emanating from an external source. As the electric force acting on the ions is increased, the current increases and approaches the saturation value, which is attained when all the ions of one sign generated by the rays are collected on the electrode of opposite sign. For a range of forces above a certain value X_1 the current differs only by a small percentage from the constant maximum value, and no appreciable increase takes place until much larger forces are applied. The force X_2 required to produce a definite increase above the saturation current depends on the pressure. When the gas is at atmospheric pressure and the electrodes are separated by a distance of the order of one centimetre, the force X_2 is very large; but as the pressure is reduced X_2 diminishes, and with a suitable pressure a considerable increase may be obtained in the current with a potential difference of fifty volts between the electrodes. The connection between the current and the electric force is then shown by the part *BC* of the curve figure 1, and currents much greater than the saturation currents may be obtained, while the rays from the external source are acting. The first experimental observation of this phenomenon was made by Stoletow,* who used ultra-violet light to pro-

* See Section 211.

duce the initial ionization, and subsequently corresponding effects were obtained by Kreusler,* and von Schweidler † who used Röntgen rays.

It has already been mentioned that the increase in the conductivity corresponding to this part of the curve may be explained on the hypothesis that new ions are produced by the collisions of ions with molecules. At first the negative ions alone generate others as they move through the gas, but as the force increases and the sparking potential is approached, the positive ions also acquire the property of producing others to an appreciable extent.

In order to test this theory, experiments ‡ were made on the conductivity produced by Röntgen rays in a gas between parallel-plate electrodes at various distances apart, the electric force being maintained constant. The distances used were 1 and 2 centimetres, and it was found that the effect of doubling the distance was to produce a large increase in the current, as was indicated by the theory. The experiments in fact showed that it is necessary to attribute to ions moving with comparatively small velocities the property of producing ions from molecules of the gas which is possessed by particles moving with very high velocities, such as the cathode rays and the β rays from radio-active substances.

It became obvious from these investigations that in the earlier experiments on the conductivity of gases at low pressures, in which potentials of the order of 100 volts were used, a considerable proportion of the ions were generated by collisions. Thus a number of investigations of the currents between parallel plates were made by Stoletow in which the conductivity was produced by the action of ultra-violet light on the negative electrode. In one series of experiments the currents were measured with different distances between the plates, the electric force being maintained constant. Unfortunately, the distances

* H. Kreusler, Verh. Phys. Ges. Berlin, **17**, p. 86, 1898.
† E. von Schweidler, Wien. Berichte, **18**, p. 273, 1899.
‡ Nature, **62**, p. 340, Aug. 9, 1900.

used were very small, ranging from ·25 millimetre to 1·08 millimetre, and, with the exception of the results obtained with one of the pressures, the increase in the current with the distance was not very large and the numbers obtained did not suggest the simple law that the currents increase in geometrical progression as the distance increases in arithmetical progression.

183. Thomson's theory of surface layers. An explanation of the phenomena observed by Stoletow was given by Thomson,* which depended on the hypothesis that the ions were all generated by the action of the light in a layer at or near the surface of the negative electrode. If all the ions were generated exactly at the metal surface an increase in the distance between the plates would not produce an increase in the current when the force at the surface is constant. Hence in order to explain some of the experiments it is necessary to suppose that the layer in which the ions are generated extends into the gas, and while the thickness of the layer exceeds the distance between the plates an increase in the distance between the plates will produce an increase in the current.†

The theory thus provides a possible explanation of the increase of current with the increase of electric force, since the greater the force the greater the number of ions drawn from the layer; it is also conceivable that the current would depend on the pressure of the gas. It is difficult, however, to accept the explanation of the increase of the current with the distance between the electrodes, as it is very improbable that a layer of gas one millimetre thick near the surface of the metal should possess the peculiar electrical property that when ions are generated in it by the light an electric force in the negative direction produces a large negative current, and no current is produced when the direction of the force is reversed.

* J. J. Thomson, The Discharge of Electricity through Gases (Archibald Constable and Co., 1898).

† J. J. Thomson, Phil. Mag. (5) **48**, p. 552, Dec. 1899.

184. Application of collision theory to Stoletow's experiments. The theory of ionization by collision, on the other hand, provides quite a satisfactory explanation of the larger variations in the currents in Stoletow's experiments. For although the currents obtained with various distances, in many cases, are not in good agreement with those calculated on the theory, more recent experiments with similar forces and pressures but with larger distances between the plates show that the currents obtained with ultra-violet light agree accurately with the theory.

An objection was raised to the collision theory on the ground that a potential difference of 50 volts between the electrodes is so small that the velocity acquired by the ions would not be sufficiently great to produce ionization by collision. The maximum kinetic energy of the ions would in that case be $50e/300$ whereas it has been found by Rutherford and McClung* that the energy required to ionize a molecule was $175e/300$. It was supposed, from this result, that ionization by collision could not take place until the potential difference between the electrodes was 175 volts.

The value of the energy obtained by Rutherford and McClung was deduced from measurements of the number of ions generated by a beam of rays in traversing a given length of a gas, and the diminution produced in the intensity of the rays. The estimate thus found represents the energy required to ionize a molecule by that particular method. It cannot be assumed that the energy would be the same in all cases, and other methods might be more efficient. When the rays traverse the gas the ionization of molecules is not necessarily the only effect that is produced which contributes to the reduction in the intensity of the rays, and the proportion of the total loss of energy

* E. Rutherford and R. K. McClung, Phil. Trans. A, **196**, p. 25, 1901. (A numerical error appears to have been made in the number given as the final result of this research. The energy required to ionize a molecule by this method is that corresponding to a potential fall of 87 volts instead of 175.)

to be attributed to the process of ionization would differ according to the method used to produce the ions.

185. Effect of collisions of ions generated in a gas between parallel plates. The process by which ions are generated in a gas by the action of forces smaller than those required to maintain a discharge may be investigated by considering the currents obtained between parallel-plate electrodes when the initial ionization is produced by Röntgen rays or Becquerel rays, or when negative ions are generated by ultra-violet light at the surface of the negative electrode.

In the former case both positive and negative ions are generated in the gas initially, and from the experiments with plane electrodes it is uncertain whether the additional ions are due to the action of positive or negative ions. This point has been decided by experiments with electrodes of different shapes, and it was found that for the smaller forces the effects are mainly due to the action of the negative ions. If this be admitted, it is easy to find a formula for the currents between parallel plates at various distances apart, since the electric force is constant throughout the field.

At distance x from the positive electrode let a number n_0 of negative ions be generated initially by the rays: then if they generate others by collisions with molecules, and if the new ions possess the same property, the number that arrive at the positive electrode is $n_0\epsilon^{\alpha x}$, α being the number generated by one ion per centimetre of its path.

When the gas is ionized uniformly by the rays throughout the whole distance between the electrodes, the number of negative ions generated between two parallel planes at distances x and $x+dx$ from the positive electrode being $n_0\,dx$, the total number that arrive at the electrode is

$$\int_0^a n_0\,\epsilon^{\alpha x}\,dx = n_0\,\frac{\epsilon^{\alpha a}-1}{\alpha},$$

a being the distance between the plates.

The saturation current q_0, which is obtained before ionization by collision takes place, is $n_0 a$, hence the current q for the larger forces is

$$q = q_0 \frac{\epsilon^{\alpha a} - 1}{\alpha a},$$

which gives the value of α for any force when the ratio $\frac{q}{q_0}$ is determined experimentally.

The quantity α depends on the force X and pressure p of the gas, and its values for different forces may be obtained from an ordinary current-electric-force curve when there is a definite length of the curve corresponding to the saturation current from which the current q_0 may be obtained. Such curves are easily obtained when the initial ionization is produced by Röntgen rays or by Becquerel rays except when the gas is at a very low pressure.

186. Effect of initial distribution between the plates. In the above formula for the determination of α the initial distribution n_0 is assumed to be uniform. When the rays do not fall on the electrodes this condition is satisfied approximately, but at low pressures it is more convenient to allow the rays to pass through one electrode and to fall normally on the other, so as to get the advantage of the large ionization produced by the secondary rays. A series of experiments was made with an apparatus in which the primary rays after passing through an electrode of thin aluminium fell on a brass plate which gives off intense secondary radiation. In this case a greater number of ions were generated near the brass electrodes than in other parts of the field, since some of the secondary rays did not penetrate the whole distance between the plates even when the pressure of the gas was of the order of one millimetre. The average distance traversed by the negative ions was therefore larger, and more ions were generated by collisions, when the aluminium plate was the positive electrode than when it was negative. Hence the currents between the electrodes were not the same in both directions, on account of the unsymmetrical nature of the distribution, but the mean value corresponds approximately to the current that would have been obtained if the initial ionization had been uniform.

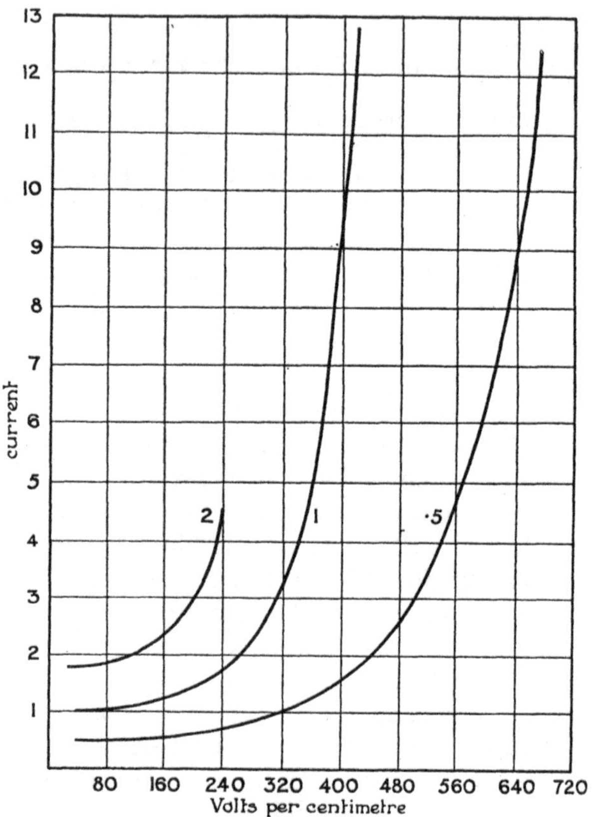

Figure 45. Current-electric-force curves for air at 4·13 millimetres pressure; initial ionization produced by Röntgen rays. Distances between the plates 2, 1, and ·5 centimetres.

187. Currents depending on the distance between the electrodes. Determination of α, the number of ions generated by an electron per centimetre. The curves, figures * 45 and 46, give the connection between the current and the electric force for air at the pressures 4·13 millimetres and 1·10 millimetres. Three sets of experiments were made at each

* Phil. Mag. (6) **1**, p. 198, Feb. 1901.

pressure with plates at distances 2, 1, and ·5 centimetres apart, and the curves show how the effects obtained by collisions are increased by increasing the distance between the plates.

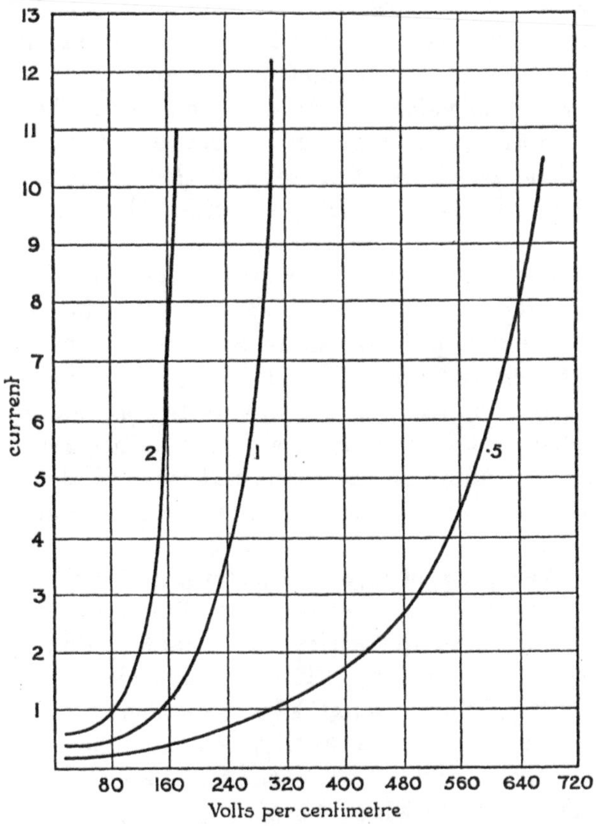

Figure 46. Current-electric-force curves for air at 1·1 millimetre pressure; initial ionization produced by Röntgen rays. Distances between the plates 2, 1, and ·5 centimetres.

In order to see to what extent the relative positions of the curves are in accordance with the theory, the value of α may be found for any force and pressure by the formula $q = q_0 \frac{\epsilon^{\alpha a} - 1}{\alpha a}$, from the experiments with different distances a between the electrodes. Thus for the pressure of 4·13 millimetres and the force 200 volts per centimetre the value $\alpha = \cdot 50$ is found from the currents obtained with the plates at 2 centimetres apart, and the value ·51 from the currents with the plates at 1 centimetre apart. Similarly, at the same pressure when the force is 320 volts per centimetre, the value $\alpha = 2 \cdot 1$ is found from the curve corresponding to the distance 1 centimetre between the plates, and the value 2·2 from the curve for the distance ·5 centimetre. The agreement between these numbers shows that the effect of increasing the distance between the plates is satisfactorily explained by the collision theory.

At the pressure 1·10 millimetres larger values of α are obtained when the same forces are acting. Thus when the force is 200 volts per centimetre $\alpha = 2 \cdot 8$, and for the force 320 volts per centimetre $\alpha = 5 \cdot 4$. The values of α do not however continue to increase indefinitely as the pressure is diminished, but a maximum value is obtained for a certain pressure depending on the force, and when further diminutions are made in the pressure the value of α diminishes.

188. Variation of α with the force and pressure. The quantity α is a function of the force X and pressure p, and varies with these quantities in a manner that is easily explained by the theory. The negative electrons traverse free paths of different lengths between their collisions with molecules, and the velocities they acquire under the electric force depend on the length of the free paths. At first it is only the molecules which terminate the longer paths that are ionized by the collisions, but as the electric force increases the velocities are increased and a larger number of molecules are ionized. Hence when the pressure is constant the quantity α is negligible until the force exceeds a certain value, then it increases with the force and approaches a maximum value, which is obtained when ionization takes place at each collision.

The value of α also depends on the pressure. When the force is constant, α is zero for large pressures, since the free paths of the electrons are then very short. As the pressure is reduced the lengths of the free paths are increased, and along some of them the electrons acquire sufficient velocity to cause ionization when they collide with molecules. The reduction of the pressure, however, causes a reduction in the total number of molecules encountered by an electron in passing through a centimetre of the gas in the direction of the force. When the pressure is very low, ionization may take place at each collision, but the total number of collisions becomes so small that very few ions are generated.

Hence when the force is constant the value of α at first increases as the pressure is reduced; for a certain pressure, depending on the force, α attains a maximum value, and finally diminishes with the pressure.

189. Values of α in hydrogen and carbonic acid. Experiments * were also made with hydrogen and carbonic acid, the initial ionization being produced by Röntgen rays. The currents obtained with different forces X and pressures p are given in the following tables:

Currents in hydrogen between plates ·53 centimetre apart; X in volts per centimetre; p in millimetres of mercury.

X	$p = 34$	$p = 23·5$	$p = 14·5$	$p = 9·5$
72	19·1	11·0	6.25	12·5
215	—	—	—	13·5
287	—	—	—	16·1
430	—	—	—	31·5
574	—	14	15·7	103
717	—	—	39·7	234
789	24·7	23·5	80·5	—

* J. S. Townsend and P. J. Kirkby, Phil. Mag. (6) 1, p. 630, June 1901.

Currents in carbonic acid between plates ·53 centimetre apart; X in volts per centimetre; p in millimetres of mercury.

X	$p = 18·3$	$p = 8·8$	$p = 3·95$	$p = 1·4$	$p = ·68$
76	140	35	19·6	7·6	5·25
152	141	35·5	—	10·8	8·1
304	—	—	33	30·8	20
380	—	—	48	54·4	29·3
456	—	—	75	95·7	40·2
532	153	64	126	162	52·9
608	—	88	216	250	72·0
684	—	129	420	411	99·0
760	198	195	778	660	161
836	232	302	1460	—	—
972	337	850	4330	—	—

The same intensity of rays was not used in the experiments at different pressures, so that the numbers in a horizontal line do not give the variations with the pressure that would be obtained with a constant source of radiation. All that is required is that the intensity of the rays should be constant while experiments are made at a given pressure with different forces.

The saturation current is given by the current obtained with the lowest force, when the larger pressures are used. For the smaller pressures the saturation current is smaller than the current obtained with the lowest force, and in order to obtain the values of α for the lower pressures it is necessary to take into consideration the ions produced by collisions, when the force 76 volts per centimetre is acting. It is not, however, possible to obtain very accurate determinations from the experiments, and it is unnecessary to consider the best method of correcting for this effect, since the values of α for low pressures may be obtained to a higher degree of accuracy by experiments on the conductivity produced by ultra-violet light.

A comparison of the results obtained with the two gases shows that additional ions are obtained by collisions at higher pressures in hydrogen than in carbonic acid when the same forces are acting.

190. Comparison of the effects of electrons and positive ions. In order to show that, as the force increases, the additional ionization which is first obtained is due to the action of the negative ions, the currents between electrodes of various shapes were investigated.* If the gas is contained inside a large sphere that acts as one electrode, the other electrode being a small sphere at the centre, the intensity of the electric force varies inversely as the square of the distance from the centre, so that when the potential difference between the electrodes is gradually increased, ionization by collision takes place at first in the neighbourhood of the inner electrode. When the outer sphere is the negative electrode, all the negative ions generated by the rays pass through the field of strong electric force before they reach the inner electrode. But when the direction of the force is reversed, the negative ions move outwards from the centre and none pass through the strong field, except those generated initially in the neighbourhood of the small sphere. If the additional ions which are generated in the gas as the force increases are due to the collisions of negative ions with molecules, the conductivity should therefore be much greater when the large sphere is negative than when it is positive. It was found experimentally that, as the force was increased, currents much greater than the saturation current were obtained when the outer electrode was negative, and when the electric force was reversed the currents only slightly exceeded the saturation current.

191. Effect of pressure on the currents between co-axial cylinders. A convenient form of apparatus was used by Kirkby to investigate, more completely, the currents in a field in which the electric force is not uniform. Two co-axial cylinders were used as electrodes, so that the electric force in the space between them varied inversely as the distance from the axis, and a large force was obtained at the surface of the inner cylinder, which was of small diameter. The arrangement of the apparatus

* Nature, **62**, p. 340, August 9, 1900.

is shown in figure 47. The outer cylinder A (4·15 cm. in diameter), containing the gas, was of aluminium, in order that the Röntgen rays, which were used to produce the initial ionization, might pass through it. The inner cylinder was a copper wire W (·206 mm. in diameter). The ends of the larger cylinder were cemented into grooves turned in two ebonite plates E_1 and E_2, and two brass tubes T_1 and T_2 were fitted into holes bored through the centres of the plates. These tubes were maintained at zero potential, so that no leakage could take place from the outer aluminium cylinder to the concentric wire except through the gas. The wire was stretched between two brass rods passing through ebonite plugs P_1 and P_2 in the ends of the tubes T_1 and T_2. The wire being thus insulated, the charge

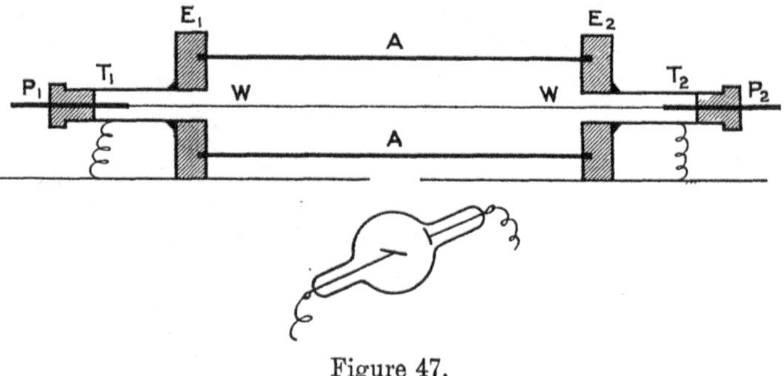

Figure 47.

which it acquired in a given time while the rays were acting was measured by a sensitive electrometer. With air at atmospheric pressure inside the cylinder, the conductivity is independent of the direction of the electric force, and no appreciable increase above the saturation current is obtained with potentials of the order of 200 volts.

At low pressures, when the smaller electric forces are acting, the currents in the two directions are constant but are less than those obtained at the higher pressures, since the total number of ions generated by the rays diminishes as the pressure is reduced. Also, at low pressures a difference may be observed in the saturation currents in the two directions, since the number E of electrons in the secondary radiation emitted by the copper wire is appreciable as compared with the number of positive or negative ions N generated in the gas. The saturation current $N+E$, obtained when the outer cylinder is positive, is therefore greater than the saturation current $N-E$ in the opposite direction, and this difference becomes more marked as the pressure is reduced. Electrons are also emitted from the surface of the large cylinder, but they nearly all impinge again on the cylinder at another point, and very few are discharged by the thin copper wire. As the force increases, the currents obtained with the outer cylinder negatively charged are much greater than those obtained when it is positively charged.

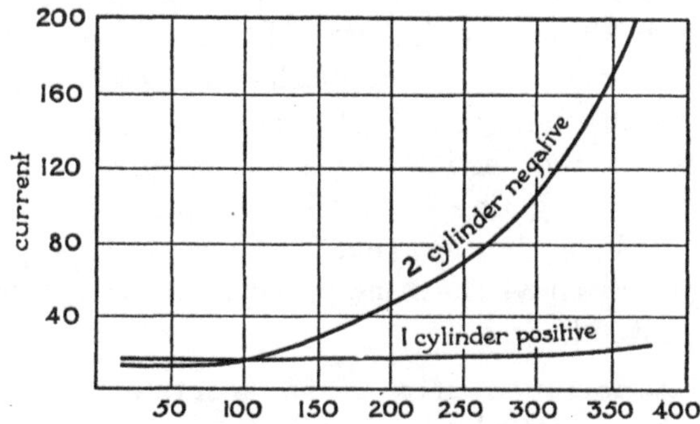

Figure 48. Potential difference between cylinder and wire in volts. Currents in air at 3·53 millimetres pressure, between a cylinder 4·15 centimetres diameter and a wire ·206 millimetre diameter.

192. Currents depending on the direction of the force between co-axial cylinders. The curves *, figure 48, for air at 3·53 mm. pressure, are examples of the current-electric-force curves obtained when the air is ionized by Röntgen rays. Curve 1 gives the currents when the aluminium cylinder was at a positive potential. In that case the currents are not much greater than the saturation current when the higher potentials are used.

When the outer cylinder is negative, much larger currents are obtained with the higher potentials, as is shown by

* P. J. Kirkby, Phil. Mag. (6) 3, p. 212, Feb. 1902.

curve 2. With an electromotive force of 300 volts, the current is about six times as great as the saturation current.

These experiments show that as the force is increased the additional ions that are produced are due to the action of the negative ions. Kirkby has shown that the currents obtained when the outer cylinder is negative are in agreement with those calculated from the values of α by integrating the effect produced by the electrons in traversing the field between the cylinders.

The investigations of the current-electric-force curves thus afford very strong evidence in support of the theory of ionization by collision, but there are a number of points of minor importance that have been overlooked in applying the theory to these experiments. The results are nevertheless of importance, for although the values of α cannot be obtained with the same high degree of accuracy as is possible when the initial ionization originates from the surface of the negative electrode, the numbers that have been obtained clearly show that the negative ions generated by Röntgen rays have the same properties as those generated by other methods.

193. Determination of α by an accurate method. The most accurate method of determining the ionization produced by negative ions is to measure the currents between parallel plates when ions are set free initially from the surface of the negative electrode by the action of ultra-violet light. When the distance x between the plates is varied over a certain range, and the electric force is maintained constant, it has been found experimentally that the number of negative ions n that arrive at the positive electrode is given by the formula $n = n_0 \epsilon^{\alpha x}$, n_0 being a constant. The range of distance over which the currents are in accurate agreement with this formula is in many cases very large, so that several currents differing by large factors may be found, and the value of α may be obtained to a high degree of accuracy from the ratio $\frac{n_2}{n_1} = \epsilon^{\alpha(x_2 - x_1)}$

of two currents corresponding to distances x_1 and x_2 between the plates.

A form of apparatus found suitable for the measurement of these currents is shown in figure 49. The light from the spark gap in the discharge circuit of a Leyden jar entered the apparatus through a quartz plate *P*, and passing through a second plate *Q* fell on a parallel zinc plate *Z*. The plate *Q* was silvered, and a small area, one centimetre in diameter at the centre of the plate, was ruled with fine parallel lines about ·1 millimetre wide, through which the light passed. The quartz plate was 4 centimetres in

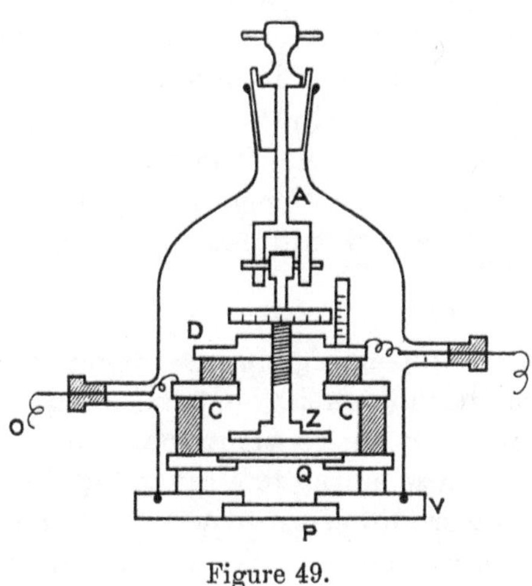

Figure 49.

diameter, and since the greatest distance between the plates *Q* and *Z* was of the order of one centimetre, the field of force was uniform in the space near the centre of the plates through which the currents passed.

The silver surface of the plate was in contact with the metal base, which was placed on insulating supports, so that it was possible to raise the potential of the plate *Q* to any desired voltage, by a battery of small accumulators.

The zinc plate *Z* was fixed to a micrometer screw that worked in a thick insulated metal plate *D*. It was found necessary to provide some form of split insulation for the upper part of the apparatus, to prevent any leakage from the base over the insulating supports to the plate *D*, which was in connection with the electrometer. In some experiments the potential of the plate *Q* was raised to several hundred volts, and unless the micrometer apparatus is carefully insulated irregular electrometer disturbances may take place. The method shown in the figure was adopted in some of the later experiments. The metal disc *C* was supported on ebonite pillars and was maintained at zero potential. The brass plate *D* was insulated from the disc *C* by the ebonite ring, so that no charge could pass from the base to the zinc plate except through the gas. The currents between the plates were measured by an electrostatic induction balance, as described in Section 20. The distance between the plates was adjusted to any required value by turning the forked axle *A*. The bearing of the axle in the brass stopper of the cover-glass was arranged as illustrated, so that no air should leak in when the axle was rotated.

194. Currents depending on the distance between the electrodes, the initial ionization being produced by ultra-violet light on the negative electrode. In order to obtain the value of α corresponding to a given force X and pressure p, the currents were determined with different distances x between the plates, the potential of the lower plate being adjusted to the value xX. Since the ions generated by the ultra-violet light do not acquire the average velocity corresponding to the electric force X until they have passed a small distance through the gas, the currents were not determined for distances less than a millimetre between the plates. For the same reason very low pressures were not used, as the velocity of the negative ions does not approximate to the final velocity corresponding to the force unless the mean free path of the electrons between collisions with molecules is less than the shortest distance between the electrodes. When these precautions are observed, the currents n_1, n_2, n_3, ..., obtained with the distances x_1, x_2, x_3, ..., between the plates, are in a constant ratio when equal increments x_2-x_1, x_3-x_2, ..., are made in the distance between the plates, and from the determination of the ratio the value of α may be obtained, since

$$\frac{n_2}{n_1}=\frac{n_3}{n_2}=\ldots=\epsilon^{\alpha(x_2-x_1)}.$$

The ratio between the successive currents remains constant for a considerable range of distance, but as x increases a point is reached when the currents increase more rapidly than the value of n, as given by the formula $n=n_0\epsilon^{\alpha x}$. This shows that some other form of ionization, in addition to the action of the negative ions by collisions, begins to produce an appreciable effect when there is a large distance between the plates. For simplicity, therefore, the investigation may at first be confined to the results obtained with the shorter distances in which the ionization of the gas is due to the action of negative ions only.

195. Illustration of the method used to determine α. The following is an example of an experiment in which the currents through air at one millimetre pressure were determined for various distances between the plates, the electric force being maintained constant at 350 volts per centimetre.* The distance a between the plates in centimetres is given in the first line of the table. The currents q are given in the second line. It is convenient, in order to compare the experiments with the theory, to choose the units so that the smallest current q_1 may have the value given by the equation $\frac{q_1}{1}=\frac{q_2}{q_1}$. The currents determined experimentally then give the total number of ions that would be produced if one negative ion, starting from the negative electrode, travels with the average velocity corresponding to the force X for the whole distance between the plates. Taking α as 5·26, the values of n given by the formula $n=\epsilon^{\alpha a}$ represent the currents for the shorter distances, but for the longer distances it is necessary to adopt a different

* Phil. Mag. (6) 6, p. 598, Nov. 1903.

expression for n, obtained by considering the effect that may be produced by the positive ions. This has been done for the numbers in the last line of the table.

CURRENTS IN AIR AT 1 MILLIMETRE PRESSURE.

a	0	·2	·4	·6	·8	1·0	1·1
q	—	2·86	8·3	24·2	81·0	373	2250
$\epsilon^{\alpha a}$	1	2·86	8·2	23·4	66·5	190	322
$\dfrac{(\alpha-\beta)\epsilon^{(\alpha-\beta)a}}{\alpha-\beta\epsilon^{(\alpha-\beta)a}}$	1	2·87	8·3	24·6	80	380	2150

196. Currents obtained in helium with small forces. With an apparatus such as that described above, the values of α may easily be measured for most gases when the pressure ranges from 1 to 10 or 20 millimetres, and the ratio X/p of the force to the pressure is from 60 to 1000, the force X being expressed in volts per centimetre, and the pressure in millimetres of mercury.

A notable exception occurs in the case of helium, as large effects are obtained in this gas by the collision of ions with molecules when comparatively small forces are used. The following are examples of experiments made by Gill and Pidduck,* in which the values of X/p were 5, 10, and 20. The corresponding values of α, deduced from the experiments, are given in the last column.

CURRENTS IN HELIUM.

a	·4	·6	·8	α
Pressure 8 mm.				
$X = 80$	2·4	3·6	6·1	2·20
$X = 160$	6·7	17·5	49·0	4·77
Pressure 10 mm.				
$X = 200$	9·4	35·7	—	5·60
Pressure 25 mm.				
$X = 125$	6·7	12·7	—	3·18
$X = 250$	72	295	—	7·13

197. Experiments in which the action of positive ions is negligible. This method of finding the effect of the negative ions is more accurate than the method in which the values of α are deduced from the current-electric-force curve. The

* E. W. B. Gill and F. B. Pidduck, Phil. Mag. (6) **23**, p. 837, 1912.

number of ions set free initially by the light is the same in the experiments at various distances between the plates when the electric force and pressure are constant, and the effect produced by collisions while the ions move with the same mean velocity over a measured distance may be found. Also, it is possible to observe when the positive ions produce an appreciable effect, since the ratio of the successive currents obtained by making equal increments in the distance between the plates is then no longer constant, but increases with the distance. With a current-electric-force curve it is impossible to decide to what extent the positive ions contribute to the ionization, and in order to ensure that too high a value is not attributed to α with the higher forces, it is necessary to make determinations from curves corresponding to different distances between the plates. In some of the earlier determinations this precaution was not observed, and the values of α that were given were too high.

198. α/p as a function of X/p. The values found for α for various forces and pressures may be recorded in a simple manner, as it was observed that when the points whose co-ordinates are $\frac{\alpha}{p}$ and $\frac{X}{p}$ are marked on a diagram they all lie on one curve. The variables α, X, and p are therefore connected by an equation of the form $\frac{\alpha}{p}=f\left(\frac{X}{p}\right)$. This result was first obtained from the determinations of α deduced from the current-electric-force curves when the conductivity was produced by the action of Röntgen rays,* and it was subsequently confirmed by the more accurate determinations made with ultra-violet light. In some of these experiments † the pressures and forces were chosen in order that this relation should appear by inspection, as for example, the following determinations of α in hydrogen:

* Phil. Mag. (6) 1, p. 198, Feb. 1901.

† The researches published in the Philosophical Magazine (6) **6**, p. 598, Nov. 1903, and **8**, p. 738, Dec. 1904, contain many examples of such experiments.

$p = 8$ mm. $X = 1{,}050$ volts per cm. $\alpha = 14{\cdot}8$
$p = 4$ mm. $X = 525$ volts per cm. $\alpha = 7{\cdot}4$
$p = 2$ mm. $X = 262$ volts per cm. $\alpha = 3{\cdot}7$.

This connection between the three variables, which was deduced from the numbers found experimentally, affords the strongest evidence in support of the theory of ionization by collision. For if the number of molecules α ionized by an electron, or negative ion, per centimetre of its path depends on the velocity with which it collides with the molecule, it may easily be seen that a relation of the above form must hold between the quantities α, X, and p.

When a negative ion traverses a centimetre of the gas in the direction of an electric force, it collides with several molecules, some of which become ionized. According to the theory, the chance of producing a new ion on collision depends on the velocity at impact, and in general it is only those molecules which terminate the longer free paths that are ionized. The velocities on impact depend on the product of the force X and the lengths of the free paths of the negative ion, so that if the force X be increased by the factor z all the velocities will be increased. If the pressure p be also increased to zp, then all the free paths will be reduced in the ratio 1 to z, and consequently the velocities of the negative ions on colliding with the molecules will be restored to their original values, and the same proportion of collisions will result in ionization as with the original force X and pressure p. Since the number of collisions per centimetre is increased in the same ratio as the pressure, the value of α is thus increased to $z\alpha$ when X and p become zX and zp. This relation * may be expressed mathematically by the equation $\alpha = pf\left(\frac{X}{p}\right)$.

* A similar equation in a different notation was subsequently given by Thomson, who took the view that in a certain fraction of the number of collisions the effect of a collision is to leave the electron clinging to the molecule with which it collided and thus to end its career as an ionizing agent. (J. J. Thomson, Phil. Mag. (6) 1, p. 369, April 1901.)

The following table gives the values of α/p in terms of X/p for the different gases.*

TABLE OF VALUES OF α/p.

$\frac{X}{p}=$	1000	900	800	700	600	500	400	300	200	100
Air	10·5	10·0	9·3	8·7	7·9	7·0	5·82	4·4	2·6	·72
Hydrochloric acid	15·4	14·8	14·0	13·0	11·9	10·5	8·9	6·8	4·1	1·21
Carbon dioxide	12·6	11·9	11·0	10·3	9·1	7·8	6·4	4·8	2·8	·82
Water vapour	9·7	9·4	9·0	8·5	7·95	7·2	6·35	5·2	3·6	1·31
Nitrogen	—	—	—	—	7·0	6·2	5·2	3·95	2·3	·42
Hydrogen	—	—	—	—	—	—	3·7	3·3	2·62	1·36
Argon	—	—	—	—	9·2	8·5	7·5	6·2	4·4	2·0
Helium	—	—	—	—	—	—	—	—	2·37	2·0

The results of the experiments may also be conveniently represented by means of curves such as those given in figures 50 and 51.

199. Nature of ions generated by collision in different gases. The results of the experiments on the ionization by collisions lead to many conclusions of theoretical interest. In the first place it will be observed that all the effects of the negative ions are explained on the hypothesis that after travelling a given distance x the original number is multiplied by the factor $\epsilon^{\alpha x}$.

The experiments show that all the negative ions have the property of producing others by collisions to exactly the

* The determinations of $\frac{\alpha}{p}$ in terms of $\frac{X}{p}$ for air, hydrogen, carbonic acid, water vapour, and hydrochloric acid, are given in the following papers published in the Philosophical Magazine (6) **3**, p. 557, 1902; **5**, p. 389, 1903; **6**, p. 598, 1903; **8**, p. 738, 1904. Further experiments on carbonic acid and nitrogen were made by Hurst, who also collaborated in some of the earlier researches (H. E. Hurst, Phil. Mag. (6) **11**, p. 535, 1906). The experiments on argon and helium were made by Gill and Pidduck (E. W. B. Gill and F. B. Pidduck, Phil. Mag. (6) **16**, p. 280, 1908, and **23**, p. 837, 1912). The numbers given in Chapter IX for $\frac{\beta}{p}$ in terms of $\frac{X}{p}$, and the experimental verifications of the theory of sparking potential mentioned in Chapter X, are also to be found in these publications.

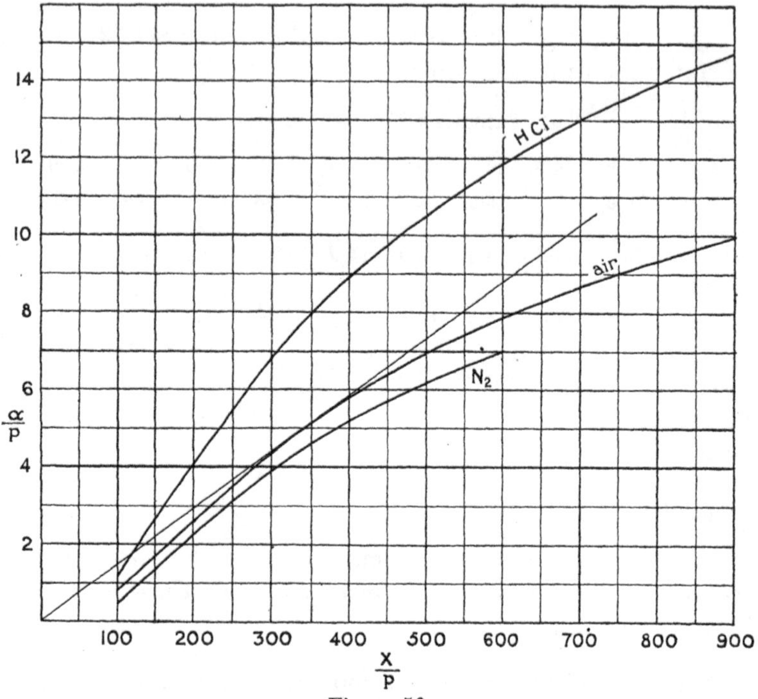

Figure 50.

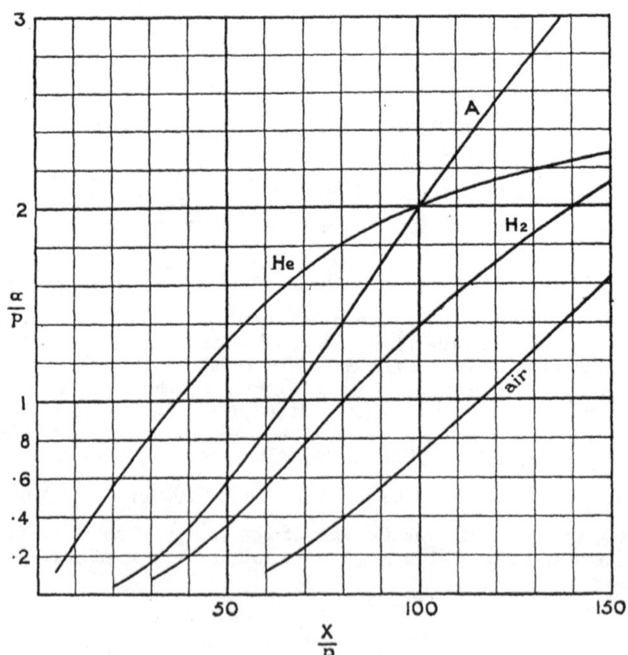

Figure 51.

same extent. At the beginning they consist principally of the ions generated initially either by rays or by ultra-violet light, but after travelling a certain distance a large number of additional ions may be formed, so that the number n consists principally of ions generated by collisions in the distance x. The same equation $dn = \alpha n dx$ holds throughout the motion when the force is constant, which shows that α must be the same for the ions generated initially as for the ions generated by collisions.

Considering the experiments with ultra-violet light, in which the ions initially present are those set free from a metal plate by the action of the light, it follows that the negative ions generated by collisions in any gas have precisely the same properties as the ions set free by the action of ultra-violet light from a metal surface. It follows from this that the negative ions generated in the different gases are all the same, being identical with the ions set free by ultra-violet light from the same metal surface.

200. Nature of ions generated by Röntgen rays. The negative ions generated by Röntgen rays from the molecules of a gas are the same as other negative ions. The values of α which have been determined from the conductivity produced by Röntgen rays, although not so accurate as those made with ultra-violet light, are in very good agreement with the latter, and were shown to satisfy the relation $\frac{\alpha}{p} = f\left(\frac{X}{p}\right)$ before that property was confirmed by the more accurate experiments. The principal discrepancies occur with the larger values of $\frac{X}{p}$ which, as has already been explained, is due to the fact that an appreciable effect was produced by the positive ions in the earlier experiments, which was not recognized at the time, and consequently the values of α that were obtained were too large. The best agreement is to be found in the case of carbonic acid, since the positive ions have a much smaller effect as compared with the negative ions in this gas than is the case with air or hydrogen. The following are the values of $\frac{\alpha}{p}$ deduced from the different experiments with carbonic acid.

VALUES OF $\frac{\alpha}{p}$ IN CARBONIC ACID, OBTAINED BY DIFFERENT METHODS.

$\frac{X}{p}$ =	100	200	300	400	500	600	700
Ionization produced initially by ultra-violet light	·82	3·7	4·8	6·4	7·8	9·1	10·2
Ionization produced initially by Röntgen rays	·80	3·8	5·0	6·8	8·3	9·6	10·4

The general agreement between the values of $\frac{\alpha}{p}$ and $\frac{X}{p}$, deduced from experiments in which the initial ionization is produced by different methods, shows that the conductivity obtained in air, carbonic acid, and hydrogen must be due to the same effect, and that the negative ions generated from the molecules of the different gases by the action of Röntgen rays are the same as the negative ions generated by the action of ultra-violet light from a metal plate.

201. Electronic state of negative ions. In order to account for the large difference between the positive and negative ions it is necessary to assume that the negative ions are in the electronic state, and that the property of generating others by collisions is due to the high velocity they acquire while moving along their free paths. More recent experiments have led to results in accordance with this conclusion, as it has been found that the velocity of the negative ions under an electric force is much greater than the velocity of positive ions, and also their rate of diffusion corresponds to that of bodies of very small mass. What is remarkable is that the electronic state* begins to be prevalent for values of X/p that are much smaller than the values at

* See Sections 132 and 133.

which the ionization by collisions becomes appreciable. For example, in air when $\frac{X}{p}$ is 60, $\frac{\alpha}{p}$ is about ·12 and becomes very small for lower values of X/p. The experiments on the velocity of ions show that when X/p is ·05 the velocity of the negative ions in dry air is about 180 centimetres per second, or three times as great as the velocity of the positive ions. This increase in the velocity of the negative ions must be due to a diminution in the group of molecules associated with the negative electrons when X/p is very small. The average size of the negative ions continues to diminish as X/p increases, and when $X = 2$ volts per centimetre and $p = 3{\cdot}8$ millimetres, or $X/p = {\cdot}5$ approximately, the velocity as found experimentally is 10^6 centimetres per second, which, as has been shown in Section 132, corresponds to the value $5{\cdot}8 \times 10^{17}$ for the ratio $\frac{e}{m}$, e being the charge and m the mass of a negative ion. This is of the same order as quantity $5{\cdot}3 \times 10^{17}$ found by recent accurate experiments as the value of $\frac{e}{m}$ for electrons. Since $\frac{e}{m}$ is 10^{13} when m is the mass of a molecule of air, it follows from these investigations that when X/p has the comparatively small value ·5, the electronic state is prevalent, and it is even more probable that when $\frac{X}{p}$ has attained the value 60, at which $\frac{\alpha}{p} = {\cdot}12$, the electrons move freely and are not associated with molecules of the gas.

202. Free electrons in gases at several millimetres pressure. The view held by Thomson, that in an appreciable number of collisions the electrons adhere to molecules and subsequently cease to act as ionizing agents, is contrary to the above conclusion. There does not appear to be any experimental evidence to show that an appreciable number of molecules become permanently associated with negative electrons when the forces are sufficiently great to produce ionization by collision. The measurements of the conductivity are in accurate agreement with the formulae obtained on the hypothesis that when the negative ions act as ionizing agents, their activity remains constant as they move through the gas. The connection which an electron may have with a molecule, if there is any, must be of very short duration. The experiments from which the value of e/m were deduced are also conclusive on this point.* Thus, for example, the magnetic deflection of a stream of negative ions was observed after the ions had traversed 6 centimetres of air at a pressure of 3·8 millimetres, under an electric force of 2 volts per centimetre. Assuming the ions to be in the electronic state, their mean free path in air at that pressure would be about $8{\cdot}4 \times 10^{-3}$ centimetre, so that if they travelled in straight lines along the distance of 6 centimetres, each ion would, on an average, collide with 700 molecules. The number of collisions is, however, much larger than this, for while the ions move with a velocity of 10^6 centimetres per second in the direction of the electric force, their velocity of agitation is of the order $2{\cdot}4 \times 10^7$ centimetres per second, so that the length of the zigzag path they traverse in the gas is twenty-four times as great as the distance they travel in the direction of the force. The number of molecules encountered by each ion, before the value of $\frac{e}{m}$ was observed, must therefore have been of the order 2×10^4. It follows from this that an electron may make several thousands of collisions without becoming permanently connected with a molecule of gas at low pressures, when the electric force X is not less than $\cdot 5p$, X being expressed in volts per centimetre and p in millimetres of mercury.

203. Effect of impurities on the electronic state. A point that should be noticed, which is of theoretical interest, is that in the experiments in which different pressures were used and the ratio X/p was constant, the values of α were

* See Sections 132 and 133.

accurately proportional to the pressure. This indicates that the velocity of the electrons is a function of X/p only, and the law probably holds for much wider ranges of pressures than those over which it has been verified experimentally.

With the lower values of X/p, at which ionization by collision is negligible, the velocity was found to be approximately a function of the ratio X/p in dry gases, but the agreement with this simple law was not very accurate in some cases.* In these cases a small percentage of water vapour produced a large effect on the motion of the negative ions, the velocity in the moist gas being very much less than in the dry gas. With these values of X/p, ·05 to ·5, water vapour tends to increase the mass associated with the negative ions, and they do not move as free electrons. At the higher values of X/p, when ionization by collision begins, this effect disappears and electrons move freely, even in pure water vapour, and, for the values of X/p of the order of 100, ionization by collision takes place in the vapour to as great an extent as in air or other gases.

Similarly with other impurities; it has been observed, for instance, that the velocities of the ions in argon and nitrogen are greatly reduced by small quantities of oxygen, the experiments being made with forces much smaller than those required to produce ionization by collision. When larger forces were used and ions are generated by collisions, impurities produce no very remarkable effects, and the velocity is accurately a function of X/p, since the value of α satisfies the relation $\alpha = pf\left(\frac{X}{p}\right)$. Thus, in nitrogen the addition of small percentages of oxygen does not give rise to large changes in the effects produced by collision, as is shown by the determinations of the sparking potentials. In an experiment † in which the gas was at 4 millimetres pressure and the distance between the electrodes 7·2 millimetres, Hurst found that the sparking potential in pure

* See Sections 101 and 108.

† H. E. Hurst, Phil. Mag. (6) **11**, p. 535, 1906.

nitrogen was 507 volts, and in nitrogen containing 1·1 per cent. of oxygen the sparking potential was 516 volts.

In air the sparking potential at that pressure is about 570. In the same way with argon,* the effects produced by collision in the gas after it had been purified were practically the same as those obtained with a specimen of the gas containing a small quantity of air, of the order of 1 per cent. as measured by pressure.

Helium, which has very low sparking potentials, appears to differ in many ways from other gases. The sparking potentials are increased, and the conductivity produced before sparking takes place is greatly diminished, by small percentages of air. The investigations, however, show that these effects are chiefly due to a large reduction in the activity of the positive ions, the ionization by collisions with negative ions being much less altered by the impurities.

204. Velocity of ions as a function of X/p. It thus appears that impurities have much more effect on the velocity of the electrons when X/p is small than when X/p is large, and this may be the reason why the velocities of the order of 200 centimetres per second, as obtained experimentally, are not accurately functions of X/p.

The theory on which the general nature of these phenomena may be explained indicates that in the transition stage, when the negative ions move as if they were associated with masses of an order intermediate between the mass of a molecule and the mass of an ion, the velocity would still be a function of X/p only, in any mixture of gases, provided the partial pressures of all the constituents are altered in the same proportion when variations are made in the total pressure p. The velocity U under the electric force depends on the average † mass associated with the ion, the pressure, and the electric force, and is given by an equation of the form $U = \frac{Xe}{m} \cdot \frac{D}{p}$,

* E. W. B. Gill and F. B. Pidduck, Phil. Mag. (6) 16, p. 280, 1908.

† See Section 135.

where D involves the dimensions of the mass, or in general $U=\phi\left(m, \frac{X}{p}\right)$. The average mass m is not constant, but is a function of the ratio X/p, so that the velocity is also a function of the same quantity. In order to examine the conditions which determine the average mass of the negative ion, the electron may be supposed to move freely for a time t_1 during which it makes a certain average number of collisions c_1 with high velocities with molecules, and that state is terminated when the electron happens to collide with a small velocity and adheres to a molecule of the gas. It then remains in connection with the larger mass for a time t_2 during which c_2 collisions with molecules occur, but when it happens that the impulsive force is sufficiently great to break the connection with the molecule, the electron begins to move freely for another period t_1. An increase in the electric force prolongs the periods t_1, as it has been shown that the velocity of agitation of the electrons increases with the electric force, and therefore the occasions on which the electron adheres to a molecule and moves with a small velocity become less frequent. The average mass, therefore, diminishes as the force increases.

If, however, changes take place in which the ratio of the periods t_1 and t_2 remain unaltered, the average mass remains the same. This is the case when the force and the pressure of the gas, or of each constituent of a mixture of gases, are altered in the same proportion. Thus the direct effect of increasing the pressure is to diminish the lengths of all the free paths, and consequently, if the velocities of agitation remain unaltered, all the intervals between collisions c_1 will be reduced in the same proportion as the intervals between the collisions c_2. When the force X is increased in the same proportion as p the velocity of agitation of the electrons is unaltered, since the latter acquires a value depending upon the fall of potential along the free paths. Hence when X and p are increased in the same proportion, both the intervals t_1 and t_2 are diminished in that proportion and the ratio t_1/t_2 is unaltered. The average mass is therefore a function of X/p, since it is unchanged when X and p are multiplied by the same factor.

There is thus some reason to suppose that for any mixture of gases the velocity would be determined completely by the ratio X/p, but when a small quantity of one of the gases, such as water vapour, produces a large effect on the average mass of the ion, the experimental investigations are not in accurate agreement with this conclusion, when X/p is less than ·1. It is difficult to ensure that alterations in the total pressure should be accompanied by similar alterations in the partial pressure of the water vapour, so that it is uncertain whether in the transition stage the velocity is a function of X/p for all values of the pressure.

205. Motion of electrons when ionization by collision takes place. Some information as to the nature of the process of ionization by collision may be obtained by considering the curves representing α/p in terms of X/p for the different gases. For convenience, the pressure may be taken as 1 millimetre, which is the unit chosen in these measurements, and the curves then represent the values of α in terms of the force X. The phenomena can be explained on the hypothesis that a molecule may be ionized by the collision with an electron when the latter is travelling with a velocity exceeding a certain value. It may be assumed, for simplicity, that all collisions are of the same type and that the minimum velocity required to ionize a molecule has a definite value depending on the nature of the gas.

In order to obtain an accurate expression for the value of α/p in terms of molecular quantities, which would apply to the whole range of forces that have been used in the experiments, it would be necessary to take into consideration the velocity of agitation of the electrons. It has been shown that the velocity of an electron in air in the direction of the electric force is less than the velocity of agitation for values of X/p of the order of 100. In these cases the electrons retain after collisions a considerable portion of the energy they acquire under an electric force while traversing the free paths. The velocities might be considered to be distributed approximately according to Maxwell's law, and the mean square of the velocity of agitation would be proportional to the quantity k, which is a function of the ratio X/p.* A complete investigation would therefore be very complicated when all the motions of the ions are taken into consideration, but a simple formula may be obtained which gives the values of α/p for the larger values of X/p, if it be assumed that with the larger forces the principal motion of the electron is in the direction of the force.

206. Simple formula for α/p in terms of X/p. If the velocity acquired by an ion along its free path is reduced to an average value v_0 when it collides with a molecule, the velocity at the end of a free path of length x measured in the direction of the electric force is

$$\sqrt{v_0^2 + \frac{2xXe}{m}}.$$

Since the free paths are of different lengths, the electrons will collide with different velocities, and in general those molecules which terminate the longer free paths will be ionized. The number of collisions per centimetre depends both on the velocity of agitation and the velocity due to the electric force, except when the latter velocity is very large, and in that case the number of collisions C per centimetre is constant, and the number of paths that exceed a certain value x may easily be expressed in terms of the mean free path l. In all cases when the electrons move freely, their velocity is so great that the molecules of the gas may be considered to be at rest. Let c be the number of paths that exceed the distance x, the number of paths $-dc$ of length intermediate between x and $x+dx$ is $c\theta dx$, where θdx is the probability that a collision should take place in the length dx, θ being a constant independent of c and x.

* See Section 130.

Hence $$dc = -c\theta dx,$$
or $$c = C\epsilon^{-\theta x}.$$

The total length of all the paths being Cl, where l is the mean free path,

$$Cl = \int_0^\infty x dc = \frac{C}{\theta}.$$

Hence $$c = C\epsilon^{-\frac{x}{l}}.$$

When the electron is travelling under the force X the velocities with which it collides with molecules depend on the values of the quantity $X \cdot x$, and when $X \cdot x$ exceeds a certain potential V, new ions will be generated; that is, an electron will be set free from the molecule and the molecule or possibly an atom will be left with a positive charge. The quantity α will be equal to the number of free paths of an electron that exceed $x = \frac{V}{X}$.

Hence $$\alpha = C\epsilon^{-\frac{V}{lX}} = C\epsilon^{-\frac{CV}{X}}.$$

207. Comparison of the formula with the values obtained experimentally. This equation represents the connection between α and X, and if the constants C and V are chosen to give the right values of α for two points on the curve corresponding to high values of the force X, it will be found that the above equation will represent the values of α for a considerable range of the larger forces.

Thus, for air, taking $C = 14\cdot6$ and $V = 25\cdot0$, the following table gives the values of α found experimentally, and the values of $C\epsilon^{-\frac{CV}{X}}$ for the different forces:

X	1000	900	800	700	600	500	400	300	200	100
α determined experimentally	10·5	10·0	9·3	8·7	7·9	7·0	5·82	4·4	2·6	·72
$C\epsilon^{-\frac{CV}{X}}$	10·1	9·7	9·3	8·6	7·9	7·0	5·85	4·3	2·34	·38

Thus the formula is in agreement with the experiments for forces exceeding 300 volts per centimetre. The values of α for forces larger than those given above cannot be obtained accurately by the direct experimental method. With the large forces the action of the positive ions adds considerably to the conductivity, and it is necessary to use short distances between the plates in order to prevent a discharge taking place. Under these conditions it is difficult to obtain a reliable expression for the current from which the value α could be obtained, and the most accurate estimate of the maximum value is probably the number C, which brings the expression for α into agreement with the experimental results obtained with forces of the order 800 volts per centimetre.

208. Definitions of 'collision' adopted in physical investigations. Assuming, as in the above investigation, that the path of the electron is approximately a straight line in the direction of the electric force, the mean free path of the electron is $\frac{1}{C}$, since C represents the total number of molecules with which the electron collides in a path of one centimetre length. The mean free path thus found is longer than the mean free path obtained from the kinetic theory of gases. The mean free path of a molecule of air, as deduced from the viscosity of air, is $7{\cdot}5 \times 10^{-6}$ centimetre at atmospheric pressure, so that the mean free path of a molecule of air at one millimetre pressure is $5{\cdot}7 \times 10^{-3}$ centimetre. Since the dimensions of an electron are small compared with those of a molecule, and the velocity of the electron is large compared with the velocity of agitation of the molecules, the mean free path of an electron in air at one millimetre pressure is $5{\cdot}6 \times 5{\cdot}7 \times 10^{-3} = 3{\cdot}2 \times 10^{-2}$. Hence, according to the definition of a collision adopted in the kinetic theory of gases, the number of molecules that an electron would encounter in traversing a centimetre of air at one millimetre pressure is 31, or about twice $[C = 14{\cdot}5]$ the maximum value of α. It is not to be expected that these two numbers should be equal, as the number of collisions that may be said to occur in a path of a given length depends on the way in which a collision is specified, and the definition suggested by the theory of ionization is widely different from that which is found most convenient for the purposes of the kinetic theory of gases.

Since the dimensions of the molecules may be obtained from the determinations of the mean free paths, it is interesting to compare the radius σ of a molecule as deduced from the kinetic theory with that obtained from the mean free path $\frac{1}{C}$. If the molecules be regarded as elastic spheres, the quantity $N\sigma^2$ may be deduced from the coefficient of viscosity of a gas, N being the number of molecules per cubic centimetre at atmospheric pressure. Taking for N the value 3×10^{19}, the value of σ may be calculated from the numbers given by Jeans,* the mean radius of a molecule of air obtained on this principle being $1{\cdot}64 \times 10^{-8}$ centimetre.

The experiments on the electrical conductivity show that ionization by collision does not take place every time an electron comes within the distance σ of the centre of a molecule, however great the velocity of the electron may be. If R be the distance an electron must approach the centre in order that ionization may result from the collision, the total number C of such collisions that occur per centimetre in a gas at a millimetre pressure is

$$\frac{\pi R^2 N}{760}; \quad \text{hence } R^2 = \frac{760 C}{\pi N};$$

and for air the value of R is $1{\cdot}08 \times 10^{-8}$ centimetre.

209. Dimensions of molecules. The curves representing the values of α/p in terms of X/p for the different gases may all be represented by a formula of the same form, namely

$$\frac{\alpha}{p} = C\epsilon^{-\frac{CVp}{X}},$$

* Jeans, Dynamical Theory of Gases, p. 251.

for the larger values of α/p. The following table gives the values of C and V for the different gases, and also the values of R and σ for the radii of molecules as deduced from the theory of ionization by collision, and the theory of the viscosity of gases:

	C	V	$C \times V$	$R \times 10^8$	$\sigma \times 10^8$
Air	14·6	25	365	1·08	1·64
Nitrogen	12·4	27·6	342	1·00	1·67
Hydrogen	5·0	26	130	·63	1·18
Carbonic acid	20·0	23·3	466	1·26	2·00
Hydrochloric acid	22·2	16·5	366	1·33	1·93
Water vapour	12·9	22·4	289	1·02	1·92
Argon	13·6	17·3	235	1·04	1·60
Helium	2·8	12·3	34·4	·47	1·04

A comparison of the numbers in the last two columns shows that there is a rough agreement between both systems of measurement. As a rule, the larger values of R and σ go with the heavier molecules, helium being the most notable exception, as the dimensions of the molecule appear to be smaller than that of hydrogen.

210. Average velocity of an electron required to ionize a molecule. Minimum velocity required to ionize a molecule. The quantity Ve represents the kinetic energy of the electron when the velocity is sufficiently high to produce ions by collision, if it be supposed that the mean velocity v_0 at the beginning of each free path is small compared with that acquired under the electric force. This hypothesis may be justified by reason of the fact that with the larger forces new ions are produced in many cases, when a collision occurs. A large proportion of the kinetic energy of the electron may, under these circumstances, be transferred to each molecule with which it collides.

The velocity v corresponding to the potential V may easily be found in terms of the velocity of agitation u which a particle of the same mass would have if in thermal equilibrium with the surrounding molecules, by eliminating m from the two equations

$$mNu^2 = 3 \times 10^6, \text{ and } mv^2 = 2eV,$$

and substituting for Ne its value $1 \cdot 23 \times 10^{10}$, the ratio of the squares of the velocities becomes

$$\frac{v^2}{u^2} = 27V,$$

V being expressed in volts.

The potential V deduced from the determinations of α/p in terms of X/p gives an average value of the potential required to produce an ion, but on some few occasions ionization may take place when the electron is moving with a smaller velocity than that corresponding to the potential V. Lenard * has made a special investigation to determine the smallest velocity of electrons at which ionization occurs. The electrons were generated by the action of ultra-violet light on a metal plate A, and moved under an electric force to a gauze G at a distance 1·45 centimetres from A. Beyond the gauze, at a distance of 3 centimetres from A, an insulated ring B was set parallel to A and G. The plate A and the gauze G were maintained at potentials V_1 and V_2, which were adjusted so that the potential difference $V_2 - V_1$ was positive and less than the potential $V_2 - V_3$, which was also positive, V_3 being the potential of the ring B. Some of the electrons, which start from A with a velocity corresponding to the potential v, arrive at the gauze with the velocity $v + V_2 - V_1$, and collide with molecules in the space between G and B. The ring B received a positive charge when the molecules were ionized, and the negative ions returned to G, as the potential $V_2 - V_3$ was positive. The ring B received only an extremely small charge when $V_2 - V_1$ was small, but when $V_2 - V_1$ exceeded 8 or 9 volts, the positive charge increased rapidly. With air at the pressures ·004 and ·04 millimetre, Lenard found that the limiting velocity required to ionize the molecules was that corresponding to a potential difference of 11 volts.

Franck and Hertz † have recently made experiments on

* P. Lenard, Ann. der Phys. (4) **8**, p. 194, 1902.
† J. Franck and G. Hertz, Verh. d. D. phys. Ges. **15**, p. 34, 1913.

a similar principle with electrons emitted from a glowing wire, and found values for the minimum potentials required to generate ions in different gases. The numbers obtained are as follows: hydrogen 11 volts, oxygen 9 volts, nitrogen 7·5 volts, argon 12 volts, helium 20·5 volts, neon 16 volts.

When the potential difference between the two ends of a free path is 25 volts, the velocity acquired by an ion or an electron is about twenty-six times the normal velocity of agitation.

In the case of an electron, the normal velocity of agitation is 10^7 centimetres per second, so that the mean velocity required to produce ionization by collision is of the order $2{\cdot}6 \times 10^8$ centimetres per second.

211. Stoletow's experiments on the effect of pressure on currents produced by ultra-violet light. An interesting series of investigations has been made by Stoletow* on the variation of the conductivity produced by ultra-violet light when the potential difference and distance between the electrodes were constant, and alterations were made in the pressure of the gas. It was found that for high pressures the current was small; but it increased as the pressure was reduced, and for a certain definite pressure, depending on the intensity of the force, the current attained a maximum value. Further reductions in the pressure then produced a decrease in the current, which approached a definite value when low pressures (small fractions of a millimetre) were attained. The following table of numbers given by Stoletow affords an example of the variations in the currents which take place under these conditions. The currents i were obtained between parallel-plate electrodes, at 3·71 millimetres apart, when the potential difference between the plates was maintained by a battery of 65 Clark cells, which is equivalent to 93 volts. The force between the plates was therefore 250 volts per centimetre.

* A. Stoletow, Journal de Physique (2) **9**, p. 468, 1890.

p (mm.)	i	p (mm.)	i
754 . . .	8.46	·64 . . .	108·2
152 . . .	13·6	·52 . . .	102·4
21 . . .	26.4	·275 . . .	82·6
8·8 . . .	32·2	·105 . . .	65·8
3·3 . . .	48·9	·0147 . . .	53·8
2·48 . . .	74·7	·0047 . . .	50·7
1·01 . . .	106·8	·0031 . . .	49.5

With the higher pressures the values of X/p are so small that there is probably no effect due to ionization by collision. There are, however, large changes in the currents when the pressure is changed from 754 mm. to 8·8 mm., which must be due to a variation in the number of ions set free from the negative electrode. When the pressure is 3·3 millimetres a considerable effect is produced by collisions, as the value of X/p is then 76 and the corresponding value of α/p is ·34, so that $\alpha = 1·1$ approximately.

The ions, being set free initially from the negative electrode by the ultra-violet light, start with a velocity corresponding to a potential fall of about 2 volts.* On the average they do not begin to produce ions by collisions until a velocity corresponding to a potential fall of 25 volts is attained, so that in the first millimetre no new ions are produced by collisions. In the remainder of the distance between the electrodes, 2·71 millimetres, the number produced initially becomes multiplied by the factor $e^{1.1 \times .271} = 1·35$, or 26 per cent. of the total number that arrive at the positive electrode are produced in the gas.

The current due to the ions set free by the direct action of the light at the pressure 3·3 millimetres is therefore 36. This is greater than the currents at the higher pressures, so that the number of ions set free by the light diminishes as the pressure is increased.

At the lowest pressures the current approaches a constant value equal to the rate at which ions are set free from the

* See Section 52.

metal plate, for in these cases there is so little gas between the plates that the great majority of the ions do not collide with molecules of the gas while they traverse the distance between the electrodes. The exact effect of collisions at the lowest pressures may easily be found. Since X/p is very large, α/p has the maximum value 14·6, and at the pressure ·0047 millimetres, $\alpha = \cdot068$. Hence about 2 per cent. of the current at that pressure is due to ions generated in the distance 2·71 millimetres.

212. Photo-electric currents obtained with small forces. When similar experiments were made with smaller potential differences between the electrodes, the variations in the current were not so well marked. The current underwent smaller changes with variations in the pressure, and when the potential difference between the electrodes was of the order 30 volts, there was no maximum at any particular pressure, but the current increased continuously as the pressure was reduced from 760 millimetres, approaching a constant value when the air was at a pressure of a small fraction of a millimetre.

When higher potentials up to 200 volts were used, the currents underwent larger changes as the pressure was altered, and a maximum value was attained at a certain definite pressure p_m, depending on the force X. At the lowest pressures the current assumed the same value as was obtained with the lower forces.

These effects obtained with the larger forces may be explained on the hypothesis that the variation in the number of ions set free by the direct action of the light is comparatively small, and the large changes in the currents obtained by altering the pressure are due to the changes in the number of ions generated in the gas.

213. Pressure corresponding to the maximum current. The point at which the maximum current was obtained was examined under various conditions, and it was found that the pressure p_m for which the current is a maximum is exactly proportional to the electric force X, and is independent of the distance between the plates. The actual value of the maximum current depends on the product of the distance and the electric force, or the potential difference between the plates.

Stoletow gives the following values of the pressure corresponding to the maximum current, obtained in a series of experiments in which various forces and distances between the plates were used.

The first column of numbers $E\,(cl)$ denotes the potential difference between the electrodes, the unit being the electromotive force of a Clark cell. The distance l between the plates is given in millimetres, the pressure p_m at which the current attains a maximum value is in millimetres of mercury, and the quantity $\frac{p_m l}{E\,(cl)} \times 10^4$ given in the last column is proportional to the ratio of the pressure p_m to the electric force $\frac{E}{l}$.

$E\,(cl)$	l (mm.)	p_m (mm.)	$\frac{p_m l}{E\,(cl)} \cdot 10^4$
165	·25	25·3	383
165	·47	13·5	384
65	·47	5·3	383
100	·83	4·7	389
65	·83	3·0	383
60	·83	2·8	386
65	1·91	1·3	382
65	3·71	·67	382
40	3·60	·43	387

The numbers in the last column show that the ratio $\frac{p_m}{X}$ is constant, and when the force is expressed in volts per centimetre p_m is given by the formula

$$p_m = \frac{X}{372}.$$

214. Maximum value of α when X is constant and p varies. Since the potentials $E\,(cl)$ were too small to cause appreciable ionization by the action of the positive ions, the variations in the numbers of ions generated in the gas must be attributed to the changes in the quantity α caused by changing the pressure. The force X being constant, the value of p for which α is a maximum may be found by differentiating the equation $\frac{\alpha}{p} = f\left(\frac{X}{p}\right)$.

Thus
$$\frac{d\alpha}{dp} = f\left(\frac{X}{p}\right) - \frac{X}{p} f'\left(\frac{X}{p}\right) = 0.$$

The latter expression involves $\frac{X}{p}$ only, and if A be the root of the equation, then $\frac{X}{p_m} = A$. Since A is constant, depending on the form of the function f, it follows that the pressure p_m is proportional to X, which agrees with the result obtained by Stoletow.* It is not necessary to express the function in an algebraic form to find the value of p_m corresponding to a given force. Since $\frac{\alpha}{p}$ is given by means of a curve in terms of $\frac{X}{p}$, the above equation for $\frac{X}{p_m}$ may be written

$$\frac{\alpha}{p_m} - \frac{X}{p_m} f'\left(\frac{X}{p_m}\right) = 0, \text{ or } \frac{y}{x} = \frac{dy}{dx},$$

where x and y are the co-ordinates of points on the curve. The latter form shows that x and y are the co-ordinates of the point of contact of the tangent from the origin. The value of $\frac{X}{p_m}$ may therefore be obtained by a simple geometrical construction. If the tangent be drawn to the curve obtained with air the point of contact is near the point whose ordinates are $\frac{X}{p} = 370$, $\frac{\alpha}{p} = 5{\cdot}5$. There is

* This explanation of Stoletow's experiments appeared in the Philosophical Magazine (6) **1**, p. 198, Feb. 1901. It was subsequently adopted by J. J. Thomson, Conduction of Electricity through Gases, 1903 edition, pp. 233–4.

thus an accurate agreement with Stoletow's result that the maximum current occurs at the pressure $p_m = \frac{X}{372}$.

The maximum value of α and the pressure p_m corresponding to a given force X may also be found by differentiating, with respect to p, the equation

$$\frac{\alpha}{p} = C\epsilon^{-\frac{CVp}{X}},$$

since this formula for α is in agreement with the numbers found for a range of values of $\frac{X}{p}$ that include the point of contact of the tangent from the origin of co-ordinates to the curve $y = f(x)$.

Thus $$\frac{d\alpha}{dp} = C\epsilon^{-\frac{CVp}{X}} - \frac{p}{X}C^2V\epsilon^{-\frac{CVp}{X}} = 0.$$

Hence $$p_m = \frac{X}{CV},$$

and $\alpha_m = \frac{X}{V\epsilon}$, ϵ denoting the base of the Napierian logarithms.

For air the value of CV was found to be 365. The corresponding numbers for the other gases are given in the table of constants, Section 209.

215. Maximum current corresponding to a given force. It is more difficult to estimate the maximum value of the current to the same degree of accuracy. It has been shown that the currents between electrodes are given by the expression $n = n_0\epsilon^{\alpha x}$ when the distances x between the electrodes lie between certain limits. If the ions were started initially with a mean velocity corresponding to the force X, then ionization by collision would take place in the layers of gas near the plane $x = 0$ at the same rate as at other points between the electrodes, and the maximum value of the current would be $n_0\epsilon^{\frac{Xx}{V\epsilon}}$, n_0 being the number starting from the negative electrode. The ions, however, only start with a small velocity from the negative electrode, and acquire the property of generating others by collisions after travelling a short distance l_1. Hence if n_0 be the number that arrive at the distance l_1 from the negative electrode, the number that arrive at the positive electrode will be $n_0 \epsilon^{\frac{X(x-l_1)}{V\epsilon}}$.*

With a given potential difference v between the plates, the maximum value of α is $\frac{v}{\epsilon V x}$, and since the current depends on αx, the best conductivity is obtained with those gases for which V is small, such as helium, hydrochloric acid, and argon. Hydrochloric acid or carbonic acid would give results in accordance with the formula for a large range of potentials, since ionization by positive ions does not take place in these gases until comparatively large potentials are used.

216. Ionization by cathode rays and β rays. Another series of investigations that are of interest in connection with these results are the measurements of the conductivity produced by electrons moving with velocities much greater than those attained with the highest forces used in the above experiments. The average velocity of the electrons travelling under the electric force of 1,000 volts per centimetre in air at one millimetre pressure probably did not exceed the value 5×10^8 centimetres per second. The velocity of the β particles emitted by radio-active substances is very much higher, and for some rays the velocity approaches the value 3×10^{10}. The cathode rays have velocities intermediate between these, and the momentum which is imparted to them is sufficiently great to cause them to penetrate short distances in air at atmospheric pressure, as in Lenard's experiments. In these cases, the number of pairs of ions generated per centimetre of the path of an electron in air at one millimetre pressure is much less

* The Theory of Ionization of Gases by Collision, p. 35.

than 14·6, which is the maximum number generated by an electron travelling under an electric force.

217. Ionization produced by β rays and Lenard rays. Durack's experiments. This was first shown conclusively by Durack's* experiments, in which he measured the currents between electrodes in air at a low pressure when electrons are projected with a high velocity through the gas in a direction normal to the electrodes. In the experiments with β rays, the electrode which was traversed by the rays was a thin aluminium plate, and was made the high-potential electrode. The opposite plate was of thick lead to absorb the β particles, and was faced with a sheet of aluminium so as to reduce the secondary rays.

The charge acquired by the lead plate was due to the absorption of β particles, and also to the ions generated in the gas between the plates. Thus if n be the number of β particles that traverse the gas per second, and n' the number of pairs of ions generated in unit distance in the gas between the plates, the current is $ln' - n$ or $ln' + n$, according as the aluminium electrode is maintained at a positive or negative potential. If c_1 and c_2 be the two currents, the ratio $\frac{n'}{n}$ is obtained from the equation

$$\frac{ln'}{n} = \frac{c_1 + c_2}{c_2 - c_1}.$$

When the pressure of the air between the plates was altered, $c_1 + c_2$ increased in proportion to the pressure and $c_2 - c_1$ remained constant. For one millimetre pressure the value obtained for the ratio $\frac{n'}{n}$ was ·17. Hence each β particle from a radio-active substance ionizes on an average one molecule when traversing a path of 6 centimetres of air at one millimetre pressure. This number is very small, as it follows from the kinetic theory that the β particle would encounter 180 molecules in that distance. The electrons must, therefore, pass through most of the molecules

* J. J. E. Durack, Phil. Mag. (6) **5**, p. 550, 1903.

without causing ionization to take place when the velocity is very high, and since they traverse considerable distances through gases with very slight diminution in their velocity, the motion cannot be much retarded by passing through a molecule.

By similar methods Durack found for Lenard rays (cathode rays that have traversed a thin sheet of aluminium) that on an average each electron in the cathode stream generates ·4 pair of ions in traversing a centimetre of air at a millimetre pressure. The electrons in this case were moving with velocities much less than those of the β rays. Owing to the difficulty of estimating the effects of secondary ionization in these experiments, it is uncertain whether the difference between the final results is due to an error arising from this cause or whether the number of ions generated per centimetre increases as the velocity of the electron is reduced.

218. Ionization depending on the velocity of the rays. Glasson's experiments with cathode rays. Further experiments on this subject have recently been made by Glasson,* in which the ionizing power of cathode rays was investigated with a view to finding the effects produced by particles moving with different velocities. The apparatus used for this purpose is illustrated in figure 52.

The rays were generated in the discharge tube D, which is of a simple form containing a cathode C directly opposite the anode A. A small hole was bored in the centre of the anode, and a tube T was joined to the back of the electrode through which the cathode rays (emitted by C) passed and entered the chamber B. The latter was a brass cylinder having two small side-tubes S_1 and S_2 at points separated by an angular distance of 90°. The axis of the cylinder was perpendicular to the direction of motion of the rays, and a solenoid was wound round the cylinder so that when a current was flowing in it a uniform magnetic field H was created inside B.

* J. L. Glasson, Phil. Mag. (6) **22**, p. 647, 1911.

The side-tube S_2 opened into a brass cylinder E containing a small insulated aluminium box F with a wide opening. The cathode rays emitted by the negative electrode C entered the cylinder B, and when there was no magnetic force they impinged on the surface at a point P. When a current was sent through the solenoid the rays were deflected through different angles as they emerged from the tube T with different velocities. Their paths in the magnetic field are circles touching the line CP at T,

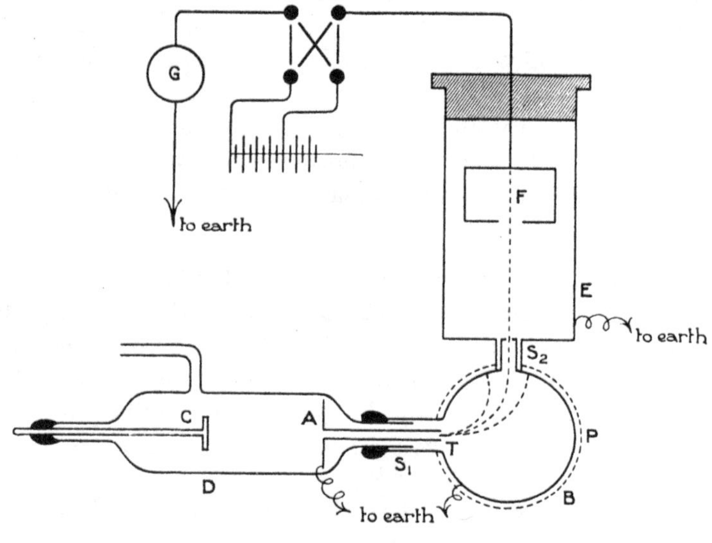

Figure 52.

and the radius R of the path of those that enter the tube S_2 is equal to the radius of the cylinder B. The velocity v of these rays is given by the equation

$$\frac{mv^2}{R} = Hev.$$

When they pass through the opening in the tube S_2 they ionize the gas, and entering the box F impinge on the surface remote from the aperture. The secondary rays, or rays irregularly reflected from the aluminium, are given off in all directions, and very few escape through the opening in F. Also, most of the ions generated inside F either recombine or are lost by diffusion to the inner surface, and only a very small proportion emerge through the aperture when the potential of the box F is raised above that of the surrounding cylinder E.

The box F was connected to one terminal of a battery, the other terminal being connected to earth through a sensitive galvanometer, since the currents obtained with cathode rays are so large it is unnecessary to use an electrometer to measure them. The cylinders E and B, and the positive electrode A of the discharge tube, were also connected to earth, so that after emerging from the tube T the rays are not affected by electric forces. The potential of F was too small to alter the motion of the rays appreciably, and was just sufficient to collect all the ions generated in the gas by the rays on the outer surface of F, which acted as one electrode, and the inner surface of E, which acted as the other electrode.

The current through the galvanometer consists of the current n carried by the cathode rays, and the current $n'l$ due to ions generated by the cathode rays while traversing the molecules in the distance l from S_2 to F. The latter current may be either positive or negative according to the potential of the cylinder F. If c_1 and c_2 be the currents through the galvanometer then $\frac{n'l}{n} = \frac{c_1 + c_2}{c_2 - c_1}$; so that the number of ions $\frac{n'}{n}$ generated per centimetre by each electron in the cathode stream was determined.

The pressure was varied within certain limits and the ratio $\frac{n'}{n}$ was found to be proportional to the pressure when the velocity of the rays was constant, that is, when the same current in the solenoid was used to deflect the rays into the tube S_2. In these experiments there was open communication between the various parts of the apparatus so that the pressure was the same throughout, and it was necessary to use small pressures so as to obtain cathode rays. The highest pressure attainable under this condition was ·2 millimetre, and the pressure was varied from ·2 to ·1 millimetre. The potential required to maintain the discharge in the tube D varies, and with low pressures it increases rapidly as the pressure is reduced. The velocity imparted to the cathode rays is therefore larger with the low pressures than with the high pressures. In this way, rays with velocities differing over a large range may be obtained.

If p be the pressure, α_v the number of ions generated per centimetre in air at one millimetre pressure by an electron projected with a velocity v, the value of α_v was deduced from the ratio $\frac{n'}{n}$, since $n' = p\alpha_v n$, and the velocity v was obtained from the intensity H of the magnetic field required to make the electron move in a circle of radius R while in the cylinder B.

The following table contains the values of α_v, corresponding to different velocities v of cathode rays, as given by Glasson:

$v \times 10^{-9}$	α_v
4·08	2·01
4·42	1·72
4·76	1·53
5·10	1·38
5·44	1·26
5·78	1·12
6·12	·99

The numbers show that with high velocities of the order 5×10^9 centimetres per second the number of molecules ionized by an electron per centimetre of its path diminishes as the velocity increases.

219. Action of rays of various types. The effect produced by an electron thus depends on the order of the velocity with which it moves through the gas. The results of the experiments with the higher velocities may be summarized as follows:

In a path of one centimetre in air at a millimetre pressure, each β particle emitted with velocities from 2×10^{10} to $2{\cdot}8 \times 10^{10}$ centimetres per second generated on an average ·17 pair of ions.

A cathode ray particle travelling with a velocity of 4×10^9 centimetres per second generates 2 pairs of ions.

An electron moving under an electric force of 1,000 volts per centimetre generates 10 pairs, which increases to a maximum of 14·6 as the force is increased.

It may be shown that the mean velocity under the electric force in the latter case is of the order 2×10^8 centimetres per second, but owing to the uncertainty of the effects of collisions on the motion of the electrons it is possible only to obtain an approximate estimate of the velocity from theoretical considerations.

When the velocities are large, the effect of a collision on the motion of an electron probably varies in a manner that corresponds to the effect produced on a molecule, so that the gas may offer more resistance to the motion when the velocity is of the order 10^8 than when it is 10^{10} centimetres per second.

It will also be remarked that with the higher velocities the momentum of the electron is sufficient to carry it along its path, but with the lower velocities an electric field is necessary to maintain the motion. In the latter case the velocities on collision exceed the mean velocity under an electric force.

220. Velocity of electrons before ionization by collision takes place. The formula $U = \frac{Xe}{m} \cdot \frac{l}{u}$, given in Chapter III for the velocity of an ion or electron in an electric field, applies only when the velocity of agitation u is large compared with the velocity U in the direction of the electric force. The latter velocity is proportional to the force provided the velocity of agitation and the mass m are constant, as is the case for ions moving in gases at high pressures under the action of small electric forces.

When large forces are used and the pressure is reduced, the average mass m and the velocity of agitation depend on the force and the pressure, since the ions are in the transition stage and the electronic state becomes more prevalent as $\frac{X}{p}$ increases. With these forces and pressures the velocity in the direction of the electric force increases rapidly with the force.

When the ratio $\frac{X}{p}$ is sufficiently great (of the order ·5 for dry air) the electrons move freely with a constant mass m which is very small, so that the velocity of agitation is large. If the motion were in thermal equilibrium with the molecules of the gas this velocity would be inversely proportional to the square root of the mass of the particles, and since the ratio of the mass of a molecule of air to the mass of an electron is $5{\cdot}3 \times 10^{17} \div 10^{13}$, the velocity of agitation of the electrons would be 230 times the velocity of agitation of a molecule of air at 0° C., or 10^7 centimetres per second. The actual velocity of agitation of the electrons, however, exceeds the value 10^7 by a factor $\sqrt{k}$ which depends on the ratio $\frac{X}{p}$.* When the force is one volt per cm., and the pressure one millimetre, $k = 11{\cdot}5$ approximately, so that the velocity of agitation is $3{\cdot}4 \times 10^7$ centimetres per second. The velocity under the electric force is still small compared with the velocity of agitation, being $1{\cdot}25 \times 10^6$ when $X/p = 1$.

221. Motion of electrons when X/p becomes large. With large forces a point is reached when the velocity in the direction of the electric force exceeds the velocity of agitation, and in that case the above formula cannot be used to determine U. That stage may be reached when $\frac{X}{p}$ is 1,000, since ionization then takes place in a large percentage

* See Sections 131 and 132.

of the total number of collisions, and the velocity of the electron may be considerably reduced at each collision that it makes with a molecule.

If the velocity of an electron after a collision be neglected in comparison with the velocity acquired in moving along a free path under the action of the force, the mean velocity U through the gas may be easily found.

Let x_1, x_2 ... be the lengths of the free paths; t_1, t_2 ... the intervals between collisions, then

$$x_1 = \frac{Xe}{2m} \cdot t_1^2, \quad x_2 = \frac{Xe}{2m} \cdot t_2^2, \ldots$$

and $$U = \frac{x_1 + x_2 + \ldots}{t_1 + t_2 + \ldots} = \sqrt{\frac{Xe}{2m}} \cdot \frac{x_1 + x_2 + \ldots}{\sqrt{x_1} + \sqrt{x_2} + \ldots}.$$

If a large number n of paths are traversed, the number whose lengths are intermediate between x and $x + dx$ is

$$\frac{n}{l} \epsilon^{-\frac{x}{l}} dx,$$

and the denominator $\sqrt{x_1} + \sqrt{x_2} + \ldots$ of the above fraction becomes

$$\frac{n}{l} \int_0^\infty \sqrt{x} \epsilon^{-\frac{x}{l}} dx$$

$$= n \int_0^\infty y d\left(\epsilon^{-\frac{y^2}{l}}\right)$$

$$= n \int_0^\infty \epsilon^{-\frac{y^2}{l}} dy = \frac{n\sqrt{l\pi}}{2}.$$

Hence $U = \sqrt{\frac{2Xel}{m\pi}}$, since the numerator of the fraction $x_1 + x_2 + \ldots = nl$.

The velocity in this case is proportional to $\sqrt{\frac{X}{p}}$, and when $X = 1{,}000$ volts per cm. and $p = 1$ mm., $U = 2 \times 10^8$.*

This number is probably a low estimate of the velocity,

* Taking $l = 3{\cdot}2 \times 10^{-2}$ as in Section 133.

and as the force increases the effect of each collision on the motion of the electron diminishes. The whole character of the motion then changes, and under a high electric force there is an acceleration of the electron along its whole path, and the velocity does not tend to attain a final value as when particles move under a force in a highly resisting medium.

CHAPTER IX

IONIZATION BY COLLISIONS OF POSITIVE IONS

222. Effect of ionization by positive ions on the currents between parallel plates. It has been explained in the preceding chapter how the current between parallel plates depends on the ionization produced by the motion of negative ions in the gas. When the electric force is constant, the currents for a certain range of distance between the plates are given by the formula $n = n_0 \epsilon^{\alpha x}$ for the conductivity produced by ultra-violet light falling on the negative electrode. For larger distances the currents increase with the distance x more rapidly than the exponential term $\epsilon^{\alpha x}$. It appears from this that the effect of some other process of ionization begins to be appreciable, and the investigations show that the conductivity may then be explained on the hypothesis that positive ions also generate others in the gas.

In order to calculate the currents that would be produced if both positive and negative ions generate others by collisions with molecules, the simple case of the conductivity between parallel plates may be taken, and the ions may be considered to move with a definite average velocity over the whole distance between the plates. This assumption introduces a small error in estimating the relative values of the currents for the different distances between the plates, but when the distance between the plates is large compared with the mean free path of the ion, this error is probably much less than the experimental error involved in measuring the currents.

When a number n_0 of negative ions are generated initially at the negative electrode, the formula for the number n that eventually reach the positive electrode may be found as follows: Let α be the number of pairs of ions generated by a negative ion in moving through a centimetre of the gas and β the number generated by a positive ion.

Let p be the number of pairs of ions generated by both processes in the layer of gas between the negative electrode and the parallel plane at the distance x from the electrode, and q the number of pairs generated in the layer of thickness $a-x$, a being the distance between the electrodes.

The number of negative ions n that arrive at the positive electrode is $n = n_0 + p + q$.

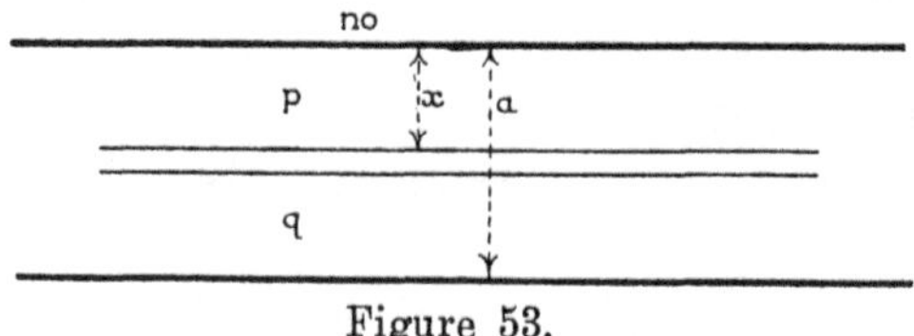

Figure 53.

The number of pairs of ions generated in the space between the two planes x and $x+dx$ is

$$(n_0 + p)\,\alpha\, dx + q\,\beta\, dx,$$

since the number of negative ions that pass through the layer in one direction is (n_0+p), and the number of positive ions that pass through in the opposite direction is q.

Hence $$dp = (n_0 + p)\,\alpha\, dx + q\,\beta\, dx.$$

Substituting for q its value $n - n_0 - p$, the differential equation for p becomes

$$\frac{dp}{dx} = (n_0 + p)(\alpha - \beta) + n\beta,$$

of which the solution is

$$n_0 + p = A\,\epsilon^{(\alpha-\beta)x} - \frac{n\beta}{\alpha-\beta}.$$

The constant of integration A is found from the condition $p = 0$ when $x = 0$. Hence $A = n_0 + \dfrac{n\beta}{\alpha-\beta}$, and since $n_0 + p = n$ when $x = a$, the following relation between n and n_0 is obtained,

$$n = \left(n_0 + \frac{n\beta}{\alpha-\beta}\right)\cdot \epsilon^{(\alpha-\beta)a} - \frac{n\beta}{\alpha-\beta};$$

or $$n = n_0 \frac{(\alpha - \beta)\, \epsilon^{(\alpha-\beta)a}}{\alpha - \beta\, \epsilon^{(\alpha-\beta)a}}.*$$

223. Experimental evidence of the action of positive ions. When the pressure and electric force are constant, the currents obtained by the action of ultra-violet light on the negative electrode are in agreement with this formula. The currents for the different distances between the plates depend on the two constants α and β which may be found from two ratios $\frac{n_1}{n_2}$ and $\frac{n_2}{n_3}$ of currents at different distances. Experiments have been made to test the effect of a variation in the intensity of the light, and it was found that the currents at the different distances were altered in the same proportion. The ratios $\frac{n_1}{n_2}$ and $\frac{n_2}{n_3}$ are therefore independent of the number of ions initially produced at the negative electrode. The table of numbers given in Section 195 shows to what extent the currents q in air at one millimetre pressure may be represented by the formulae $\epsilon^{\alpha a}$ and $\frac{(\alpha-\beta)\, \epsilon^{(\alpha-\beta)a}}{\alpha - \beta\, \epsilon^{(\alpha-\beta)a}}$. The following tables † give examples of similar results obtained with air at the pressures 4, 6, and 8 millimetres, the electric force being proportional to the pressure in each case. It will be observed that the currents are in agreement with both formulae for the smaller

* This equation, and the theory of sparking to which it leads, was first given by the author in the *Electrician*, April 3, 1903, after several cases had been verified experimentally by the methods described above.

The same explanation of the currents between parallel plates was subsequently given by Sir J. J. Thomson in the 1906 edition of his treatise on the 'Conduction of Electricity through Gases', pp. 270–4, but it is difficult to find evidence in support of the theory from the researches which are there quoted.

Another theory of the sparking potential was given by Sir J. J. Thomson in the 1903 edition of his treatise which is unsatisfactory, as it implies that the sparking potential for parallel plates is the same as the potential required to maintain a current when the normal cathode fall of potential is established. (See Phil. Mag. (6) **6**, p. 358, 1903, and (6) **8**, p. 750, Dec. 1904.)

† See researches by the author and H. E. Hurst, Phil. Mag. (6) **8**, p. 738, 1904.

distances a. This shows that the positive ions may be acting all the time, but the effect they produce does not become noticeable until the larger distances are reached, since β is so small compared with α.

I. Currents q in air at 4 millimetres pressure under a force $X = 700$ volts per centimetre, the distance a between the plates being expressed in centimetres : $\alpha = 8{\cdot}16$, $\beta = {\cdot}0067$.

a in cm.	·2	·3	·4	·5	·6	·7	·8
q	5·12	11·4	26·7	61	148	401	1,500
$\epsilon^{\alpha a}$	5·11	11·6	26·1	59	133	301	680
$\dfrac{(\alpha-\beta)\epsilon^{(\alpha-\beta)a}}{\alpha-\beta\epsilon^{(\alpha-\beta)a}}$	5·11	11·6	26·5	62	149	399	1,544

$S = {\cdot}871$, $S \times X = 609$, $V = 615$.

II. Currents q in air at 6 millimetres pressure under a force of 1,050 volts per centimetre: $\alpha = 12{\cdot}42$, $\beta = {\cdot}0108$.

a in cm.	·1	·2	·3	·4	·5
q	3·47	12·1	42·5	163	880
$\epsilon^{\alpha a}$	3·46	12·0	41·3	143	495
$\dfrac{(\alpha-\beta)\epsilon^{(\alpha-\beta)a}}{\alpha-\beta\epsilon^{(\alpha-\beta)a}}$	3·47	12·1	42·8	163	872

$S = {\cdot}527$, $S \times X = 601$, $V = 604$.

III. Currents in air at 8 millimetres pressure under a force of 1,400 volts per centimetre : $\alpha = 16{\cdot}47$, $\beta = {\cdot}013$.

a in cm.	·1	·2	·3	·4
q	5·2	27·5	158	1,840
$\epsilon^{\alpha a}$	5·19	26·9	140	724
$\dfrac{(\alpha-\beta)\epsilon^{(\alpha-\beta)a}}{\alpha-\beta\epsilon^{(\alpha-\beta)a}}$	5·2	27·6	158	1,850

$S = {\cdot}431$, $S \times X = 603$, $V = 603$.

The value of $\frac{X}{p}$ in the three experiments was 175, and the values of $\frac{\alpha}{p}$ as deduced from the observations are 2·04, 2·07, and 2·06, the values of $\frac{\beta}{p}$ being ·00167, ·0018, and ·0016. Hence $\frac{\alpha}{p}$ and $\frac{\beta}{p}$ are practically the same at different pressures when the force is proportional to the pressure.

The following tables contain some examples of experiments in hydrogen and carbonic acid. In hydrogen β is larger than in air for the same forces and pressures; in carbonic acid β is very small. The currents q obtained experimentally are given in arbitrary units, and the quantities n were calculated from the formula $n = \frac{(\alpha - \beta)\, \epsilon^{(\alpha - \beta)\, a}}{\alpha - \beta\, \epsilon^{(\alpha - \beta)\, a}}$, α and β having the values given in the first column of each table.

HYDROGEN, 16 MILLIMETRES PRESSURE.

a		·2	3	·4
$X = 1050$,	q	8·2	26·7	126
$\alpha = 10·4$, $\beta = ·086$	n	8·4	26·7	124

$S = ·463$, $S \times X = 485$, $V = 490$.

HYDROGEN, 8 MILLIMETRES PRESSURE.

a		·2	·3	·4	·5	·7	·8
$X = 525$	q	—	4·87	—	14·4	51·5	122
$\alpha = 5·2$, $\beta = ·043$	n	—	4·85	—	14·6	53	124
$X = 700$	q	6·05	15·2	44·4	174	—	—
$\alpha = 8·86$, $\beta = ·059$	n	6·05	15·3	44·2	176	—	—

$X = 525$, $S = ·927$, $S \times X = 486$, $V = 487$.
$X = 700$, $S = ·57$, $S \times X = 398$, $V = 385$.

HYDROGEN, 4 MILLIMETRES PRESSURE.

a		·1	·2	·3	·4	·6	·8	1·0
$X = 350$	q	—	2·5	—	6·05	14·9	43·7	177
$\alpha = 4·43$, $\beta = ·0295$	n	—	2·44	—	6·05	15·3	44·2	176
$X = 700$	q	2·65	7·25	25·3	960	—	—	—
$\alpha = 9·6$, $\beta = ·214$	n	2·65	7·5	26	870	—	—	—

$X = 350$, $S = 1·14$, $S \times X = 398$, $V = 385$.
$X = 700$, $S = ·405$, $S \times X = 283$, $V = 282$.

HYDROGEN, 2 MILLIMETRES PRESSURE.

a		·2	·4	·6	·7	·8
$X = 350$	q	2·65	7·32	24·6	68	822
$\alpha = 4·8$, $\beta = ·107$	n	2·65	7·5	26	65	870

$S = ·81$, $S \times X = 283$, $V = 287$.

CARBON DIOXIDE,* 2 MILLIMETRES PRESSURE.

a		·1	·2	·3
$X = 1400$	q	8·32	70·7	747
$\alpha = 21·21$, $\beta = ·0085$	n	8·32	71·7	755

$S = ·369$, $S \times X = 516$, $V = 517$.

* These determinations of α and β in carbonic acid were made by Hurst. H. E. Hurst, Phil. Mag. (6) **2**, p. 535, 1906.

CARBON DIOXIDE, 1 MILLIMETRE PRESSURE.

a		·2	·3	·4	·5	·6
$X = 700$	q	8·54	24·7	73·1	226	787
$\alpha = 10{\cdot}68$, $\beta = {\cdot}0041$	n	8·54	24·8	73·8	228	790
$X = 1050$	q	13	51	289	—	—
$\alpha = 12{\cdot}68$, $\beta = {\cdot}0353$	n	13	50·4	278	—	—

$X = 700$, $S = {\cdot}736$, $S \times X = 515$, $V = 509$.
$X = 1050$, $S = {\cdot}465$, $S \times X = 488$, $V = 491$.

CARBON DIOXIDE, ·5 MILLIMETRE PRESSURE.

a		·2	·4	·6	·8
$X = 525$	q	3·64	13·3	53·1	303
$\alpha = 6{\cdot}41$, $\beta = {\cdot}0174$	n	3·64	13·3	52·9	301

$X = 525$, $S = {\cdot}929$, $S \times X = 488$, $V = 485$.

The currents q are in good agreement with the expression found for n provided the pressure of the gas is not reduced to a very low value. The free paths of the ions between collisions with molecules increase as the pressure is reduced, and according as they become comparable with the distance between the electrodes the formula for n becomes less accurate.

In air at a millimetre pressure the mean free path of a positive ion is of the order ·06 millimetre.

These experiments were as a rule only applied to cases in which the product of the pressure p and the distance a exceeded the value of pa that corresponds to the minimum sparking potential.

224. Comparison of the values of α and β. The value of β is connected with the force and pressure by a relation similar to that which holds for the quantity α; that is, $\frac{\beta}{p}$ is a function of the ratio $\frac{X}{p}$ only.

The curves, figure 54, represent $\frac{\beta}{p}$ in terms of $\frac{X}{p}$ for the different gases. It will be noticed that for the same force

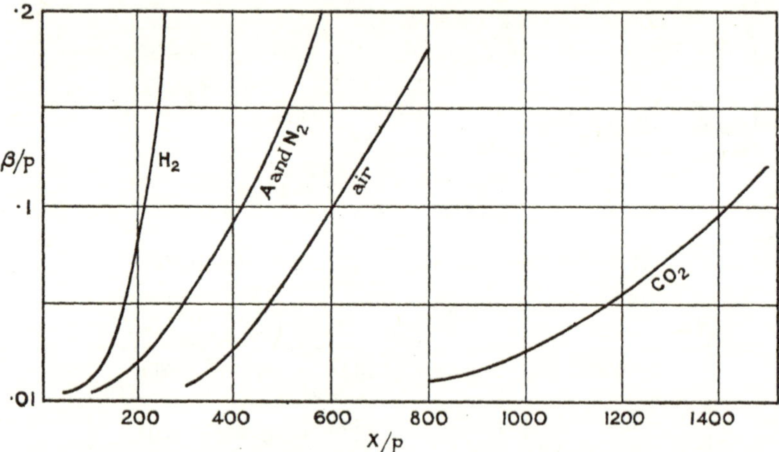

Figure 54. X in volts per centimetre, p in millimetres of mercury.

and pressure β is much less than α for any gas. The values of $\frac{\beta}{p}$ for argon are practically the same as for nitrogen, but with helium the values of β are much larger than for other gases.

The numbers recently obtained for pure helium are given in the following table, which shows that as compared with hydrogen the positive ions in helium produce much more ionization under the action of small electric forces.

HELIUM.

$\frac{X}{p}$	$\frac{\alpha}{p}$	$\frac{\beta}{p}$
5	·126	—
10	·27	—
20	·57	—
38	1·03	·0135
40	1·08	·0164
80	1·83	·040
120	2·1	·070
200	2·37	·165

By comparing the values of α and β it may be seen that the positive ions are much less efficient than the negative ions in producing new ions by collisions with molecules when both have the same kinetic energy.

Since the negative ions in these cases are free electrons, the positive ions are large compared with them, and if it be assumed that the mass of a positive ion is of the same order of magnitude as that of a molecule of the gas from which it is derived, each positive ion would make four times as many collisions with molecules as a negative ion travelling the same distance in the gas. The free paths of an electron in a gas at pressure p will therefore be the same as the free paths of a positive ion in a gas at pressure $\frac{p}{4}$, and if the same electric force be acting in each case the energy acquired between the collisions will also be the same. The relative ionizing powers may, therefore, be found by comparing the value of α in a gas at any pressure with the values of β in the gas at a quarter of the pressure when the same force is acting.

The case of hydrogen may be taken as an example. With the force $X = 50$ volts per centimetre and $p = 1$ millimetre, $\alpha = \cdot 35$. The curves representing the effects of positive ions give $\frac{\beta}{p} = \cdot 090$ when $\frac{X}{p} = 200$, so that $\beta = \cdot 022$ when $X = 50$ and $p = \cdot 25$. Hence when a positive and negative ion collide an equal number of times with molecules and acquire equal kinetic energies along their free paths, the negative ion generates sixteen times as many ions as the positive ion. The ratio of the activities of the positive and negative ions, as estimated in this way, varies with the electric force X. Under a force of 75 volts per centimetre, $\alpha = \cdot 9$ when $p = 1$, and $\beta = \cdot 065$ when $p = \cdot 25$, so that the effect of the negative ion is fourteen times that of the positive ion in this case.

If similar calculations are made for the other gases it will be found that the action of the positive ions as compared with that of the negative ions diminishes as the density of the gas increases. Thus, with carbonic acid,

$$\alpha = 3{\cdot}2 \quad \text{when } X = 215 \text{ and } p = 1,$$
$$\beta = {\cdot}003 \text{ when } X = 215 \text{ and } p = {\cdot}25.$$

Hence the effects of the positive and negative ions are in the ratio 1 : 1070.

The relative ionizing power of positive and negative ions depends on their velocities.

In the initial stages of ionization by collision, when the ions are moving under small electric forces and a given kinetic energy is imparted to ions of different masses, the smaller the mass of the ions the greater the probability of new ions being generated when a collision with a molecule occurs.

When ions are moving with very high velocities, as when 'α' and 'β' particles are emitted by radio-active substances, the relative effects of positive and negative ions are quite different, the number generated by the positive ion per centimetre being much greater than the number generated by a negative ion.

225. Theory of sparking. The theory of ionization by collision, which affords an accurate explanation of the currents obtained by the aid of ultra-violet light, also explains how sparking takes place. When the distance a between the electrodes does not exceed a certain definite value (which depends on the electric force and the pressure p), the current q only lasts so long as the light continues to act on the negative electrode. For distances somewhat larger than the values of a given in the preceding tables, a spark takes place between the electrodes and a large current flows through the gas. The point at which the spark is obtained may be deduced from the formula

$$n = n_0 \frac{(\alpha - \beta)\, \epsilon^{(\alpha-\beta)a}}{\alpha - \beta\, \epsilon^{(\alpha-\beta)a}}.$$

The denominator of this expression diminishes as a increases, and vanishes for a certain value of a. The quantity n then becomes infinite, and a continuous current flows between the electrodes after the light ceases to act on the electrodes. The critical distance S at which n becomes infinite is given by the equation

$$\alpha = \beta \epsilon^{(\alpha-\beta) S},$$

and the theoretical value of the sparking potential at the distance S is $S \times X$.

When the values of α and β had been determined, another series of experiments was made to determine the sparking potentials. The plates were set at the distance S given by the formula

$$S = \frac{1}{\alpha-\beta} \log\left(\frac{\alpha}{\beta}\right),$$

and the potential between the plates was gradually raised until a spark took place. The least potential V at which the sparking occurred is given at the foot of each of the tables in Section 223, also the value of S and the theoretical potential $S \times X$. There is a close agreement between the values of V and the product $S \times X$.

The following tables * give the results of the experiments with different gases that have been made in order to test the accuracy of the theory, as applied to a uniform field. The electric force X is given in volts per centimetre, and the sparking distance S in centimetres. The product $S \times X$ is the theoretical value of the sparking potential, and V the potential obtained experimentally.

AIR, SPARKING POTENTIALS.

X	p	S	$S \times X$	V	$p \times S$
1050	8	·765	803	803	6·12
1400	8	·431	601	603	3·45
1050	6	·572	601	604	3·43
700	4	·871	610	615	3·48
1050	4	·454	477	480	1·82
525	2	·91	481	488	1·82
700	2	·575	403	407	1·15
350	1	1·13	395	398	1·13
437	1	·832	364	365	·83
350	·66	·965	338	340	·64
437	·66	·766	335	336	·505

* The numbers in these tables are taken from researches already quoted in Section 198.

Hydrogen, Sparking Potentials.

X	p	S	$S \times X$	V	$p \times S$
1050	20	·643	675	675	12·9
875	16	·710	621	619	11·4
1050	16	·463	485	490	7·4
700	12	·791	555	561	9·5
1050	12	·353	370	389	4·23
525	8	·927	486	487	7·42
700	8	·57	399	395	4·56
1050	8	·306	322	322	2·45
350	4	1·14	399	395	4·56
525	4	·613	322	323	2·45
700	4	·405	283	282	1·62
350	2	·810	283	287	1·62
525	2	·501	269	273	1·00
350	1	·806	282	289	·81

Nitrogen, Sparking Potentials.

X	p	S	$S \times X$	V	$p \times S$
700	4	·72	502	507	2·88
—	4	·60	—	460	2·40
—	2	·85	—	382	1·70
525	2	·665	349	344	1·33
—	2	·65	—	339	1·30
—	2	·58	—	332	1·16
—	1	1·00	—	310	1·00
—	1	·868	—	300	·868
—	1	·601	—	298	·601
525	1	·580	304	300	·580
—	1	·500	—	310	·500
—	1	·400	—	330	·400

Carbon Dioxide, Sparking Potentials.

X	p	S	$S \times X$	V	$p \times S$
1400	2	·369	516	517	·738
700	1	·736	516	509	·736
875	1	·571	500	495	·571
1050	1	·465	488	491	·465
525	·5	·929	488	485	·464
700	·5	·706	494	497	·353
875	·5	·613	537	530	·306
525	·25	1·06	558	564	·265

ARGON, SPARKING POTENTIALS.

X	p	S	$S \times X$	V	$p \times S$
505	10	1·087	549	549	10·87
610	4	·471	287	276	1·88
410	4	·947	388	380	3·79
618	2	·378	234	233	·76
405	2	·618	248	245	1·24
615	1	·399	245	248	·399
517	1	·482	249	244	·482
401	1	·580	233	235	·580
309	1	·770	248	237	·770
402	·66	·599	241	248	·395
301	·66	·802	242	238	·528

226. Effect of impurities in helium. The experiments which were made with helium, to test the theory, were conducted on a slightly different principle from that adopted in the other cases. The distance a between the electrodes was increased by small amounts at a time and the potential difference aX was applied, X being constant for each distance. A certain distance S' was thus found experimentally for which a spark took place. The quantities α and β having been also found by previous experiments for the force X and the pressure p, it was possible to compare S' with the theoretical sparking distance $S = \dfrac{\log \alpha/\beta}{\alpha - \beta}$. The following table gives S and S' for certain forces X and pressures p, and also the sparking potential V corresponding to the distance S':

HELIUM, SPARKING POTENTIALS.

X	p	S	S'	V	$p \times S'$
240	2	·84	·81	194	1·68
240	3	·71	·71	170	2·13
200	5	·79	·78	156	3·95
198	5·2	·82	·83	164	4·26

The following values of the sparking potential V were also observed, but without obtaining the values of α and β:

$p \times S$	1·87	4·98	16·0	22·4	35·2
V	175	162	210	240	280

The numbers given in the last table are taken from the experiments recently made by Gill and Pidduck,* in which special precautions were taken to remove impurities from the gas. It was found that the values of β in helium were affected to a great extent by the presence of small quantities of air, and the sparking potentials were increased in a corresponding manner. The values of α were also affected, but to a smaller extent. In order to remove traces of air, the gas was kept in a long vertical tube containing a quantity of charcoal, which absorbs all the impurities when the tube is surrounded with liquid air. The electrical properties of the gas were determined immediately after it had been taken from the charcoal tube, and in some cases experiments were made with the charcoal tube in connection with the apparatus. The effect of small impurities was observed by keeping the gas in the vessel containing the parallel plates and making determinations on successive days with the same force and pressure. There was a small leak in this apparatus, and when the helium was at 5 millimetres pressure ·25 per cent. of air got into the gas during the day.

The following table gives the results of the experiments in which $X = 200$ volts per centimetre and $p = 5$ millimetres:

	α	β	S	V
Pure helium	5·4	·082	·78	156
Helium + ·25 per cent. air	5·1	·069	·85	170
Helium + ·50 per cent. air	5·0	·065	·88	176
Helium + ·75 per cent. air	5·0	·051	·93	187

Thus small quantities of air have a large effect on the conductivity of helium, which accounts for the fact that the sparking potentials obtained in previous investigations were much higher than those just given. The large effect of impurities on the sparking potential in helium was first observed by Strutt,† who obtained values of the sparking potential as low as 261 volts, but the method of

* E. W. B. Gill and F. B. Pidduck, Phil. Mag. (6) **23**, p. 837, 1912.
† Hon. R. J. Strutt, Phil. Trans. **193**, p. 377, 1900.

freezing out impurities by surrounding a charcoal tube with liquid air was not available at the time he made his experiments, so that the purification was probably not very complete. It will be noticed that, as far as testing the theory is concerned, it matters little whether the gas is impure or not, since the potentials change in accordance with the quantities α and β, and the theory is applicable to mixtures of gases such as air as well as to pure gases.

All the determinations of the sparking potentials with the different gases given in the above tables were made by observing the point at which sparking took place when ultra-violet light of small intensity acted on the negative electrode. The potentials thus obtained were quite definite and showed no irregularities depending on the length of time the gas is subjected to the electric force.* They were in most cases about two volts below the potential required to spark when the light was not acting.

In general, sparking in gases is due to the multiplication of the few ions which are present under normal conditions in the gas. Theoretically a small number of ions would be sufficient to initiate a discharge when the plates are at the proper sparking distance corresponding to any electric force, and it has been shown by Elster and Geitel and C. T. R. Wilson † that a few ions are being continually generated in a gas even when it is contained in a closed vessel.

227. Sparking potential as a function of the number of molecules between parallel plates. The above tables show that the sparking potential depends only on the product of the pressure and the distance between the plates. For example, in air at 8, 6, and 4 millimetres pressure, the sparking potentials, for the distances ·431, ·572, and ·871 respectively, are the same. Thus the sparking potential for parallel plates depends only on the number of molecules between the plates. This property of a gas was first observed by de la Rue and Müller,‡ who determined the

* See Section 235.
† See Sections 11 and 12.
‡ W. de la Rue and H. W. Müller, Phil. Trans. **171**, p. 109, 1880.

sparking potentials for flat surfaces when the pressure of the gas and the distance between the plates were varied over large ranges.

They found that the potentials obtained by varying the pressure when the distance between the plates was constant were the same as the potentials obtained by making similar changes in the distance when the pressure was constant. Paschen * made a complete investigation of the sparking potential for spherical electrodes, and found that the potential was independent of the curvature provided that the distance between the spheres was somewhat smaller than the radius of curvature. He found that in these cases also the sparking potential depended on the product pS only.

It may easily be seen that the sparking potential, as given by the theory, is a function of the product pS. For if V denotes the potential SX, and m the product pS, then $\alpha = pf\left(\frac{X}{p}\right) = pf\left(\frac{V}{m}\right)$, and similarly $\beta = p\phi\left(\frac{V}{m}\right)$. If these expressions for α and β be substituted in the equation for S,

$$(\alpha - \beta)\, S = \log\left(\frac{\alpha}{\beta}\right),$$

the equation for V becomes

$$m\left[f\left(\frac{V}{m}\right) - \phi\left(\frac{V}{m}\right)\right] = \log\left[f\left(\frac{V}{m}\right) / \phi\left(\frac{V}{m}\right)\right],$$

which shows that V is a function of m only.

228. Sparking potentials at low pressures. Villard's investigation of the development of cathode rays. The curves, figure 55, represent the sparking potentials in terms of the product pS for the different gases, for the points that have been found to be in agreement with the theory.

In general, the potential difference between two electrodes required to initiate a discharge depends on the shape of the electrodes, their distance apart, and the pressure of the gas. The phenomena are simplified when the gas is

* See Section 241.

contained between two metal plates, since the sparking potential depends only on the product pS, and in order to investigate the changes that take place, the distances may be fixed and the pressure reduced from 760 millimetres to a small fraction of a millimetre. If the distance S is one centimetre, the sparking potential in air at atmospheric pressure is about 30,000 volts, at 6 millimetres it is about

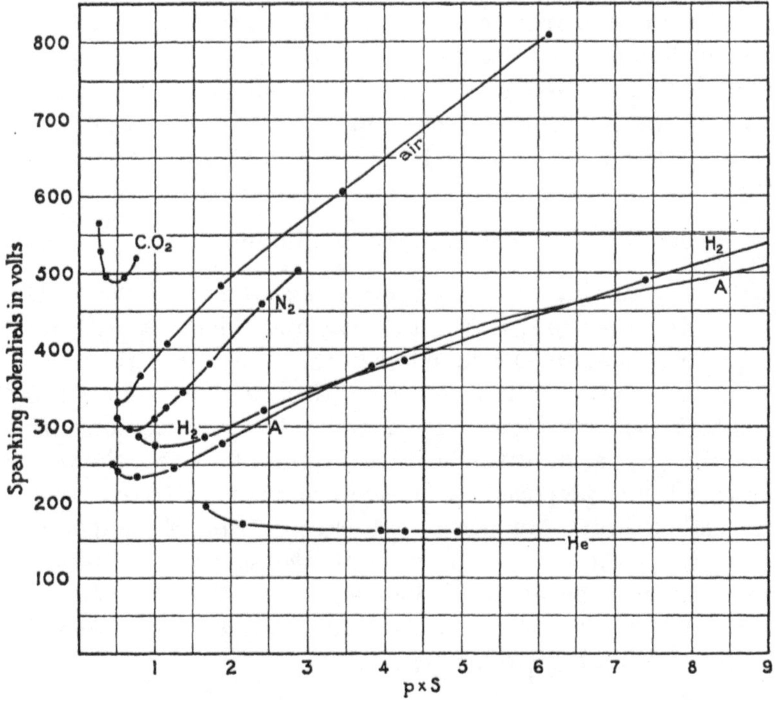

Figure 55. p the millimetres of mercury, S the spark length in centimetres.

800 volts, and at ·6 millimetre the minimum sparking potential is reached, which is about 340 volts. When further reductions are made in the pressure the potential rises rapidly. The sparking potential soon becomes much greater than the potential corresponding to atmospheric pressure.

It is necessary to consider for what range of values of the product pS the discharges may be attributed to the ionization of the molecules of the gas by the collisions of positive and negative ions. The experiments show that this hypothesis leads to satisfactory conclusions when the product pS varies from the value corresponding to the minimum sparking potential to values ten or twelve times as great. The theory probably applies to any case in which pS is large, but when the gas is at low pressures it is necessary to consider other effects that take place.

Thus, in discharge tubes, when the pressure is very low, one of the principal features of the discharge is the stream of cathode rays composed of electrons emitted by the negative electrode. These cathode rays are observed when the ratio of the electric force to the pressure throughout the tube, or at the cathode, is much larger than the ratio $\frac{X}{p} = 660$ that corresponds to the minimum sparking potential. Several interesting experiments were made by Villard * in order to discover the way in which the cathode rays originate, and he found that these rays are given off from the parts of the negative electrode on which the positive ions impinge. Even when the electric force at the surface of the cathode is very great, or when there is a large fall of potential in the layer of gas near the cathode, no cathode rays are given out except from those points at which positive ions collide with the surface.† There is thus direct experimental evidence to show that negative ions are set free when positive ions collide with a metal surface with a high velocity.

229. Effect of electrons set free from the negative electrode. In the experiments described in the first sections of this chapter, where the negative ions are generated initially by the action of ultra-violet light, the whole effect of the positive ions has been attributed to ions generated by the collisions with molecules of the gas. If, instead of that process of ionization, the positive ions were supposed

* P. Villard, Journal de Physique (3) **8**, p. 5 and p. 148, 1899.

† See Section 266.

to act by setting free negative ions when they collide with the negative electrode, the equation for the currents would be nearly the same as that obtained on the hypothesis that was adopted. Let γN be the number of negative ions set free when a number N of positive ions collide with the electrode. The factor γ must depend on the velocity of the positive ions, and is therefore a function of the ratio X/p.

When the negative ions are generated initially by the action of the light, the first group $n_0\, \epsilon^{\alpha x}$ that arrive at the positive electrode includes those generated by the direct action of the light, so that the number of pairs of ions generated in the gas is $n_0\,(\epsilon^{\alpha x}-1)$. On the arrival of the positive ions at the negative electrode, a number of negative ions $\gamma n_0 . (\epsilon^{\alpha x}-1)$ are set free, and these act in the same way as those generated initially by the light. The total number of ions that arrive at the positive electrode is therefore

$$n = n_0\, \epsilon^{\alpha x}\left[1+\gamma\,(\epsilon^{\alpha x}-1)+\gamma^2\,(\epsilon^{\alpha x}-1)^2+\ldots\right]$$

$$= \frac{n_0\, \epsilon^{\alpha x}}{1-\gamma\,(\epsilon^{\alpha x}-1)}.$$

If $\gamma = \beta'/(\alpha-\beta')$, the equation for n becomes

$$n = \frac{n_0\,(\alpha-\beta')\,\epsilon^{\alpha x}}{\alpha-\beta'\,\epsilon^{\alpha x}}.$$

The two hypotheses thus lead to equations for the currents which are nearly the same, and it would be difficult to decide which of the formulae are in best agreement with the experimental results, as the difference between them is about of the same order as the experimental error.

230. Relative importance of the effects of positive ions at different pressures. Other experiments, however, show that at the higher pressures (exceeding the pressure corresponding to the minimum sparking potential) the effect produced by the positive ions in ionizing molecules of the gas is of more importance than the effect produced by setting free electrons from the negative electrode. If, for instance, the case of the discharges from points or small spherical electrodes be considered, it will be seen that the discharge from a negative point may be explained on either hypothesis. The region in which the ions acquire high velocities is confined to the neighbourhood of the point, and the positive ions, generated in the gas by negative ions moving outwards, move towards the point and the current would be maintained if they generated a certain number of ions by collisions with molecules of the gas, or if they set free electrons from the surface of the point.

If the force is reversed the positive ions move away from the positively charged point, and their velocity diminishes and may be very small when they reach distant conductors. In this case electrons would not be set free from the negative electrode, and unless the positive ions generate ions in moving through the gas near the point, it would be impossible to explain the discharge on these principles.

The most satisfactory evidence on this subject may be obtained from a consideration of the potentials required to produce a discharge between a cylinder and a concentric wire, since the field of force is quite definite in this case.*

Another experimental result, which indicates that the negative ions set free from the negative electrode by the collisions of positive ions have only a small effect at the higher pressures, is that the sparking potentials have almost universally been found to be independent of the metal of which the electrodes are composed. If the discharge depended on the number of ions coming from the electrode, there would probably be considerable variations in the sparking potentials, as it is improbable that the quantity γ, which is involved in the above equations, would be the same for all metals or that it would remain constant if the electrode tended to become oxidized. In other cases in which negative ions are set free from metal surfaces the number that are produced depend to a large extent on the nature of the surface.

With parallel plates, therefore, the most satisfactory

* See Section 253.

hypothesis is that for the larger values of the product ap the principal effect of the positive ions arises from ions generated in the gas. The relative importance of the ions set free from the negative electrode increases as the pressure diminishes, and when ap is very small the ions generated in the gas by the positive ions are probably few in number compared with the number that are set free from the negative electrode.

There are no reliable experiments to show the relative values of these two effects in a discharge for a given value of the product pS, but for simplicity it may be assumed that the gas effect predominates when pS exceeds the value corresponding to the minimum sparking potential.

231. Ions set free from hot metals. It is necessary to observe that positive ions are not set free from the positive electrode, unless it is at a high temperature. In this respect they differ from the electrons which may be set free from metal surfaces under various conditions.

Some of the properties of the ions generated at the surface of a hot metal in gases at atmospheric pressure have been explained in Section 60. As the pressure is reduced and the electric force is increased, changes take place in the proportion of positive and negative ions that are set free, and the ions acquire the same properties as those generated in gases at low pressures.

At the higher pressures the number of positive ions generated at the surface of the metal is much larger than the number of negative ions, except when the temperature is very high (exceeding 400° C.), in which cases the numbers of the positive and negative ions become approximately equal. At low pressures the positive currents become very small, and the negative ions are greatly in excess even at moderately high temperatures. The currents that may be obtained from hot metals vary by very large factors, which depend on different circumstances. The number of ions set free initially depends on the temperature of the wire, the nature and pressure of the surrounding gas, the length of time the metal is maintained at a high temperature, and on the state of the surface.

The variation of the current with the electric force when the gas is at low pressures is much the same as the variations of current obtained by other methods, and may be illustrated by some experiments made by McClelland * on the currents between an incandescent wire and a large concentric cylinder.

The wire was heated by a current from an insulated battery of a few cells. A point on the circuit was raised to a high potential by a battery of small accumulators, so as to establish an electromotive force between the wire and the outer cylinder. The cylinder was insulated, and the current from the wire to the cylinder was measured by a quadrant electrometer in the ordinary way. The following table gives the currents obtained through air at ·66 millimetre pressure when the wire is charged negatively:

Potential of the wire in volts.	Negative current in arbitrary units.
40	19
80	38
160	61
240	78
280	99
320	135

If a definite number of ions are generated initially at the surface of the wire the fraction of these that reach the outer cylinder increases with the electric force, and if all the ions reached the outer cylinder with a comparatively small potential difference between the wire and the cylinder the current would attain a saturation value, and would be independent of the force until ionization by collision set in. This effect is not easily obtained with negative ions at low pressures, since their rate of diffusion is abnormally great, and a large number of them become discharged by the wire. A large force may be required before the majority of the ions generated near the surface move away from the

* J. A. McClelland, Proc. Camb. Phil. Soc. 11, p. 296, 1901.

wire, and ionization by collision may begin before the current approaches a constant value.

The numbers given in the above table show that the rate of increase of the current with the force is least when the potential changes from 160 to 240 volts. This may be due to the fact that at 160 volts nearly all the ions generated at the surface of the wire reach the outer cylinder, and that ionization by collision begins to produce an appreciable effect when that potential is reached.

With the wire at the same temperature, the currents obtained when the wire is positively charged are much smaller and the variations of the current with the electric force are given by the following numbers, the air, as in the previous experiments, being at ·66 millimetre pressure:

Potential of wire in volts.	Positive current in arbitrary units.
8	19
18	22
40	24
120	25
240	27
280	29
320	37
360	50

From 40 to 240 volts there is only a small variation of the current, so that saturation is nearly reached with a potential of 40 volts. The increase of the current for the higher voltages is much less than the corresponding increase when the wire is negative, and it is most probably due to a small effect produced by the collisions of positive ions with molecules of the gas.

Of particular interest are the cases in which the wire is coated with salts.* Thus Wehnelt † observed that when a small particle of calcium or barium oxide is heated on a strip of platinum foil, the number of electrons set free is so large that the effect may be compared with the cathode rays obtained in an exhausted tube. If the strip of platinum is

* See H. A. Wilson, The Electrical Properties of Flames and of Incandescent Solids.

† A. Wehnelt, Ann. der Phys. (4) **14**, p. 425, 1904.

set up in a tube with the particle of oxide opposite the centre of a metal plate with a small hole in it, an intense stream of electrons falls on the metal plate when a potential difference of about 200 or 300 volts is established between the hot cathode and the plate. If the pressure of the gas be reduced to a very low value, a narrow beam of cathode rays passes through the aperture in the positive electrode, and the beam is marked by a bright line in the gas. The path of the rays may thus be seen very distinctly and the deflections produced by transverse electric or magnetic fields may be easily observed.

CHAPTER X

DISCHARGES BETWEEN CONDUCTORS OF VARIOUS SHAPES

232. Classification of discharges. The electrical discharges that take place between conductors of different shapes and sizes have been the subject of numerous investigations ever since the principal features of the phenomena were first discovered.

The discharges have generally been classified according to their appearances, as sparks, brushes, glows, or dark discharges. The gas surrounding a conductor at ordinary temperatures insulates well for a large range of potentials, but if the potential difference between two conductors is gradually increased, a point is reached when the gas begins to conduct. The change from the insulating to the conducting state usually occurs very abruptly, and it is easy to find a definite value of the potential when a discharge begins to take place from the conductor into the surrounding gas.

The different types of discharge have distinctive features, but it is difficult to draw a definite line between them or to specify the exact conditions under which each occurs, as the discharges may alter either gradually or abruptly from one type to another. The form in which the discharge takes place depends on the shape and size of the electrodes, their distance apart, the intensity of the current, the pressure of the gas, and the capacity of the electrodes.

233. Spark discharges. In air at atmospheric pressure only one type of discharge—the spark discharge—takes place between two large conductors with smooth surfaces when their distance apart is small compared with their linear dimensions, that is, when the field of force between the two conductors is fairly uniform. Thus when two metal spheres are separated by a distance which is not very large compared with the diameter of either sphere, and the potential difference between the spheres is raised gradually, a point is reached when a spark discharge takes place along a path which is indicated by a bright line extending from one sphere to the other. If the supply of electricity to the electrodes is maintained by an electrostatic machine the discharge takes place almost instantaneously, and while it lasts a large current flows between the conductors. The sparking potential V required to produce the discharge is quite definite; when the gas is insulating, a potential slightly smaller than V produces no appreciable current that can be detected by an ordinary galvanometer. After the spark has taken place the spheres are sometimes completely discharged, so that in a spark discharge the current after it is started continues to flow when the potential difference between the electrodes is less than the sparking potential.

If the inside and outside coatings of a Leyden jar are connected to the spheres the sparks are more brilliant, and a larger quantity of electricity passes in each discharge. In this case the discharge is in general oscillatory, owing to the self-induction of the condenser circuit, but the potential V required to produce the discharge does not depend either on the capacity of the condenser or the self-induction of the circuit.

In order to start a discharge it is necessary that there should be some ions in the gas initially, and when the force in the field between the electrodes is raised to a certain point these are multiplied by collision to a sufficient extent to maintain the discharge. In some cases ions are not present in sufficient numbers to start the discharge immediately when the potential difference V is established between the electrodes, but if ultra-violet light falls on the negative electrode, or if the gas is ionized by Röntgen rays or Becquerel rays, this difficulty is removed. Under these conditions it is impossible to raise the potential difference between the electrodes above the value V. If the rate of rotation of the electrostatic machine is increased the only effect that is produced is an increase in the number of sparks that pass per second.

234. Brush discharges. If the distance between the electrodes is increased, so that there are large variations of the force in the field between the electrodes, the discharge takes place in a different form. At points of the electrodes where the intensity of the force is greatest short brush-discharges are first produced, but the path of the discharge is not luminous for the whole distance between the electrodes. The brush consists of small fine sparks branching out from the conductor and terminating in the gas. The appearance of the brush is different in positive and negative discharges, being bluish purple when the brush is formed at a positive surface and red when it is formed at a negative surface.

The negative brushes are usually shorter than the positive, and the latter type may be obtained from 10 to 15 centimetres in length from a charged conductor when the surrounding conductors are at long distances from it. It is difficult to obtain brush discharges from very large spheres, but with spheres of about one centimetre in diameter and 4 or 5 centimetres apart, brush discharges are easily obtained when the electrostatic machine is rotating slowly. The brush may be converted into a spark discharge by increasing the rate of rotation of the machine.

When the distance between the electrodes is not very great, the potential difference V required to start the brush discharge does not differ much from the potential required to convert the brush into a spark. As an example, the following experiment quoted by Gaugain * may be mentioned: When two spheres one centimetre in diameter and 2·2 centimetres apart are used as electrodes, a brush discharge is obtained when the electrostatic machine rotates

* J. M. Gaugain, Ann. de Chim. et de Phys. (4) 8, p. 75, 1866.

once in four seconds, the potential difference between the electrodes being V. When the period of revolution was reduced to two-thirds of a second sparks passed between the electrodes, the potential required to produce them being practically the same as V, but after each spark had passed the potential dropped below V.

For large distances between the electrodes brush discharges are obtained at potentials considerably less than the potential required to produce a spark.

The tendency of the discharge to assume the form of a spark is increased by adding capacity to the electrodes, which accounts to some extent for the discrepancies between the results obtained by different observers.

The brush discharge resembles the spark discharge in many respects. It consists of several small branches which may be gradually reduced to two or three, and then it is not much different from the spark. Also Wheatstone* found that the brush discharge is intermittent by observing the images of the brush in a rotating mirror. The principal difference between the brush and the spark discharge is that during the brush discharge the potential of the conductor remains practically constant, but when sparks take place between two conductors the potential drops after each spark to a value that is small compared with the potential required to produce a spark.

235. Discharges between spheres. Baille's experiments. When the distance between two spherical electrodes is less than the diameter of one of the spheres, the potential V required to produce a spark depends on their distance apart and is practically independent of the diameter. In air, at atmospheric pressure, the potential V increases with the spark length S, and the electric force V/S diminishes as the distance between the spheres is increased.

If the distance S is fixed and the diameter $2R$ of the electrodes is reduced, then V is constant for the large values of R, but diminishes as R diminishes when the

* C. Wheatstone, Phil. Trans., **124**, p. 586, 1834.

diameter becomes less than the spark length S. When the electrodes are in the shape of small cylinders with rounded or pointed ends, the potential required to produce a discharge is much smaller than the potential required to send a spark across the same distance between two large spheres. Thus the latter potential is about 3×10^4 volts when $S = 1$, but if the end of a wire of half a millimetre diameter is fixed at one centimetre from a plate or large spherical electrode, the potential required to start a discharge is $4{\cdot}4 \times 10^3$ volts when the end of the wire is spherical, and is less if the end of the wire is pointed.

The following table given by Baille* illustrates these properties of the discharge. The numbers in the horizontal lines are the equivalent distances for the different spheres, or the distances at which a given potential produces a discharge. With the small electrodes the discharges were in the form of the brush, and the numbers in the last line give the distances between the electrodes above which the discharge is of that type. The exact arrangements of the experiments are not described, but it may be mentioned that other observers obtained brush discharges for shorter distances.

Diameter of the electrodes in centimetres	∞	6	3	1·7	1·0	·6	·35	·1	Points.	
									90°	10°
Equivalent lengths of sparks or brushes in centimetres	·5 1·0 1·5	·485 ·993 1·495	·48 1·07 1·556	·480 1·43 1·86	·498 1·50 3·5	·670 3·20 7·2	1·30 4·5 10·0	2·10 6·3 15·4	2·25 5·8 15·0	2·70 7·2 17·5
Distance for which the spark is replaced by a brush	[Brushes were not formed in these cases.]			13·8	10·6	8·07	3·0	1·20	1·80	·50

* J. B. Baille, Ann. de Chim. et de Phys. (5) **25**, p. 486, 1882.

236. Discharge from a point to a plane. Töpler's investigations of point discharges. If one of the electrodes is a sphere and the other a plane at a fixed distance from it, the spark is preceded by a brush discharge when the radius of the sphere is below a certain value. A further reduction in the radius of the sphere is necessary in order to obtain a glow discharge.

According to Töpler,* who has investigated the various conditions under which the different types of discharge are obtained, a glow does not occur in the discharge from a sphere to a plane surface when the diameter of the sphere exceeds 1·5 centimetres. The glow covers the surface of the conductor where the electric force is large, and does not extend normally in bright lines from isolated points as in the case of a brush discharge.

The glow discharge is associated principally with positively charged surfaces, and in the discharge from a rounded end of a wire to a plane this type is obtained when the current does not exceed 2×10^{-4} ampère, and the distance between the point and the plane is greater than 1·5 centimetres.†

An increase in the current converts the glow either into a brush or a series of sparks. The particular form assumed by the current depends on the capacity of the electrodes and the distance of the point from the plate.

When the capacity is very small, the glow is converted by an increase of current into a brush, and that form of discharge is maintained until the discharge assumes the appearance of an arc.

When a condenser of small capacity is connected to the point, the glow tends to be converted into a spark discharge as the current increases, but for the larger distances a brush intervenes between the glow and spark discharges. Thus, when the discharge takes place from the end of a wire ·5 millimetre in diameter and the wire is connected to a condenser of 50 cm. capacity, a glow discharge takes place

* M. Töpler, Ann. der Phys. **7**, p. 477, 1902.
† Ibid., **2**, p. 560, 1900.

when the current is less than 2×10^{-4} ampère, which becomes converted into a series of sparks when the distance from the point to the plane is between 2 and 7 centimetres. For larger distances the glow is first converted into a brush and the brush subsequently into a series of sparks, when the current increases.

When the wire is negatively charged a glow surrounding the point may be obtained, but only with very small currents. The discharge in this case tends to assume the form of a brush when the currents exceed a certain value, which is of a much smaller order than 10^{-4} ampère. The negative brush may develop into a continuous arc or a series of sparks as the current increases, according as the capacity connected to the point is small or large.

A glow also occurs over the surface of a charged wire or cylinder when it is at a large distance from other conductors, or when the cylinder is surrounded by a large coaxial cylinder. The latter case is of interest, as the force at any point of the field between the two conductors may be expressed by a simple formula when the current is small, and it may be shown that in this case, as in the case of the discharge between parallel plates, the conductivity is produced by the action of positive and negative ions on molecules of the gas.

When a discharge takes place from a sharp point one or two centimetres from a plate or large sphere, a small continuous current passes through the gas, which in its initial stages is not accompanied by any luminous effect in the gas. The air in this case begins to conduct when the potential difference between the point and plane is comparatively small, between 1,000 and 2,000 volts. The sharper the point the less the potential required to produce the discharge, and it is remarkable that the potential necessary to start the discharge is much less when the point is negatively charged than when it is positively charged. If the point is blunt, such as the rounded end of a wire ·5 millimetre diameter, the initial potential is the same for positive and negative discharges. With larger wires the initial potential for positive discharges is less than for negative discharges.

The dark discharge may be converted into a glow and a glow into a brush discharge by increasing the current.

237. Determination of sparking potentials. In order to obtain a consistent series of observations of sparking potentials, there are some points of practical importance that should be observed. Two conditions are necessary in order to produce a spark between two conductors: the potential difference must not be less than a definite value V, and it is necessary to have some ions in the gas initially by means of which a large number may be generated by collisions with the molecules. The importance of the initial ionization has been recognized comparatively recently, and the first indication of this necessary condition was given by the discovery, made by Hertz, that sparking is facilitated when ultra-violet light falls on the electrodes. Several experiments that have been made on this and similar effects * gradually led to the conclusion, definitely stated by Warburg,† that in general discharges are immediately preceded by a small current in the gas. Warburg was unable to detect this small current by means of a sensitive electrometer, but he quotes experiments on the effect of a magnetic field on the sparking potential which led him to form this conclusion.

The investigations show that under ordinary conditions, when a potential difference is established between two electrodes which is sufficiently great to produce a spark. the discharge may not pass immediately, but if the potential is kept constant a discharge eventually takes place at a time which may be as much as several minutes after the potential is established. The interval between charging the electrodes and the passage of the discharge is denoted by Warburg as the retardation. The sparking potential V may be defined as the least potential required to produce

* See G. Jaumann, Wied. Ann. **55**, p. 656, 1895.

† E. Warburg, Wied. Ann. **62**, p. 385, 1897; Sitz. der Akad. der Wiss. zu Berlin, **10**, p. 128, 1897.

a spark when some of the gas is ionized initially. When a potential only slightly greater than V is applied, the retardation may be very long if the initial ionization is very small. The average value of the retardation, under given conditions, diminishes as the applied potential exceeds the sparking potential. When no light is falling on the electrodes, Warburg found that on many occasions no spark passed when a potential difference four or five times as great as the sparking potential was established between the electrodes for a short period of the order 10^{-3} second. The tendency of the spark to pass as soon as the potential is applied increases when the electrodes are exposed to daylight. When ultra-violet light acts on the negative electrode, or when Röntgen rays traverse the gas, the spark passes almost as soon as the electromotive force is applied. In these cases the smallest potential required to produce the spark is very definite and only varies by a small percentage when large variations are made in the intensity of the light.

Under ordinary conditions, when the electrodes are charged the potential cannot be assumed to be less than the sparking potential if a spark does not pass immediately; it is necessary to maintain the potential for some minutes before any conclusion can be arrived at. When this precaution is observed the sparking potentials obtained in ordinary daylight are practically the same as those found when Röntgen rays or ultra-violet light traverse the gas.

238. Causes which influence the accuracy of determinations of the sparking potential. When no special precautions are taken to provide initial ionization it has been found that, after the first spark has passed between the electrodes, the potential required to produce the second spark is smaller than that required to produce the first. It is probable in these cases that the difference between the observed potentials would have disappeared if, in the first instance, the effect of the charge had been observed for a longer time or if additional ions had been produced between the electrodes by rays from some external source.

The reason which is sometimes given to explain why the second spark passes easier than the first is that some of the ions generated in the first discharge remain in the gas, and provide the initial ionization which causes the second spark to pass more readily.

It is probable that the effect arises in some way from an increase of the initial ionization, but it cannot be due to ions left in the gas from the first discharge. They disappear very quickly by diffusion to the surface of the electrodes, but even if there were no diffusion, the way in which the experiments are conducted precludes the possibility of any effect arising from the residual ions, for when the potential is gradually raised, any ions that may be left between the electrodes are removed almost immediately by the smaller forces, and when the sparking potential is approached there would be no ions left in the gas that were generated in the first discharge.

Two possible causes may facilitate the passage of the second spark; the surface of the negative electrode may be altered by the first discharge and the small effect due to ordinary daylight increased, or the gas may be altered by the action of the large current that passes in the first discharge. The modifications of the properties of gases, produced by discharges, have recently been studied by Strutt,* who found that nitrogen continues to glow for a short time after an electric discharge has been passed through it, and that the glowing gas conducts well even under an electromotive force of a few volts. Some small effect of this kind may prevail for some minutes after a discharge has passed, which could only be detected by a very sensitive electrometer, but would make a great difference in the initial ionization that facilitates the second spark.

239. First determination of sparking potentials by Kelvin. The earliest measurements of sparking potentials in absolute units were made in 1860 by Kelvin † shortly after he

* Hon. R. J. Strutt, Proc. Roy. Soc. **87**, p. 183, 1912.
† W. Thomson, Proc. Roy. Soc. **10**, p. 326, 1860.

had perfected his well-known electrostatic voltmeters. These were used to determine the electromotive force required to produce sparks between two flat surfaces in air at atmospheric pressure. The surfaces were not quite plane, but were very slightly convex in order that the edges of the electrodes should be further apart than the central portion. The curvature was so slight that the electric force was practically uniform in the air between the central parts of the two opposing surfaces where the sparking took place. The differences of potentials producing sparks across different thicknesses of air were measured first with the absolute electrometer and afterwards by the portable torsion electrometer. The absolute electrometer referred to was the attracted disc electrometer, and was arranged to measure the maximum electric forces across the stratum of air; whereas the measurements by the portable electrometer were of electromotive forces producing sparks. The latter determinations are presumably mean values of the potential while a continuous series of sparks passes between the electrodes, and would be much less than the maximum. The measurements with the attracted disc electrometer are the sparking potentials or the 'electrostatic force preceding the spark'. These latter have been given in absolute units and when reduced to volts the values of the sparking potentials in air at atmospheric pressure between flat surfaces at different distances as given by Kelvin are:—

Distances in centimetres.	Sparking potentials in volts.
·0086	690
·0127	978
·0152	1190
·0190	1278
·0281	1692
·0408	1854
·0563	2433
·0584	2445
·0688	2907
·0904	3660
·1056	4185
·1325	5208

These values are smaller than those obtained in subsequent investigations. This is probably due to the cause suggested by Baille, who attributed the low values to the increase of temperature of the air produced by the continuous series of sparks, as it appears from the description Kelvin gives of his experiments that the measurements were taken while a number of sparks were passing.

In order to test this point, Baille * made some experiments to determine the maximum potential between plane electrodes, with an attracted disc electrometer, while sparks were passing at various rates. He found that the potentials were constant so long as the sparks did not pass more frequently than once in five seconds, when the sparks passed at the rate of two a second the potential fell about 7 per cent., and with continuous sparking the reduction was 10 per cent.

240. De la Rue and Müller's investigations. Among the earlier experiments, those made by de la Rue and Müller † are of considerable importance. The sparking potentials for air were determined when the electrodes were at different distances apart and also when the electrodes were at a fixed distance and the pressure was reduced. These determinations were made with the large battery of 10,000 silver chloride cells that were used in the interesting series of investigations on various subjects connected with discharges which are published in the Transactions of the Royal Society.

The investigations of the sparking potentials between a given pair of parallel plates led to the conclusion that the sparking potential depended only on the number of molecules between the plates.

One series of values of sparking potentials was obtained for discharges passing between spherical surfaces at distances from ·0126 to ·21 centimetre, in air at atmospheric pressure. The electrodes were convex discs 3·8 centimetres

* J. B. Baille, Ann. de Chim. et de Phys. (5) **25**, p. 486, 1882.
† W. de la Rue and H. W. Müller, Phil. Trans. **171**, p. 109, 1880.

in diameter, the radius of curvature being 7·5 centimetres. At the centre of the discs the field of force was practically uniform, as the spark lengths were so short compared with the radius of curvature. The mean values of the observations for the different distances are given, and if a curve * be drawn through the points whose ordinates are the potentials, and abscissae the products of the distances S between the electrodes and the pressure $p = 760$, the potential corresponding to any particular value of the product pS may be found.

The potentials thus obtained may be compared with those found for parallel plates when the distance was constant at 3·3 millimetres and the pressure of the air was reduced from 709 millimetres to 9·7 millimetres. The following table gives the potentials obtained in these experiments when the temperature was 15·2° C. The pressures p are given in the first column, the product pS in the second column, and the potentials V when $S = \cdot 33$ centimetre in the third column. The potentials V for the same values of the product pS given in the fourth column are obtained from the curve drawn from the observations at different distances when the pressure was constant at 760 mm.

p	pS	V in volts, $S = \cdot 33$.	V in volts, $p = 760$.
709·6	234	11,300	
636·5	210	10,200	
548·7	180	9,100	9,100
456·0	151	8,000	7,900
358·7	118	6,480	6,400
244·6	81	4,940	4,800
188·5	62	3,710	3,900
109·1	36	2,470	2,620
38·3	12·6	1,240	1,320
9·7	3·2	618	

The agreement between the numbers in the two last columns shows that the sparking potential depends only on the product $p \times S$. The importance of this result was noticed by de la Rue and Müller, and they concluded that: 'The potential necessary to produce a discharge between parallel

* A part of this curve is given in figure 57.

flat surfaces at a constant distance and various pressures, or at a constant pressure and various distances may be represented by hyperbolic curves. The resistance to the discharge between parallel flat surfaces being as the number of molecules intervening between the plates.'

They also point out that this law does not apply to point discharges.

241. Sparking potentials determined by Baille and Paschen. Baille also measured the potential required to produce sparks in air at atmospheric pressure between flat surfaces and spheres (from 6 centimetres to 1 millimetre in diameter). His results show that the sparking potentials are practically the same for plane electrodes and spherical electrodes, provided that the radius of the sphere is not less than the spark length, and there is but little difference between the sparking potentials for spherical electrodes, of diameters between 6 centimetres and 3·5 millimetres when the distance between the electrodes does not exceed the diameter of the sphere. The potential required to produce a spark of given length S diminishes with the diameter D of the electrode when D is less than S.

Similar experiments were made by Paschen,* who obtained results of the same general character and consistent among themselves. The absolute values of the potentials corresponding to the larger values of pS are in good agreement with those obtained in previous investigations, as is shown by the following table (given by Paschen) of the potentials required to produce sparks in air at atmospheric pressure between spheres of one centimetre radius, as obtained by Baille,† Quincke,‡ and Czermak.§

* F. Paschen, Wied. Ann. **37**, p. 69, 1889.

† J. B. Baille, Ann. de Chim. et de Phys. (5) **25**, p. 486, 1882.

‡ G. Quincke, Wied. Ann. **19**, p. 545, 1883.

§ P. Czermak, Wien. Berichte, **97** (II^a), p. 307, 1888.

Sparking Potentials in Electrostatic Units.

Distance between spheres in centimetres.	Baille.	Quincke.	Czermak.	Paschen	
				from observations made in winter.	from observations made in summer.
·1	15·12	14·78	15·14	16·08	15·84
·2	26·37	26·39	26·57	27·75	26·74
·3	36·96	37·31	37·50	38·85	37·88
·4	47·20	46·69	47·69	49·41	48·26

The difference between the numbers in the last two columns is due to a temperature change of about 8° C. For smaller values of the product of pS, the potentials given by Paschen are considerably larger than those found by Baille.

Paschen also investigated the effect of changes in pressure, in order to find if the potential V is a function of the product pS. His experiments cover a large range of pressures and spark lengths, and they verify this relation in a remarkable manner.

242. Sparking potentials determined by Carr. More recently a series of experiments have been made by Carr, corresponding to cases in which the product pS has smaller values than were taken in previous investigations. When p is measured in millimetres of mercury and S in centimetres, the sparking potential for parallel plates in air attains a minimum value when pS is about ·6. When the pressure is reduced below the value $·6/S$ the potential increases rapidly as the product pS diminishes. The potentials investigated by Carr were in the neighbourhood of the minimum sparking potential and corresponded to values of the product pS between the limits ·1 and 15. The potentials were established by means of a battery of small accumulators having a maximum potential of 1,800 volts. These experiments were made after the retardation had been investigated, and Carr was careful to maintain the potential difference between the electrodes for a considerable time, as in some cases the discharge did not take place until ten or fifteen minutes after connecting the electrodes to the battery.

When the pressure is reduced below the value corresponding to the minimum sparking potential, and the electrodes are set up inside a large air-tight receiver, the discharge does not take place across the shortest distance between the electrodes, but tends to start from the edges and to follow a longer path, for which the value of pS is more nearly equal to ·6.

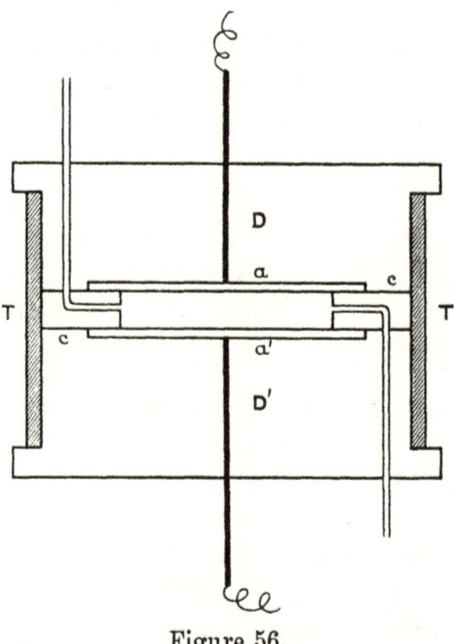

Figure 56.

In order to avoid this difficulty Carr used plane electrodes a, a', resting on an ebonite ring c, figure 56. The electrodes were embedded in ebonite discs D and D' fitting into the glass tube T and the joints were made air-tight with sealing wax. Experiments were made with iron, zinc, aluminium, and brass electrodes, but the sparking potentials were found to be independent of the metal that was used. The gas was led into the space between the electrodes through narrow channels leading through the ring c. With this arrangement the spark can only take place across the space where the field of force is uniform and the spark length S is equal to the thickness of the ebonite ring c. For the range of values of the product pS between ·1 and 15, Carr found that the sparking potentials in air depended only on the number of molecules between the plates, the same result as was previously found by de la Rue and Müller and by Paschen.

243. Sparking potentials independent of the temperature when the density of the gas is constant. Hence when the temperature is constant the sparking potentials corresponding to different pressures p and lengths S for parallel-plate electrodes, or for spherical electrodes when the radius of curvature exceeds the spark length, may be expressed by means of a curve of which the ordinates are the potentials and the abscissae the product pS.

The curves corresponding to ordinary temperatures may be extended to other temperatures if the abscissae be taken as the density of the gas. The effect of an increase of temperature on the sparking between electrodes was first clearly shown by Harris,* who found that when air is contained in an air-tight receiver, so that the density remains constant, the potential difference between two electrodes required to produce a discharge was unaltered when the temperature varied from 50° to 300° Fahrenheit. When the gas was allowed to expand by heating, the potential was reduced to the same extent as when the temperature is constant and the change in density is made by reducing the pressure.

The same result has been obtained more recently by Cardani† over a larger range of temperature. The sparking potential, in these investigations, was found to be independent of temperature changes up to 300° C., when the density of the gas is constant.

Hence if the curves representing the sparking potentials

* W. Snow Harris, Phil. Trans. **124**, p. 230, 1834.

† M. Cardani, Rendiconti della R. Acc. dei Lincei, **6**, p. 44, 1888.

in terms of the products pS be taken as giving the potentials for gases at 15° C., the sparking potential at any temperature t° C. and pressure p may be found by taking the quantity $pS\frac{288}{273+t}$ as the abscissa of the point.

244. Values of the potential required to produce a spark through air between parallel plates expressed in terms of the product pS. The sparking potentials for parallel

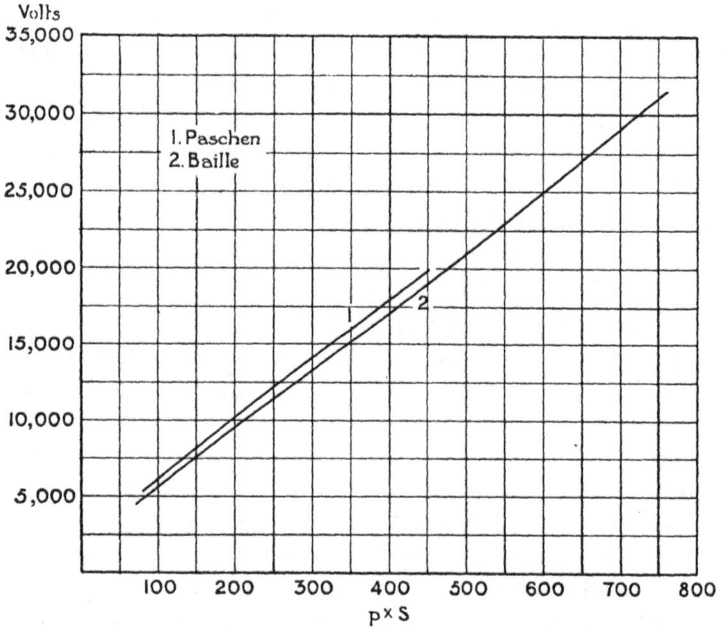

Figure 57. p in millimetres of mercury, S in centimetres.

plates obtained by the different physicists are represented by the curves given in the figures 57, 58, 59, and 60, which correspond to different ranges of the product pS.

The results differ by amounts which exceed ordinary experimental errors. The high values obtained by Baille and Paschen may be due to the fact that at the time their investigations were made it was not realized that a spark may not pass until some minutes after the electrodes are charged. In Kelvin's experiments where sparks passed in more rapid succession, the retardation did not produce a serious effect, but, as Baille has shown, the sparking potentials obtained when sparks pass continuously are too low, owing to an increase of temperature, which reduces the number of molecules between the plates when the pressure is constant.

It will be noticed that as the values of pS increase there is better agreement between the different observers.

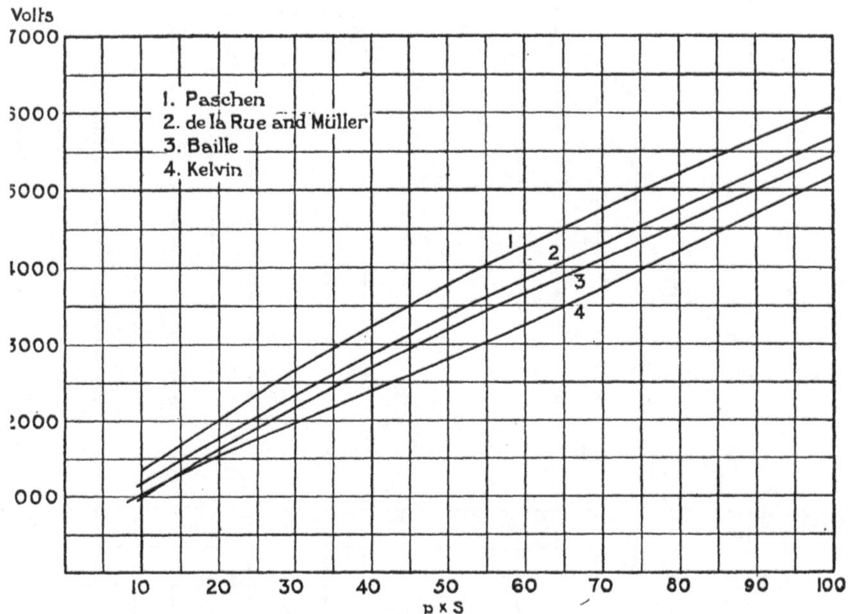

Figure 58. p in millimetres of mercury, S in centimetres.

The initial ionization, which develops into a discharge under the action of the electric force, arises from various causes, the most efficient being the ultra-violet effect, as has been shown by Warburg. The initial ionization may, therefore, be considerably increased as the distance S is increased and more light falls on the electrodes. The initial ionization arising from other causes also increases with the number of molecules between the plates, so that

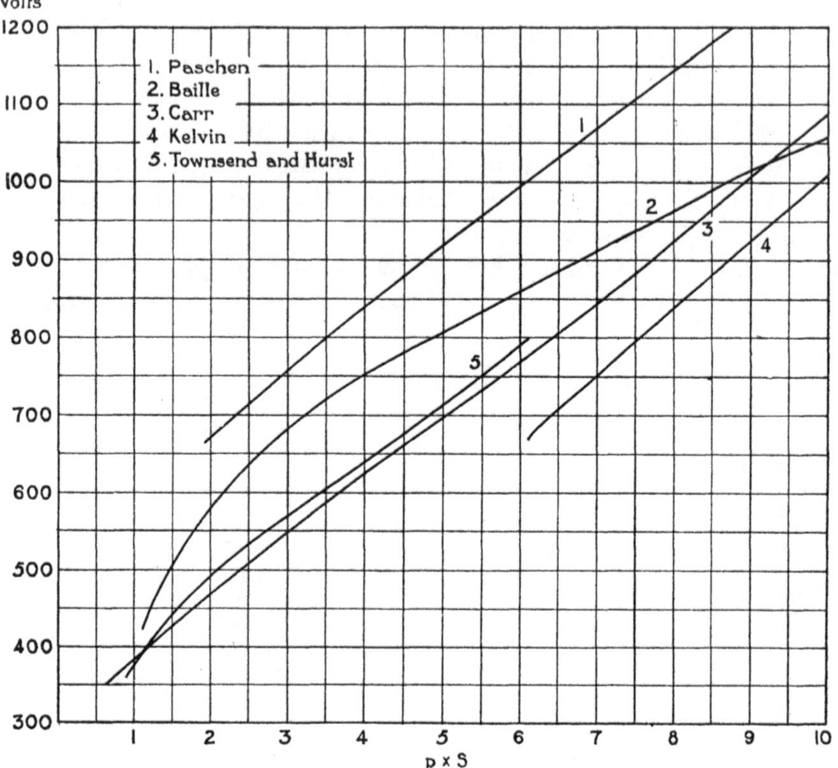

Figure 59. p in millimetres of mercury, S in centimetres.

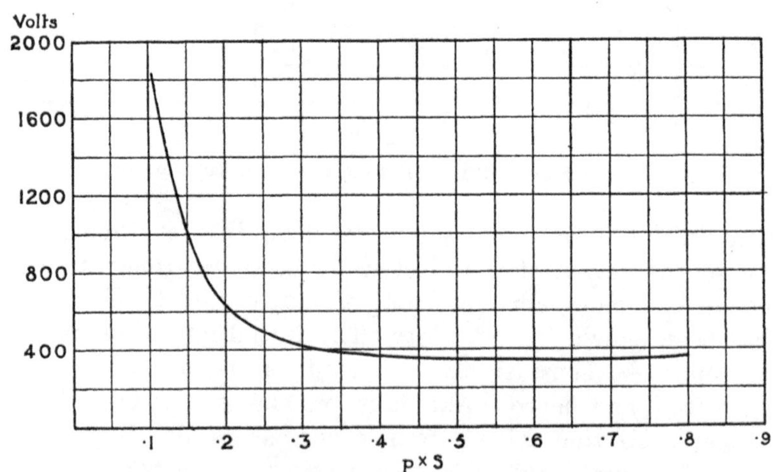

Figure 60. p in millimetres of mercury, S in centimetres.

errors are less likely to occur with the higher values of pS, when the effect of the retardation is not taken into consideration.

The sparking potentials, given in Chapter IX, which are the values of the potentials obtained when ultra-violet light falls on the negative electrode, are represented by the curve 5, figure 59, and are not much different from those found by Carr.

245. Sparking potentials in different gases at high pressures. For the larger potentials the curves are practically straight lines and, taking the numbers given by Baille, it may be seen that for values of the product of pS above 100 the potential is given by the equation

$$V = 39\,pS + 1700,$$

and the corresponding electric forces between the electrodes are given by the equation

$$\frac{X}{p} = 39 + \frac{1700}{pS}.$$

The expression shows that the ratio of X/p, corresponding to the sparking potential, diminishes continuously as pS increases from the smallest to the largest values. When $pS = 760$, $X/p = 41{\cdot}2$, and the above equation shows that the value of X/p when sparking takes place in a uniform field does not fall below the value 39.

Since the sparking depends upon the actions of the positive and negative ions in generating others by collisions, neither of these effects can vanish for values of X/p greater than 39 or 40; but since β is less than α, and sparking does not take place when X/p is less than 39, it may be assumed for all practical purposes that β is zero when X/p is less than 39. This fact is of importance in the theory of the discharge from the surface of a wire.

Similar results may be obtained for other gases by taking the values found by Wolf for the sparking potentials at high pressures, the potential being a linear function of the pressure when the distance between the electrodes is constant. The following table gives the potentials V in electrostatic units required to produce sparks through different gases between two spheres of 5 centimetres radii when the spark length is one millimetre, P being the pressure in atmospheres.*

In the experiments with hydrogen the pressure was increased from 1 to 9 atmospheres, and with the other gases the pressure was increased from 1 to 5 atmospheres. The potentials may also be expressed in terms of the product pS, since they depend only on the numbers of molecules between the electrodes. The sparking potentials in volts corresponding to values of the product pS exceeding 76 (p being measured in millimetres of mercury and S in centimetres), are also given in the following table:—

	V in electrostatic units, P in atmospheres, $S = \cdot 1$.	V in volts in terms of pS.
Hydrogen	$6{\cdot}509\ P + 6{\cdot}2$	$25{\cdot}6\ pS + 1860$
Oxygen	$9{\cdot}60\ P + 4{\cdot}4$	$38{\cdot}0\ pS + 1320$
Air	$10{\cdot}7\ P + 3{\cdot}9$	$42{\cdot}1\ pS + 1170$
Nitrogen	$12{\cdot}08\ P + 5{\cdot}0$	$47{\cdot}6\ pS + 1500$
Carbonic acid	$10{\cdot}22\ P + 7{\cdot}2$	$40{\cdot}4\ pS + 2160$

The numerical coefficients of the product pS in the second column represent the values of X/p in the different gases for which the value of β vanishes.

246. Sparking potentials in different gases at low pressures. Several determinations have been made of the sparkling potentials in different gases. Among the earliest are the investigations made by de la Rue and Müller, who found the potentials required to produce sparks through hydrogen and carbonic acid between flat surfaces for different pressures of the gases. With hydrogen the plates were set at 5·88 millimetres apart and the pressure was varied from 760 to 15·4 millimetres of mercury. With carbonic acid the plates were at 3·096 millimetres apart and the pressure varied from 760 to 31·3 millimetres of mercury.

The sparking potentials in these gases have also been investigated by Paschen, and as in air it was found that for

* M. Wolf, Wied. Ann. **37**, p. 306, 1889.

uniform electric fields the potential depended only on the product of the pressure and the spark length. The electrodes were spheres of one centimetre radius, and the distances varied from a millimetre to 6 or 8 millimetres, but in no case did they exceed a centimetre. Under these conditions the potentials obtained in these gases, as in air, are practically the same as if the electrodes were plane surfaces. The pressure in Paschen's experiments was varied from 2 centimetres to 76 centimetres.

The following are the sparking potentials in electrostatic units (300 volts = 1 E.S. unit) for hydrogen and carbonic acid as given by Paschen. The table also contains the potentials which were deduced from the numbers given by de la Rue and Müller.

$p \times S$.	Hydrogen.		Carbonic Acid.	
	De la Rue and Müller.	Paschen.	De la Rue and Müller.	Paschen.
·2	—	1·52	—	2·91
·4	—	1·89	—	3·70
·6	—	2·20	—	4·27
·8	—	2·43	—	4·83
1·0	2·2	2·82	4·3	5·33
1·2	—	3·13	—	5·79
1·5	—	3·51	—	6·46
2·0	3·3	4·09	6·0	7·57
2·5	—	4·65	—	8·61
3·0	4·7	5·23	8·0	9·68
3·5	—	5·76	—	10·63
4·0	5·7	6·28	9·7	11·56
4·5	—	6·84	—	12·43
5·0	6·7	7·29	11·3	13·27
5·5	—	7·80	—	14·06
6·0	7·3	8·27	12·7	14·81
6·5	—	8·75	—	15·57
7·0	8·7	9·19	14·3	14·34
7·5	—	9·68	—	17·37
8·0	9·3	9·96	16·0	17·92
9·0	10·3	10·86	18·0	19·36
10·0	11·3	11·69	19·0	20·66
12·0	13·3	13·32	22·0	23·31
15·0	16·0	15·69	27·0	27·27
20·0	20·0	19·45	33·3	33·39
25·0	24·0	23·09	—	39·19
30·0	29·0	26·68	—	45·09
35·0	33·0	30·18	—	51·07
42·0	39·0	34·75	—	59·09
45·0	—	36·71	—	62·35

Carr * has also investigated the potentials required to produce sparks through various gases between parallel plates, when the pressures are low, so as to include cases in which the values of the product pS are above and below the value corresponding to the minimum sparking potential.

In all cases it was found that the potential depended only on the product of the pressure and the spark length.

The following table gives the sparking potentials obtained by Carr in different gases corresponding to a fixed distance of 3 millimetres between the plates, when variations are made in the pressure. The potentials V are given in volts, and the pressure p in millimetres of mercury (see table on p. 361).

247. Discharges between co-axial cylinders. Gaugain's experiments. The discharges between plates, or spherical electrodes at short distances apart, are of importance from a theoretical point of view, since the field of force between the electrodes preceding the discharge is uniform and many of the principal features of the discharge may be explained by the theory given in Chapter IX. When electrodes of different shapes are used, the electric force varies along the path of the discharge, and in most cases it is impossible to make an accurate investigation of the field of force. The only simple example of a field of force that is not uniform, and at the same time is perfectly definite, is that between two co-axial cylinders, and the particular case in which the inner cylinder is a wire and the outer cylinder a large tube is of interest both from a theoretical and a practical point of view.

The forces required to produce discharges through air between co-axial cylinders were first investigated by Gaugain,† and by means of very simple apparatus he discovered the principal conditions on which the phenomena depend.

The discharge that takes place may either be in the form

* W. R. Carr, Phil. Trans. A, **201**, p. 403, 1903.

† J. M. Gaugain, Ann. de Chim. et de Phys. (4) **8**, p. 75, 1866.

Spark Length = 3 millimetres.

Air.		Hydrogen.		Carbon dioxide.		Nitrous oxide.		Acetylene.		Hydrogen sulphide.		Sulphur dioxide.		Oxygen by electrolysis.	
p	V	p	V	p	V	p	V	p	V	p	V	p	V	p	V
51	1480	13·6	415	8·75	674	8·64	716	14	765	9·73	617	13·5	1145	8·42	576
41·5	1275	8·54	356	5·57	563	5·00	560	8·7	633	5·29	467	7·7	842	4·85	523
31·5	1015	5·40	301	3·55	477	2·95	470	7·4	583	3·90	442	4·5	651	3·42	483
21·4	790	4·66	286	2·25	427	2·03	430	6·4	555	2·88	428	2·54	531	2·82	468
14·1	630	4·02	278	1·91	420	1·70	420	5·5	530	2·48	418	2·30	511	2·34	461
9·31	526	3·44	282	1·63	419	1·41	418	4·7	509	2·12	414	2·01	501	1·94	458
5·99	452	2·93	292	1·41	425	1·18	422	4·1	490	1·83	414	1·83	486	1·60	455
3·84	405	2·52	310	1·20	432	·982	430	3·5	480	1·57	432	1·61	471	1·33	456
2·51	371	2·15	356	1·02	449	·816	450	3·0	474	1·39	457	1·40	466	1·10	463
2·18	361	1·85	440	·875	487	·670	500	2·55	468	1·20	472	1·23	461	·910	493
1·89	356	1·59	564	·758	542	·560	581	2·2	468	·86	543	1·20	459	·759	533
1·64	358	1·35	780	·651	599	·466	706	1·9	471	·73	592	1·04	457	·625	631
1·42	364	1·16	1054	·558	699	·389	866	1·66	476	·63	668	·92	459	·517	766
1·22	375	1·00	1382	·482	815	·326	1061	1·42	483	·54	817	·80	465	·428	1001
1·06	397	·861	1789	·420	971	·277	1370	1·21	490	·46	1013	·70	481	·354	1250
·928	441			·362	1162	·251	1830	1·04	498	·39	1286	·62	498	·294	1578
·804	494			·314	1445			·89	513	·33	1650	·55	531	·268	1802
·710	576			·274	1756			·65	588			·43	621		
·616	691							·47	754			·27	1000		
·536	863							·34	1064			·23	1590		
·465	1092							·28	1302						
·411	1395							·24	1650						
·357	1786														

of a spark or of a brush, but in either case the potential difference between the electrodes is independent of the length of the cylinders provided the length is large compared with the diameter of the outer cylinder. The inner cylinder, in his experiments, was a long wire or rod which projected through the open ends of the outer cylinder, and was either maintained at zero potential, or connected to a sensitive electroscope arranged so that the leaves should touch an earthed conductor when they diverged through a small angle. The current through the gas was measured by the number of times the leaves touched the earthed conductor in a minute. The outer cylinder was a thick metal tube with bell-shaped ends so as to increase the distance between the inner cylinder and the edges of the tube. The discharge then passed between the central parts of the cylinders and there was no tendency for brushes to take place from the edges of the outer cylinder. The outer cylinder was charged by an electrostatic machine, and the potential to which it was raised was measured by a graduated electroscope which became charged by induction to a potential proportional to the potential of the cylinder. The electric force at a point on the path of the discharge is given by the equation $X = 2\,E/r$ where E is the charge per unit length of the inner cylinder and r the distance of the point from the axis.

248. Force at the surface of the inner cylinder required to produce a discharge. The potential differences V required to produce discharges were measured in a series of experiments in which outer cylinders of various diameters $2A$ were used with the same inner cylinder of constant diameter $2a$. The potential V was found to increase with the diameter of the outer cylinder in such a way that the force at the surface of the inner cylinder was independent of the radius A. The charge per unit length of the inner cylinder being $E = V/(2 \log A/a)$ the force X_1 at the surface of the inner cylinder is $X_1 = V/(a \log A/a)$.

The following numbers give an example of the experiments made by Gaugain when the inner cylinder was one centimetre in diameter:

Radius of inner cylinder $a = \cdot 5$ centimetre.

V in electrostatic units.	A in centimetres.	$\dfrac{V}{a \log A/a}$.
43	1	125
85·4	2	123
113·2	3	126

When the outer cylinder was of one or two centimetres radius the discharge always passed in the form of a spark. With the larger cylinder, sometimes a brush discharge and sometimes a spark discharge took place, but there was no appreciable difference between the potentials.

When the inner cylinder was a wire one millimetre in diameter a brush discharge took place in each case, and the potential required to maintain the discharge increased as the current increased. The force at the surface of the wire required to produce a small current was also in this case found to be independent of the radius A of the outer cylinder when A varies from 1 to 5 centimetres. The force X_1 at the surface of the inner cylinder is thus independent of the radius A, where A exceeds a certain value, but X_1 increases as the radius a of the inner cylinder diminishes.

249. Recent investigations of the critical force for high potential wires. The force $X_1 = V/(a \log A/a)$ required to produce a discharge from the surface of a cylinder is usually called the critical force. The values of this force, in terms of the radius of the cylinder, are given by the following table, X_1 being measured in electrostatic units:

$2a$ in millimetres.	X_1 Gaugain.	X_1 Watson.
30	95	—
10	125	133
5	145	155
1	243	223
·27	390	—
·20	425	—
·11	520	—

The numbers given in the third column are the values of

X_1 obtained recently by Watson. The agreement between the two sets of figures is very satisfactory considering the kind of apparatus that was used in the earlier experiments.

The phenomena connected with the discharge of electricity from high-potential wires in air have been the subject of many investigations during late years, as a considerable loss of energy may result from the glow discharge when wires of high potentials are used to convey electric energy between stations at a long distance apart. In addition to the points investigated by the earlier physicists, several measurements have also been made recently of the currents from the surface of the wire when the potential exceeds the minimum value required to initiate the discharge.

250. Watson's determination of the critical forces. A very complete investigation has been made by Watson* for wires increasing in diameter from ·7 millimetre to 12·76 millimetres. Measurements were made of the critical

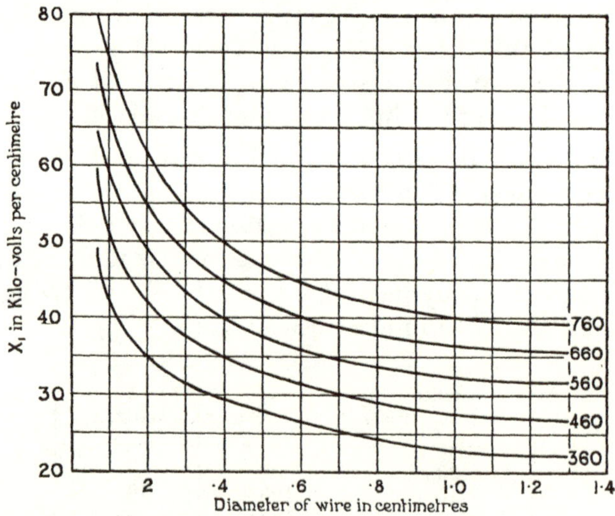

Figure 61. The figures on the curves give the air pressure in millimetres.

stress, or the force at the surface of the wire, $X_1 = V/(a \log A/a)$, when the discharge begins. The wire in these experiments

* E. A. Watson, The Electrician, Feb. 11, 1910.

was stretched along the axis of a galvanized iron cylinder 180 centimetres long and 20 centimetres in diameter. The ends of the large cylinder were closed and the wire was insulated somewhat after the manner illustrated in figure 47. When the wire was quite clean, both positive and negative discharges required very nearly the same voltage to start them.* The critical forces X_1 for wires of different diameter are given by the curves figure 61. The five curves correspond to the five different pressures of the air at which the experiments were made.

251. General theorem relating to the sparking potentials. It is interesting to examine these curves, as it may be seen that the corona or glow discharge from the wire is produced by the action of the positive and negative ions on the molecules of the gas, in the same way as the discharge between parallel plates is produced. An exact relation between the critical forces corresponding to different pressures may be obtained on this hypothesis, and an approximate formula which gives the values of the critical force in terms of the radius of the wire may also be found from the potentials required to produce sparking between parallel plates. The critical force X_1, the pressure p, and the radius of the wire a, are connected by an equation of the form $X_1 a = F(ap)$, which is a particular example of a more general relation that applies to conductors of any shape. The theorem may be stated as follows:

If V be the potential difference required to produce a discharge through a gas at pressure p between two conductors A and B, the same potential difference will produce a discharge through a gas at a lower pressure $p' = \frac{p}{k}$ between two conductors A' and B' of the same shape and in the same relative position, but with all the linear dimensions in-

* It was found by other observers that the starting potential is less for a positive discharge than for a negative discharge with the larger values of ap, but the negative discharge starts for a smaller potential than the positive with the smaller values of ap. A similar effect is obtained with point discharges. [See Phil. Mag. (6) **27**, p. 789 1914; also note at end of this chapter.]

creased in size, so that the distance between points on A' and B' exceeds the distance between the corresponding points on A and B in the ratio $k:1$.*

Let σ be the surface distribution per unit area of the conductors A and B, A being at zero potential and B at the potential V. The potential at any point P in the field between A and B is $\int \frac{\sigma dS}{r}$. If the surfaces A' and B' are charged with the distribution $\frac{\sigma}{k}$ per unit area, the potential at the point P' in the field between A' and B' corresponding to the point P, is $\int \frac{\sigma dS'}{kr'}$, which is equal to the potential at P, since $dS' = k^2 dS$, and $r' = kr$. Hence A' and B' will be equipotential surfaces, their potentials being zero and V respectively.

The number of ions generated by a negative ion in traversing a distance ds between two points P and Q on the path of the discharge between A and B is

$$\alpha ds = p \cdot f\left(\frac{X}{p}\right) ds = p \cdot f\left(\frac{1}{p} \cdot \frac{dv}{ds}\right) ds;$$

dv being the fall of potential between P and Q.

The number of ions generated by a negative ion in traversing the distance $ds' = k \cdot ds$ between the two corresponding points P' and Q' is

$$\alpha' ds' = p' \cdot f\left(\frac{1}{p'} \cdot \frac{dv'}{ds'}\right) ds' = \alpha ds,$$

since $p \cdot ds = p' \cdot ds'$, and the fall of potential dv' between P' and Q' is the same as the fall between P and Q.

Hence if ions are generated by positive and negative ions along a line of force of length S between A and B in sufficient numbers to produce a discharge, a similar effect will take place on the corresponding line S' from A' to B' when A' and B' differ in potential by the same amount as A and B, the pressures being inversely proportional to the lengths S and S'.

* See The Electrician, 71, p. 348, June 1913.

The connection between the sparking potential and the product of the pressure and spark length, discovered by de la Rue and Müller, for the discharge in a uniform field between parallel plates, is the simplest example of this general property of discharges between conductors.

252. Relation connecting the critical force with pressure of the gas, and the diameter of a wire ; $aX_1 = F(pa)$. The preceding theorem, applied to cylinders, shows that if V is the potential difference required to produce a discharge through a gas at pressure p between a pair of coaxial cylinders of radii a and A, V' the corresponding potential for the gas at pressure p' between a pair of coaxial cylinders of radii a' and A' proportional to a and A, then V and V' will be the same if $pa = p'a'$. Since the force at the surface of the inner cylinder is $X_1 = V/(a \log A/a)$, it follows that $aX_1 = a'X_1'$ if $pa = p'a'$. This relation holds in all cases when the ratio A/a is constant.

In the particular case in which the inner cylinder is a wire of small diameter, all the ionization takes place near the surface of the wire. If the outer cylinder is sufficiently large to extend beyond the volume of gas in which ionization is taking place, the number of ions generated by an ion along the path of the discharge is then independent of the radius A, and the quantity aX_1 depends only on the product pa.

This conclusion may be tested by means of the numbers obtained for the critical forces at various pressures. The following table gives some of the values of the forces X_1 in kilovolts per centimetre, found by Watson for different values of the radius a and pressure p, and the two last columns show that X_1a is constant when pa is constant.*

* Examples of the application of the theory to discharges at lower pressures for values of ap from 20 to ·05 are given by the author and Mr. P. J. Edmunds, Phil. Mag. (6) **27**, p. 789, 1914.

p	$2a$	X_1	$2a \cdot p$	$2a \cdot X_1$
760	·1	75	76	7·5
560	·136	55	76	7·5
360	·211	34	76	7·2
760	·2	61	152	12·2
560	·272	44·5	152	12·5
360	·422	28·5	152	12·0
760	·5	46·5	380	23·2
560	·68	35·0	380	23·7
360	1·055	22·0	380	23·2

253. Expression for the critical force in terms of the diameter of the wire. An approximate formula for the critical force X_1 corresponding to wires of various diameters may be obtained by considering the force required to produce a spark of length S in a uniform field. The force V/S is comparatively large when S is small, but diminishes as S increases and approaches the value 30,000 volts per centimetre for long sparks in air at atmospheric pressure. For forces exceeding 30 kilovolts per centimetre both positive and negative ions generate others by collisions with molecules, and when the distance between the plates is sufficiently great a discharge takes place, but with smaller forces it is impossible to produce a spark, so that the effect of the positive ions is practically zero. Hence when the air at the surface of a positively charged wire conducts, the ionization does not take place at points so far from the axis that the force is less than 30 kilovolts per centimetre, and all the ions are generated inside a cylinder of a certain radius c given by the formula $\frac{2E}{c} = 30$, E being the charge per unit length of the wire. The length of the path in which ionization takes place is $c - a = \frac{X_1 a}{30} - a$, X_1 being the critical force at the surface of the wire. The distances $c - a$ are very small, and for this reason the forces required to produce a discharge are independent of the radius A of the outer cylinder, as in most of the experiments A is greater than c. Taking the values given by Watson for the critical force X_1 at atmospheric pressure, the following are the distances $c-a$ in centimetres for wires of radius a:

a	·6	·5	·4	·3	·2	·1	·05
$c-a$	·18	·166	·153	·140	·133	·103	·075
X_1 in kilovolts, found experimentally	39	40	41·5	44	50	61	75
$X_1 = 30 + \frac{9}{\sqrt{a}}$	42	43	44	46	50	58·5	71

Hence the distance in which the ionization takes place increases from ·75 millimetres for a wire of half a millimetre radius, to 1·8 millimetres for a wire of 6 millimetres radius.

The mean value of the force in the distance $c-a$ is $(X_1+30)/2$, and if this force acted throughout the distance $c-a$ it would produce approximately the same effect as the forces in the field near the inner cylinder. Hence the force $(X_1+30)/2$ is approximately equal to the force V/S required to produce a discharge in the uniform field between two plates at the distance $c-a$ apart. A simple expression for the force V/S may be obtained from the determinations of the sparking potentials for plane electrodes. According to the results obtained by Baille, the potentials V in kilovolts for different spark lengths S in air at atmospheric pressure are given by the equation $V = 30\ S + 1{\cdot}35$ for distances of the order of one millimetre, so that the electric force in kilovolts per centimetre is

$$\frac{V}{S} = 30 + \frac{1{\cdot}35}{S}.$$

Hence $$\frac{X_1+30}{2} = 30 + \frac{1{\cdot}35}{c-a} = 30 + \frac{1{\cdot}35}{a\left(\frac{X_1}{30}-1\right)}.$$

This equation reduces to $a\,(X_1 - 30)^2 = 81$.

Hence the critical strength X_1 for a wire of radius a in air at atmospheric pressure is

$$X_1 = 30 + \frac{9}{\sqrt{a}}.$$

The numbers found by this formula for wires from ·6 to ·05 centimetre in radius are given in the last line of the preceding table, and may be compared with the numbers found experimentally, which are given in the third line.

As it has been shown that for any pressure the product $X_1 a$ is a function of the product ap, the critical force for a wire of radius a in air at pressure P is given by the formula

$$X_1 = P\left(30 + \frac{9}{\sqrt{(aP)}}\right)$$

for values of aP between ·05 and ·6, a being measured in centimetres and P in atmospheres.

254. Whitehead's experiments with alternating forces. The forces required to start a discharge under various conditions, when alternating potential differences are established between two coaxial cylinders, have been investigated by Whitehead.* The results are of the same general character as those obtained with continuous forces.

Any irregularity on the surface of the wire lowers the critical voltage. When the wire is clean and smooth, the discharge starts when the amplitude (or maximum value) of the alternating force at the surface of the wire is approximately equal to the critical force X_1 obtained by charging the wire gradually with an electrostatic machine. The relation between the diameter of the wire d and the force X_1 at which the glow starts was found to be given accurately by an empirical formula $X_1 = 32 + \frac{13 \cdot 4}{\sqrt{a}}$, for diameters from 5 millimetres to 1 millimetre, which is of the same form as the theoretical formula obtained in the

* J. B. Whitehead, American Institute of Electrical Engineers, June 1911.

preceding section, but gives somewhat larger values of the forces.

The critical force X_1 diminishes as the frequency of the alternating force increases, the reduction in the force being 2 per cent. when the frequency changes from 25 cycles per second to 60, and 6 per cent. when the frequency is increased to 90.

The effect of surface irregularities may be seen from the results of the experiments with stranded wires. It was found that the critical voltage of a stranded wire was less than that of a round wire of the same diameter as the over-all diameter of the stranded wire. Thus if the over-all diameter is ·349 centimetre, the critical voltage of a wire made of three strands was less than that of a round wire in the ratio 183 : 215. With five strands and an over-all diameter of ·45 centimetre the critical force was less than that of a round wire in the ratio 224 : 244. In these experiments the diameter of each strand was ·162 centimetre and the diameter of the outer cylinder was 9·52 centimetres.

255. Considerations bearing on the theory of discharges. These experiments are of interest as they show definitely that the glow discharge from wires must originate from the same processes of ionization as those which cause the discharge to take the form of a spark between parallel plates. They also afford a method of deciding between the two theories on which it is possible to explain the large currents that are developed between parallel-plate electrodes in gases at the higher pressures, when the electric force approaches the value required to produce a spark.

It has been mentioned in sections 229 and 230 that these currents may be explained by two methods which differ in respect to the action that is attributed to the positive ions. Thus the phenomena of sparking between parallel plates may be attributed to the continuous production of ions by the action of negative ions on the molecules of the gas, combined with a similar but smaller effect of the positive ions also on the molecules of the gas. Or it may be supposed that the conductivity arises from the action of the negative ions on the molecules of the gas, combined with an action of the positive ions on the negative electrode which consists in setting free negative ions when they collide with a high velocity (which depends on the ratio X/p). This latter effect of positive ions is undoubtedly large when they move under a high force in a gas at a very small pressure, as was shown by Villard, but it is not the predominating effect at the higher pressures. In these cases the combination of the effects of the positive and negative ions in generating ions from the molecules of the gas is the only hypothesis of this kind which affords an explanation of the discharges from cylinders. The experiments on the sparking potentials in uniform fields show that, whatever effect the positive ions produce, it must vanish when the electric force is less than 30 kilovolts per centimetre in air at atmospheric pressure. With a cylindrical electrode the critical force is therefore independent of the radius of the outer cylinder, since the positive ions cease to act at the larger distances where the force $2\,E/r$ is less than 30 kilovolts per centimetre.

When the inner cylinder is positively charged, the positive ions collide with the outer cylinder with a velocity corresponding to the force $2\,E/A$, which becomes very small when A increases. Thus in Gaugain's experiments, in which the inner cylinder was ·5 centimetre radius and the outer cylinders were 1, 2, and 3 centimetres radius, the discharges took place when the forces at the surface of the outer cylinder were 20, 10, and 5 kilovolts per centimetre. These are much below the value 30, at which it may be supposed that the positive ions cease to generate ions either from the molecules of the gas or from the negative electrode. In these experiments the only objects with which the positive ions collide with velocities sufficiently high to produce ionization are the molecules of air which are within a distance of 1·66 millimetres of the surface.

It is obvious, therefore, that in the experiments between parallel plates the predominating effect of the positive ions at the higher pressures is due to ions generated by collisions with molecules of the gas.

256. Effect of charge in the gas in determining the nature of a discharge. Spark discharges. The potential difference v between electrodes, required to maintain a current, may be greater or less than the potential V required to initiate the discharge. When the discharge takes the form of a spark v is less than V, and when a glow or brush takes place v is greater than V. Gaugain found that the potential required to maintain a current between coaxial cylinders increases with the current when the outer cylinder is much larger than the inner. Thus when $a = \cdot 5$ cm. and $A = 1$ cm., sparks take place between the cylinders, and it is impossible to raise the potential above the critical value required to initiate the discharge, but when $a = \cdot 05$ cm. and $A = 1$ cm., the discharge takes place at first in the form of a brush or a glow, and the current increases with the potential difference between the electrodes. It will be seen in the course of these investigations that the form of the discharge is determined by the effect of the electric charge in the gas on the distribution of the force along the path of the current. In the case of the spark a large quantity of electricity passes in each discharge when the supply is maintained by an electrostatic machine, so that the electricity continues to pass between the electrodes when the potential is less than the sparking potential. This is also seen when the discharge is maintained by a battery of cells connected to the electrodes through large resistances; the potential difference between the electrodes when a continuous current is passing is generally much less than the sparking potential. This is caused by the volume distribution of electricity in the field between the electrodes, which alters the distribution of the force in such a way as to increase the production of ions. A necessary condition for sparking is that ionization should take place along a large part of the distance between the electrodes.

257. Potential required to increase the current in brush discharges. Theoretical investigation of currents between coaxial cylinders. When ions are generated in a strong field of force near one of the electrodes and move through a long distance under a small force to the other electrode, a brush or glow discharge takes place. The current from a wire to a concentric cylinder, or from a point to a plane, are examples of the latter forms of discharge. The ions are generated in a small volume of the gas near the wire and the rest of the space is filled with a volume distribution of electricity of the same sign as the charge on the wire. This distribution reduces the force near the wire and thus tends to impede the formation of ions; with small currents, the increase of the force in other parts of the field near the surface of the outer cylinder does not compensate for this, as the force is not brought up to the value $40\,p$. It is necessary, therefore, in order to maintain a current, to raise the potential difference between the electrodes by such an amount as will bring the force at the surface of the wire up to the critical value of X_1.

The curves (figure 62) given by Watson represent the currents per kilometre of a wire ·7 millimetre diameter inside a coaxial cylinder of 20 cm. diameter in terms of the quantity $v/(a \log A/a)$, v being the difference of potential between the wire and the cylinder in kilovolts. This latter expression represents approximately the force at the surface of the wire when the currents are small, but it becomes inaccurate as the current increases. It is nevertheless convenient to express the currents in terms of the quantity $v/(a \log A/a)$. The five continuous curves given in the figure relate to negative currents at five different pressures of the air, the dotted curve is for a positive discharge from the same wire at atmospheric pressure.

The theory shows that the critical force X_1, or the minimum value of the force $v/(a \log A/a)$, is independent of the radius A of the outer cylinder when A is large, but the increase of potential $(v - V)$ required to maintain a given current depends to a great extent on the radius of the outer cylinder. The increase of potential required to maintain a small current may be found by considering the effect of the charge in the gas on the force at the surface of the wire.

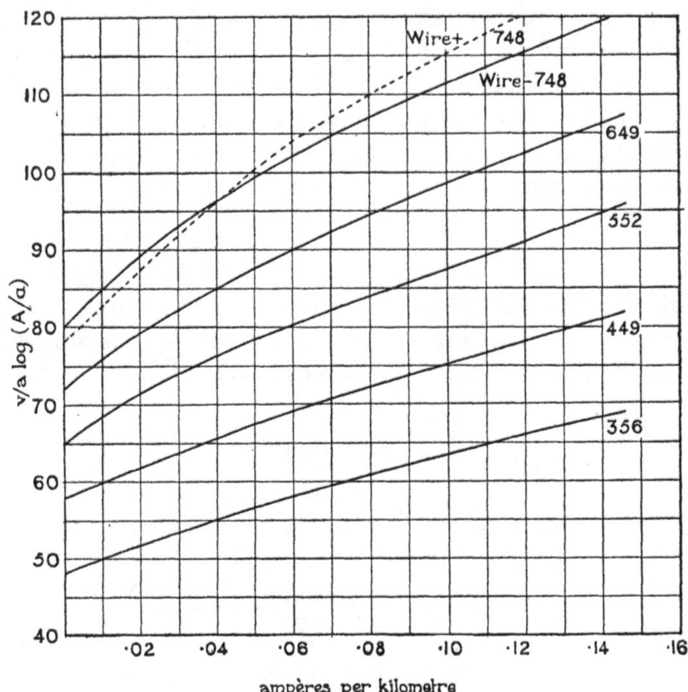

Figure 62. Wire ·07 centimetre diameter. The figures on the curves give the air pressures in millimetres.

Let I be the current per unit length of the wire, q the charge per unit volume at a point in the gas, r the distance of the point from the axis, w the potential. For simplicity, a positive discharge may be considered, since the velocity of the positive ions is proportional to the electric force for a large range of forces. The distribution q at points where r is greater than c is given by the equation

$$I = 2\pi r q k_1 \frac{dw}{dr},$$

where k_1 is the velocity of the positive ions under unit force. The force $\frac{dw}{dr}$ is approximately equal to $\frac{v}{r \log A/a}$ since q is small. Hence

$$I = \frac{2\pi q k_1 v}{\log A/a}.$$

Since I is constant, the distribution q is uniform, and the problem is reduced to finding the force X at the surface of a wire of radius a in terms of the potential difference v between the wire and the cylinder and the distribution q extending from the distance $r = c$ to $r = A$. When c is small compared with A the value of X is given by the equation

$$X = \frac{v}{a \log A/a} - \frac{\pi q A^2}{a \log A/a}.$$

In order to maintain the current by ions generated by collisions near the surface of the wire it is necessary that X should not fall below the critical value

$$X_1 = \frac{V}{a \log A/a}.$$

Hence $$v - V = \pi q A^2.$$

Substituting for q its value in terms of the current I, the relation connecting the potential v and I is obtained, namely

$$v(v - V) = \frac{IA^2 \log A/a}{2k_1}.*$$

It is difficult to apply this formula to the negative discharges, since with the large values of X/p the velocity of the negative ions depends on the moisture in the air, but the formula is in agreement with the observations, as it

* This formula applies to small currents. For larger currents the relation between $v - V$ and I is given by the equation

$$\frac{v - V}{V} \log \frac{A}{a} = (1 + \theta)^{\frac{1}{2}} - 1 + \log \frac{2}{(1 + \theta)^{\frac{1}{2}} + 1},$$

where $\theta = vIA^2/ka^2 X_1^2$. It has been shown (Phil. Mag. (6) **28**, p. 83, 1914) that the velocities k_1 and k_2 for positive and negative ions may be deduced by this formula from experimental determinations of $v - V$ and V. The theory is in agreement with the curves figure 62, also with Schaffers' experiments (V. Schaffers, Phys. Zeitschr. **14**, p. 981, 1913; **15**, p. 405, 1914).

shows that the rate of increase of the current with the potential is greater for negative discharges than for positive discharges.

For positive ions the value of k_1 at atmospheric pressure may be taken as 450 centimetres per second when the forces are expressed in electrostatic units, so that for small currents the increase of potential becomes

$$v - V = \frac{IA^2}{900aX},$$

where
$$X = \frac{v}{a \log A/a}.$$

Thus if the current between a wire ·07 centimetre diameter and a co-axial cylinder 20 centimetres diameter is ·01 ampère per kilometre, the current I per centimetre of the wire is 300 electrostatic units, and the rise of potential in electrostatic units becomes

$$v - V = \frac{950}{X}.$$

According to the curves given by Watson, the value of X in electrostatic units is $8{\cdot}3 \times 10^4/300$, so that the quantity $\frac{v - V}{a \log A/a}$ becomes 17 electrostatic units or 5·1 kilovolts per centimetre.

The value of this quantity as found experimentally for a positive discharge is given by the upper curve, figure 62, and is about 5 kilovolts per centimetre.

258. Effect of diameter of outer cylinder on the potential required to maintain a current. The currents between a wire of fixed diameter and co-axial cylinders of various diameters have been investigated by Almy,* who found that the currents through air from 20 to 80 centimetres pressure were given by the formula

$$I = cv(v - V),$$

which is similar to the expression given by Warburg for

* J. E. Almy, American Journal of Science (4) **12**, p. 175, 1901.

the currents in point discharges. The constants c and V depend on the gas and on the dimensions of the wire and cylinder. In air, the constants do not change much when the direction of the current is reversed.

The currents through hydrogen were also investigated, as in this gas more concordant results were obtained. The experimental error is considerable when large currents are used, as the passage of the discharge makes a permanent change in the conductivity, particularly in air. Even with hydrogen, the agreement between the various determinations with the same force was not very accurate. The following are the results obtained with a wire of ·0034 centimetre radius and cylinders of radii 5, 3·2, and 1·5 centimetres:

Applied potential in volts.	I_1 $A = 5$	I_2 $A = 3{\cdot}2$	I_3 $A = 1{\cdot}5$
3,500	4	16	152
3,800	—	24	200
4,000	9	36	290
4,500	19	74	480
5,000	31	118	—
5,500	35	189	—

Almy concludes from these figures that the current is given approximately by the formula

$$I = \frac{C}{A^3} v (v - V),$$

A being the radius of the cylinder, V the minimum potential that gives a discharge, and C a constant depending on the gas, the sign of the discharge and radius of the wire.

It is difficult to test the formula by these observations as the values of V are not given, but the experiments show the general character of the discharge, and it is clear that the rate of change of the current with the potential increases rapidly as the radius of the outer cylinder diminishes.

259. Discharges through gases at low pressures, between cylindrical electrodes. The potentials required to produce discharges between cylinders when the gas is at low pressures of the order of a millimetre have been investigated by Meservey.* The experiments were made with the same outer cylinder in each case, which was 4 centimetres internal diameter. The potential difference V between the two cylinders necessary to produce a discharge diminishes as the pressure is reduced, and a minimum potential V is obtained at a certain pressure p_1 depending on the radius of the inner cylinder. In air, the minimum sparking potential for a wire 3·23 millimetres in diameter was 311 volts when the wire was negatively charged, the corresponding pressure being ·35 millimetre, but the minimum for other wires was greater than 311 volts. For positive discharges the minimum potential was greater than the minimum for negative discharges. In hydrogen the lowest potential at which a discharge was obtained was 240 volts, with a wire 9 millimetres in diameter, the pressure being ·75 millimetre. At pressures lower than p_1 the potential V increases rapidly as p diminishes. It was also observed that at the larger pressures the potential V_+ required to produce a discharge from a positively charged wire was greater than the potential V_- required when the wire is negative. These results are illustrated by the curves, figure 63. The curves A_+ and A_- give the sparking potentials in volts for discharges through air at various pressures between a wire ·25 millimetre diameter and a co-axial cylinder 4 centimetres in diameter. The curves B_+ and B_- refer to a wire 16·95 millimetres in diameter.

The curves A_+ and A_- intersect at the pressure ·36 millimetre, and for pressures below this value the potential V_+ is less than the potential V_-.

The difference between the positive and negative potentials increases as the radius of the inner cylinder diminishes, but for the cylinders of large diameter both potentials approach the value corresponding to the sparking potential for parallel plates separated by the distance $A-a$.

Meservey also gives a series of curves for air and

* A. B. Meservey, Phil. Mag. (6) **21**, p. 479, 1911.

hydrogen, representing the connection between the sparking potentials for wires of various diameters from ·25 millimetre to 16 millimetres for different pressures of the gases.

260. Theory of the low-pressure discharges between cylindrical electrodes. The relation connecting the three quantities X_1, a, and p, which may be expressed in the form

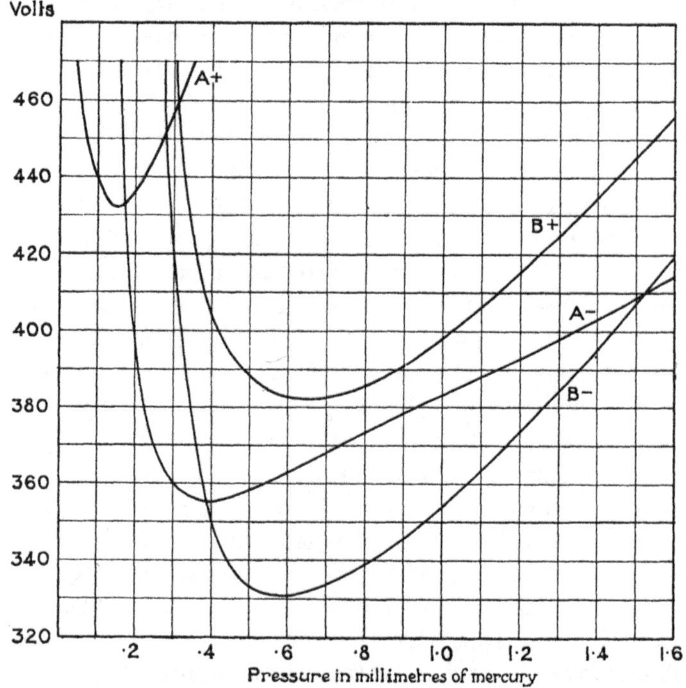

Figure 63. Curves $A+$ and $A-$ diameter of wire ·25 millimetre.
" $B+$ " $B-$ " " 15·95 "

$X_1 a = F(ap)$, applies accurately to all cases in which the radius A of the outer cylinder is proportional to the radius a of the wire or inner cylinder, or when the outer cylinder is of fixed diameter, provided it is sufficiently large so that the force at the surface does not exceed the value $39p$. In most of the experiments that have been made at the higher pressures the latter condition is satisfied, since c, the distance from the axis within which ionization takes place, is small when p is large. As the pressure is reduced the distance c increases, and if the outer cylinder is of fixed diameter c becomes equal to A, and for low pressures ionization by collision takes place in the whole space between the cylinders. This may be seen by considering the expressions for α and β in terms of X and p.

Let E be the charge per unit length of the inner cylinder required to produce a discharge at the pressure p. If E and p are reduced in the same proportion to E/k and p/k, the value of X/p at any point between the cylinders remains unchanged, but the quantities α and β representing the ionization produced by the negative and positive ions per centimetre, being proportional to the pressure,

$$\left[\alpha = pf\left(\frac{X}{p}\right) \text{ and } \beta = p\phi\left(\frac{X}{p}\right)\right],$$

are both reduced at each point to the fraction $1/k$ of their original value. A sufficient number of ions would not be produced under these conditions, so that the charge E' required to produce a discharge through the gas at pressure p/k is greater than E/k. Hence at the pressure $p' = p/k$ the distance $c' = \frac{2E'}{39p'}$ is greater than the distance $c = \frac{2E}{39p}$ corresponding to the higher pressure p.

The following table gives the critical forces X_1 and the distances c corresponding to pressures from 760 to 25 millimetres for a wire ·5 centimetre radius inside a large cylinder of radius A greater than c. The values of X_1 at the two lower pressures are deduced from the critical forces found by Watson for wires ·15 and ·035 centimetre radius when the pressure was 360 millimetres. The values of X_1 at the two higher pressures are the actual observations.

Critical forces X_1 in kilovolts per centimetre for a wire ·5 centimetre radius, the distance c being deduced from the formula

$$c = \frac{aX_1 \times 10^3}{39p}.$$

p	X_1	c
760	40	·66
360	23	·82
108	9·45	1·12
25·2	3·4	1·73

Thus c increases as p diminishes, so that when the outer cylinder is of fixed diameter and the pressure is below a certain value, the ions are stopped by the outer cylinder while they are still travelling with a velocity sufficiently great to generate other ions by collisions. When c is less than A the quantity X_1a is given by an expression of the form $X_1a = F_0(ap)$; but when A is less than c it is necessary to increase the force above the value $F_0(ap)/a$ in order to compensate for the reduction of the distance in which ionization takes place from $(c-a)$ to $(A-a)$. Hence when A is constant, the quantity X_1a should increase as p diminishes when ap is constant.

The results obtained by Meservey illustrate this effect; two examples are given in the following table. The two groups of numbers in the last column are the values of the quantity $X_1a = V_+/(\log A/a)$, when the product ap has constant values. The pressure p is expressed in millimetres of mercury and the diameter of the wire, $2a$, in centimetres.

p	$2a$	$2ap$	$V+$	X_1a
1·5	·3	·45	548	204
·75	·6	·45	434	228
·3	1·5	·45	472	480
1·5	·066	·1	587	142
·75	·132	·1	490	144
·3	·33	·1	410	164

The radius of the outer cylinder being 2 centimetres, the force X_2 at the surface is $X_1a/2$. The numbers in the last column show that X_2/p is greater than 40 in each case, so that ionization by collision takes place along the whole path of the discharge.

The effect of limiting the space by the outer cylinder, of 4 centimetres diameter, on the values of the quantity X_1a is well marked in these experiments, particularly in the case of the lowest pressure ·3 and the wire of largest diameter 1·5 when the force required to produce a discharge is more than double the value of the force $F(ap)/a$. There is therefore no simple relation connecting the three quantities X_1, a, and p when the pressure is low and the outer cylinder is small and of fixed diameter.

The sparking that took place between the cylinders when a definite potential was attained was not preceded by a brush discharge such as occurs at the higher pressures when c is less than A.

The difference between the potentials V_+ and V_- depends on the values of α and β. This effect and other phenomena arising from the same causes will be explained in Chapter XI.

261. Discharges from points. The discharge of electricity from a pointed conductor in gases at high pressures takes place somewhat in the same way as the discharge from the cylindrical surface of a wire, except that the potential required to produce a discharge from a point is generally much less than that required for conductors of other shapes. A small current passes through the gas when a sharp point is raised to a certain potential and at first no luminous effect is produced. When the potential is increased the current increases and a small glow appears at the surface of the point, which may change into a brush for the larger currents, especially when the point is not very sharp. In the latter case a spark may take place if other conductors are sufficiently near the charged point.

With sharp points the potential required to initiate a

discharge is greater for positive than for negative discharges. With blunt points the potentials are higher and the difference between the positive and negative potentials diminishes. When the discharge takes place from the rounded end of a wire in air at atmospheric pressure, and the diameter of the wire is greater than ·5 mm., the initial potential is less for positive than for negative discharges.

The intensity of the force in the electric field of a charged point cannot be expressed by a simple formula, but it may easily be shown that the electric force near the end of a charged cylinder is in general greater than at other points of the surface. Thus when a wire is surrounded by a coaxial metal cylinder which extends beyond the end of the wire, the surface distribution is greatest at the end of the wire even when it is terminated abruptly and not pointed. When the wire is charged, the values of α and β in a gas at high pressures are very much greater in a small volume near the end of the wire than at other parts of the field where α and β may be negligible, so that discharges would first take place from the end of the wire. At low pressures, when the values of α and β are considerable at all points of the field between the electrodes, the discharge does not necessarily take place more easily through that part of the gas where the force has a maximum value.

262. Zeleny's investigations of point discharges. The potentials required to produce discharges from points of definite shapes and sizes are clearly shown by the experiments recently made by Zeleny.* In these investigations the discharges were produced from the ends of straight wires of various diameters which were set up opposite the centre of a large metal plate, the axis of the wire being normal to the plate. The point and the plate were enclosed in an air-tight cylindrical vessel, to which air that had been dried by passing through calcium chloride was admitted. In one series of experiments the ends of the wire were hemispherical and in another series the ends

* J. Zeleny, Physical Review, **25**, p. 305, 1907.

were plane. The general character of the discharge was the same in both cases. For the smaller points the results obtained with the different ends were almost identical, but for the larger points the potential required to produce a given current is smaller for the wires with plane ends than for wires with hemispherical ends. This is no doubt due to the fact that in the latter case the curvature is comparatively small, and the electric force is not very great at any point of the surface.

With large distances between a point and a plane the currents are steady, but they tend to become intermittent as the distance is reduced. The limiting distance, at which the nature of the discharge changes, diminishes with the diameter of the point, and it also depends on the sign of the discharge. For a given distance between the plate and the point, steady positive discharges are obtained with points of comparatively large diameter, from which it is impossible to obtain steady negative discharges. Thus, when the point is 1·5 centimetres from the plate steady positive discharges were obtained by Zeleny from a cylindrical point with a round end when the diameter of the cylinder was 2 millimetres, but with negatively charged points the current became intermittent when the diameter of the wire exceeded ·5 millimetre. When the point was one centimetre from the plate a steady negative discharge was only obtained with wires of diameter less than ·27 millimetre.

With points of certain sizes steady currents are produced with the smaller potentials, but when the electric force increases irregularities set in, which may be detected by means of a telephone in the galvanometer circuit. The intermittence frequently seems to be some irregular effect superposed on the steady current, which may constitute such a small part of the total current as not to affect the results appreciably. When the telephone indicates an irregularity a little speck of light appears to start from the surface of the point as if a small disruptive discharge were taking place. Under these conditions the discharge tends to change from a dark discharge—or a glow discharge where the luminosity is confined to the surface of the point—to a brush discharge where the paths of the currents are indicated by short bright lines extending from the point into the gas. These small disruptive discharges or brushes increase in frequency as the current is increased, and a definite note is heard in the telephone. The pitch may become so high that the telephone and ear fail to detect it.

The potential at which the discharge from some points begins is different from that at which the current ceases, and various other irregularities occur, particularly with negative discharges. Gorton and Warburg * found that in these cases the potentials at which the current begins and ends are the same if the gas surrounding the point is exposed to certain radiations. In Zeleny's experiments a tube containing a small quantity of radium bromide was placed at the side of the cylindrical vessel in which the point and plate were set up. The presence of the radium had no effect on the larger currents, but it served to make the discharge begin more regularly at definite voltages.

263. Currents through air from points of definite shapes. Warburg's formula for the current. The following tables give the results of Zeleny's experiments, in which steady currents were obtained through air at atmospheric pressure from positively and negatively charged cylindrical points with hemispherical ends, the points being at 1.5 and 1 centimetre from the plate. The current was supplied to the point by means of a small Wimshurst machine, and the potential difference between the point and the plane was measured by an electrostatic voltmeter. The current was measured by a D'Arsonval galvanometer, which gave a deflection of one scale division for a current 10^{-8} ampère. The starting potential is the potential that produced the smallest current that could be detected by the galvanometer, which was about 2×10^{-9} ampère. All the points were made of brass wires, except the two small points of

* F. R. Gorton and E. Warburg, Ann. der Phys. **18**, p. 128, 1905.

·0244 and ·029 millimetre diameter, which were of platinum. The same wires were used in the experiments on positive and negative discharges, but in the case of the negative discharges the currents for the larger wires were all intermittent, and are not given in the tables. In a few of the experiments which are recorded in the tables the currents were partly or wholly intermittent.

POSITIVE DISCHARGE FROM CYLINDRICAL POINTS WITH HEMISPHERICAL ENDS. Distance from plate, 1·5 cm.

iameter in mm.	·0244	·039	·091	·174	·244	·50	·73	1·13	2·005
ressure in cm.	73·0	74·9	74·1	74·1	74·1	74·1	74·1	73·7	73·2
emperature in °C.	19	21·5	26·5	27·0	27·0	27·0	25·7	26·5	25·5
arting potential	1600	2000	2450	3015	3365	4600	5500	6700	9150
otentials in volts, vith corresponding urrents in 10^{-7} mpère	1750 ·28	3000 7·2	2500 ·5	3255 2·2	3500 1·1	5000 4·8	5800 4·4	7000 6·1	9250 3·3
	2000 1·5	4000 20·5	3000 4·5	3500 4·5	4000 6·3	6000 22·5	6000 9·3	8000 26·6	9500 12·8
	2500 5·2	5000 37·2	4000 16·9	4000 10·7	5000 20·8	7000 44·0	7000 28·7	9000 57·7	9750 21·1
	3000 9·6	6000 62·2	5000 33·8	5000 25·4	6000 41·7	8000 69·9	8000 52·0	10,000 99·9	10,000 30·0
	4000 23·7	7000 90·2	6000 56·6	6000 46·8	7000 65·9	9000 108·7	9000 88·4	— —	10,500 46·6
	5000 42·9	8000 124·3	7000 85·1	7000 73·2	8000 95·2	— —	10,000 131·2	— —	— —
	6000 68·9		8000 117·2	8000 102·0					
	7000 100·6		8500 139·7	8700 131·8					
	8000 135·6								

Positive Discharge from Cylindrical Points with Hemispherical Ends. Distance from plate, 1 centimetre.

Diameter in mm.	·039	·091	·174	·244	·50	1·13	2·00
Pressure in cm.	74·9	72·5	72·5	72·5	72·5	72·5	74·8
Temperature in °C.	22·5	24·2	24·2	24·2	24·2	24·2	26·0
Starting potential	1895	2300	2850	3200	4425	6350	8750
Potentials in volts with corresponding currents in 10^{-7} ampère	2000 1·5	2500 1·9	3000 1·9	3500 5·0	4500 1·3	6500 6·1	9000 13·3
	3000 12·8	3000 8·3	4000 20·0	4000 14·4	5000 13·3*	7000 22·2	9250 28·9
	4000 33·3	4000 28·9	5000 46·6	5000 38·9	6000 44·4*	7500 38·9	9500 45·0
	5000 60·5	5000 57·7	6000 38·3	6000 73·3*	7000 83·8*	8000 60·0	9750 59·9
	6000 101·0	6000 97·1	7000 127·6	7000 116·6*	8000 132·1*	8500 88·8	10,000 76·6
	7000 149·9	7000 128·8			8300 151·0*	9000 119·9	10,250 93·3
						9250 138·8	10,500 106·0

Negative Discharge from Cylindrical Points with Hemispherical Ends. Distance from plate, 1·5 centimetres.

Diameter in mm.	·0244	·039	·091	·174	·244	·50
Pressure in cm.	73·0	74·9	74·1	74·1	74·1	74·1
Temperature in °C.	19·0	21·5	26·5	27·0	27·0	27·0
Starting potential	1125	1475	1975	2775	3100	4650
Potentials in volts with corresponding current in 10^{-7} ampère	1250 ·3	2000 4·1	2000 ·3	3000 ·6	3500 7·2	5000 10·6
	1500 1·8	3000 16·5	2500 5·0	3500 1·5*	4000 15·5	6000 32·2*
	2000 5·4	4000 39·1	3000 12·5	4000 13·9*	5000 43·0	7000 77·7
	3000 20·2	5000 69·3	4000 33·9	5000 52·2	6000 81·6	7500 99·9*
	4000 44·8	6000 112·8	5000 63·8	6000 95·9	6500 113·2	
	5000 77·3	6500 135·9	6000 104·3	6250 105·5		
	6000 122·6		6500 119·9			
	6350 138·9					

* These currents showed some intermittence.

NEGATIVE DISCHARGE FROM CYLINDRICAL POINTS WITH HEMISPHERICAL ENDS. Distance from plate, 1 cm.

Diameter in mm.	·039	·091	·174	·267
Pressure in cm.	74·9	72·5	72·5	74·4
Temperature in °C.	20·5	24·2	24·2	21·5
Starting potential	1465	1850	2750	3025
Potentials in volts with corresponding currents in 10^{-7} ampère.	2000 7·7	2000 1·1	3000 3·0	4000 35·2
	3000 30·3	2500 9·9	3500 23·1	5000 89·7
	4000 71·5	3000 23·7	4000 46·2	5500 133·1
	5000 130·9	4000 64·9	5000 104·5	6000 171·6
	5500 170·5	5000 126·5	5500 150·7	

Several formulae have been given to express the current i in terms of v the potential of the point. Zeleny found that the formula given by Warburg,*

$$i = Cv(v-M),$$

represents accurately the values of the positive currents obtained in his experiments. In this expression, M is the starting potential and C a constant depending on the gas and the arrangement of the apparatus. The constant C increases with the diameter of the point, and when the point is at 1·5 centimetres from the plate the positive currents are given by the formula

$$i = 2{\cdot}58 \times 10^{-13}(1 + {\cdot}11\,d)\,v\,(v-M).$$

When the distance is one centimetre the currents are given by the formula

$$i = 4{\cdot}05 \times 10^{-13}(1 + {\cdot}235\,d)\,v\,(v-M).$$

The currents obtained from negatively charged points cannot be represented by any simple formula.

* E. Warburg, Wied. Ann. **67**, p. 72, 1899.

264. Effect of pressure on the currents through air. Zeleny also observed the effect of lowering the pressure of the air. Using as the point the rounded end of a wire ·18 millimetre diameter, at a distance of 1·5 centimetres from the plate, the starting potentials in volts at different pressures were as follows :

Pressure in millimetres.	Starting potential for positive point.	Starting potential for negative point.
736	3150	2800
634	2900	2550
518	2550	2250

From the general theorem given in Section 250 it may be concluded that the potential 2550 would produce a positive discharge in air at 736 millimetres pressure from a similar point $(\cdot 18 \times 518)/736 = \cdot 126$ millimetres in diameter at a distance $(1.5 \times 518)/736 = 1\cdot 03$ centimetres from the plate.

The experiments at atmospheric pressure in which the dimensions of the apparatus were nearest to these numbers are those with the points of ·09 and ·174 mm. diameter at a distance of one centimetre from the plate, the starting potentials being 2300 and 2850 volts. From these observations it may be concluded that the starting potential at atmospheric pressure for a wire of ·126 mm. diameter would be about 2536 volts when the distance is 1 centimetre and somewhat larger when the distance is 1·03 centimetres. This is in good agreement with the number 2550, so that as far as it is possible to judge from these experiments, the general theorem given in Section 251 may be said to apply to point discharges* as well as to the discharges between parallel planes and concentric cylinders.

265. Potentials required to start point discharges in different gases. Gorton and Warburg's experiments. The following table gives the potentials, obtained by Gorton and Warburg,† required to start point discharges when the

* See note at end of this chapter.

† F. R. Gorton and E. Warburg, Ann. der Phys. **18**, p. 139, 1905.

gases are ionized by Becquerel rays. In these experiments the discharge took place between the end of a platinum wire ·25 mm. in diameter, and a cylinder of platinum foil 4·5 centimetres long and 4·7 centimetres in diameter. The cylinder was open at both ends and the point was on the axis of the cylinder, near the centre, so that as the stream of ions from the point opened out they were received by the surrounding cylinder.

POTENTIALS REQUIRED TO START POINT DISCHARGES.

	Pressure 760 mms.			Pressure 485 mms.		
	Positive dis-charge.	Negative dis-charge.	Ratio of Poten-tials.	Positive dis-charge.	Negative dis-charge.	Ratio of Poten-tials.
Hydrogen	1370	1140	1·2	1120	1000	1·12
Nitrogen	1930	1400	1·36	1630	1200	1·36
Oxygen	2550	1950	1·31	—	—	—
Air	2250	1660	1·35	1930	1500	1·29
Chlorine	2680	1900	1·41	2400	1660	1·45
Bromine	—	—	—	2500	1700	1·47
Iodine	—	—	—	2620	1870	1·40

266. Short sparks through air. In all cases of discharges when the distance between the electrodes is not very short, it has been found that the potential required to start a discharge has a minimum value corresponding to a definite pressure, and it would appear, therefore, that the minimum sparking potential that can be obtained under any conditions is about 300 volts. In air the minimum sparking potential for parallel plates is about 340 volts, and for cylinders it may be as low as 310 volts. In the former case the minimum potential is obtained when the product pS is about ·7, p being the pressure in millimetres and S the spark length in centimetres. Hence at atmospheric pressure the spark length corresponding to the minimum potential is about 10^{-3} centimetre, and a spark will take place between any ordinary electrodes at a very short distance apart, along some path of the order 10^{-3} centimetre in length, when the potential is above 340 volts. If special precautions were taken to prevent the discharges from passing from the edges of the electrodes as in Carr's experiments, it would most likely be possible with plane electrodes to obtain a complete sparking potential curve at atmospheric pressure, showing a rapid increase of the potential above 340 volts for distances of the order 10^{-4} and 10^{-5} centimetre between the plates. If these precautions are not taken, sparking may take place from the edges, and the potential would not rise above 340 volts.

Some experiments made by Earhart * dealing with very short spark-gaps at atmospheric pressure seem to be at variance with the evidence, supported by numerous investigations, which shows that there is a minimum sparking potential of the order 300 volts, below which it is impossible to obtain a spark. Earhart found that when the distance between the electrodes is of the order 10^{-4} centimetre a great diminution in the sparking potential takes place, and currents may be obtained between the electrodes with potentials of about 30 volts. Under these conditions it is obviously very difficult to ensure that the electrodes do not actually touch at some point, or that some small particle does not form a bridge between them.

Some recent experiments made by Almy † show that the electrodes are liable to be drawn together and to come into contact, owing to the large electrostatic force between them, when they are charged to a comparatively small difference of potential. With small spherical electrodes, held firmly so that they are not displaced by the electric attraction, it was found that no discharge passed when the potential was 330 volts, and in all cases a discharge was obtained with 360 volts. The distances between the electrodes in these experiments varied from half a wave-length of light to ten wave-lengths. These results show conclusively that short discharges do not pass through air when the potential is less than about 340 volts. It also follows from these experiments that electricity does not escape from a metal even

* R. F. Earhart, Phil. Mag. (6) **1**, p. 147, 1901.
† J. E. Almy, Phil. Mag. (6) **16**, p. 456, 1908.

when the electric force at the surface is very large, of the order 3×10^7 volts per centimetre.

In discharge tubes at very low pressures the cathode rays, which are emitted when the electric force is very high, are due indirectly to the action of the force, which causes the positive ions to collide with the negative electrode with a high velocity.

NOTE TO SECTION 264.

It follows from the theorem given in Section 251 that if a be the radius of a wire with a hemispherical end at a distance d from a plate, and V the sparking or starting potential when the gas is at pressure p, V will also be the starting potential for a wire of radius ka at a distance kd from a plate when the pressure is p/k. Hence when a/d and ap are constant V should be constant. The following tables give the results of experiments made by Edmunds * to test this conclusion, the radii a of the hemispherical points and the distances d being in centimetres and the pressure p in millimetres of mercury.

STARTING POTENTIALS IN VOLTS FOR POSITIVELY CHARGED POINTS IN AIR.

	$d/a = 10$.			$d/a = 20$.			$d/a = 30$.		
ap	$a =$ ·025	$a =$ ·05	$a =$ ·075	$a =$ ·025	$a =$ ·05	$a =$ ·075	$a =$ ·025	$a =$ ·05	$a =$ ·075
2·5	1250	1260	1200	1360	1300	1320	1460	1400	1370
5·0	1700	1660	1630	1850	1760	1760	1960	1890	1860
10·0	2500	2310	2300	2760	2600	2600	2900	2840	2720
20·0	3860	3630	3600	4350	4000	3950	4650	4350	4120
30·0	—	4800	4740	—	5300	5230	—	5610	5450
37·0	—	5600	5560	—	6160	—	—	—	—

* P. J. Edmunds, Phil. Mag. (6) **28**, p. 234, 1914.

STARTING POTENTIALS IN VOLTS FOR NEGATIVELY CHARGED POINTS IN AIR.

	$d/a = 10$.			$d/a = 20$.			$d/a = 30$.		
ap	$a =$ ·025	$a =$ ·05	$a =$ ·075	$a =$ ·025	$a =$ ·05	$a =$ ·075	$a =$ ·025	$a =$ ·05	$a =$ ·075
2·5	1050	1050	1100	1160	1250	1250	1300	1300	1270
5·0	1570	1700	1630	1700	1810	1800	1860	1920	1860
10·0	2400	2520	2470	2610	2800	2800	2900	3030	2900
20·0	3720	3970	3900	4300	4450	4300	4650	4720	4510
30·0	—	5270	5150	—	5860	5610	—	6200	6000
37·0	—	6220	5950	—	6800	—	—	—	—

CHAPTER XI

DISCHARGE TUBES

267. Geissler discharges, and discharges in highly rarefied gases. The discharges that take place through rarefied gases contained in glass tubes may assume various forms, and an indefinite number of changes in the appearance of the discharges and the forces required to produce them may be obtained by altering the shape of the tube, the size and position of the electrodes, or the pressure of the gas.

In the following pages a brief account is given of the principal features of the discharges obtained in tubes of the simplest form, in order to show how some of the phenomena may be explained by taking into consideration the properties of the ions and electrons that have been described in the preceding chapters.

The various types of discharge may be obtained in a cylindrical tube of uniform section by altering the pressure of the gas. At moderately low pressures the gas becomes luminous in certain parts of the discharge, and the glow becomes very brilliant as the current increases. The tubes which exhibit these effects are generally known as Geissler tubes; when they contain air the pressure is usually greater than one-tenth of a millimetre.

At very low pressures the character of the discharge alters, the glow in the gas becomes faint, and a bright green fluorescence appears on the surface of the glass. At these pressures the cathode rays become conspicuous, and tubes of various forms have been designed in order to investigate the properties of positive ions or of electrons moving with high velocities. The Crookes tubes, Röntgen-ray tubes, and Braun tubes are well-known examples of the apparatus which has been used for these purposes.

268. Potential required to maintain a discharge in a Geissler tube. The potential difference between the electrodes required to produce or maintain a current through a discharge tube is very large when the gas is at atmospheric pressure, but the potential diminishes as the tube is exhausted, and when a certain pressure is reached the potential attains a minimum value. If further reductions are made in the pressure the potential rises, and at very low pressures it is difficult to produce a discharge through the tube. The potential also depends on the section of the tube; for a given pressure and distance between the electrodes, the potential is greater for a narrow tube than for a wide tube.

The potential difference V required to start a discharge in a Geissler tube is not very definite, since a charge on the surface of the glass may affect the field of force to a considerable extent, and there is usually some irregular distribution of electricity on the glass before the current starts. When a continuous current is passing, the charge on the glass becomes distributed in a regular manner and the potential v required to maintain a given current has a definite value. An example of the changes in conductivity that accompany changes of pressure is given in the following table, where v represents the potential difference between the electrodes required to maintain a current of 10^{-2} ampère through rarefied air at pressure p. The discharge took place in a cylindrical tube 3 centimetres in diameter, with plane aluminium electrodes at 11·5 centimetres apart.

p in millimetres of mercury	4	2·84	1·65	1·04	·66	·4	·29	·24	·17	·13
v in volts	650	620	500	470	490	530	590	630	740	800

There is thus a general resemblance between these discharges where the conducting gas is limited by a glass tube and those of the simpler type in which the discharge takes place through a space bounded almost completely by metal electrodes.* There are various differences of detail between

* See preceding chapter.

the two cases, as may be seen by comparing the potentials required to maintain a discharge between two parallel plates with the numbers given above.

When a discharge takes place in air between two large parallel plates at a short distance apart, the minimum sparking potential V is about 340 volts and is obtained when the product pS is about ·7, the pressure p being measured in millimetres of mercury, and S the distance between the plates in centimetres. The potential v required to maintain a current between parallel plates varies with the pressure in a similar manner: it diminishes as the pressure is reduced, and after reaching a minimum value increases rapidly when further reductions are made in the pressure. For the larger values of the product pS, v is less than V, and for the smaller values v is greater than V. The minimum value of v for parallel plates is somewhat less than 340 volts, and is therefore much less than the minimum potential required to maintain a current through a discharge tube.

269. Discharges obtained with a battery of 1,000 volts. The discharges through rarefied gases may be produced by an induction coil or an electrostatic machine, but in order to make accurate investigations of the various phenomena a constant electromotive force is required and it is necessary to use a battery* containing a large number of cells.

The current through the gas may be adjusted by means of a resistance in series with the battery, and in order to start the discharge it is convenient in some cases to use an induction coil. The experiments may be arranged by connecting the terminals of the battery circuit to the electrodes and placing the wires from the secondary of the coil near the ends of the discharge tube. The high potential produced by the induction coil causes the gas to conduct,

* The first batteries containing a sufficient number of cells to produce discharges through vacuum tubes were made by J. P. Gassiot (Proc. Roy. Soc. **10**, p. 36, 1859) and W. de la Rue, H. W. Müller, and W. Spottiswoode (Proc. Roy. Soc. **23**, p. 356, 1875). Interesting accounts of these early investigations are given in the Transactions of the Royal Society.

and a current is obtained between the electrodes which continues to flow after the action of the coil ceases. At certain pressures it is unnecessary to use a coil, as the discharge takes place as soon as the battery circuit is connected to the electrodes.

The discharges are usually classified according to the distribution of luminosity in the different parts of the tube, which changes with the pressure of the gas. The colour and distribution of the light may be affected considerably by impurities, and in order to obtain definite results it is necessary to use simple gases, such as hydrogen or nitrogen, which may be freed from impurities. The range of pressures at which a given type of discharge is obtained depends on the gas, but the principal features of the different types are the same in all gases and may be obtained in a tube containing air.

In the following description of the discharges in air, approximate values of the currents and potentials corresponding to different pressures are given. A high degree of accuracy cannot be obtained, since the conductivity is affected by the chemical changes produced by the discharge.

When the potential of the battery does not exceed 1,000 volts the various types of discharge may be obtained with a tube 10 or 12 centimetres long, but some of the well-known luminous effects in the gas are best obtained in longer tubes.

270. The negative glow and uniform positive column in air. In a cylindrical tube 3 centimetres in diameter, with plane aluminium electrodes 2 centimetres in diameter and 22 centimetres apart, a continuous discharge may be obtained with a battery of 1,000 volts when the pressure of the air is between 1·65 and ·05 millimetre. At the higher pressure the potential difference between the electrodes required to maintain the current is almost independent of the current, the values of V being 920, 925, and 930 volts for currents $2{\cdot}7\times10^{-2}$, $1{\cdot}6\times10^{-2}$, and 10^{-2} ampère. At the lower pressures smaller currents are obtained, which increase with the potential: thus, when $p = \cdot 11$ millimetre the potentials required to produce currents of $2\cdot 7 \times 10^{-3}$ and 10^{-2} ampère are 750 and 1010 volts respectively, and at a pressure ·04 millimetre the current obtained with the latter potential is of the order 10^{-5} ampère.

When the gas begins to conduct at the pressure 1.65 mm., a uniform column of bright orange-coloured glow extends from the positive electrode for a distance of 17 centimetres, where it ends abruptly, and the rest of the tube for the distance of 5 centimetres between the end of the positive column and the negative electrode is nearly all dark, except for a narrow glow about 2 millimetres wide that surrounds the negative electrode. A narrow dark space of uniform width, generally known as the Crookes or the Hittorf dark space, separates the negative glow from the electrode.

The negative glow, which is of a bluish colour, ends with a well-marked outline on the side near the cathode, but on the side remote from the cathode the outline of the glow is not so well defined. The Crookes dark space and the negative glow increase in width as the pressure diminishes. The dark space between the negative glow and the end of the positive column is known as the Faraday dark space.

As the negative glow expands, the end of the positive column recedes from the negative electrode, and at very low pressures the positive column disappears and the negative glow fills the greater part of the tube.

In addition to the long column of light extending from the positive electrode and the negative glow, there are narrow layers of luminous gas in contact with the surface of each electrode, the layer at the positive electrode being more brilliant than the positive column. Under certain conditions (with comparatively high pressures) the layers of gas near the electrodes are the only parts of the discharge which appear to be luminous.

271. Striated positive columns in air. The general appearance of the discharge, when the positive column and negative glow are the most prominent luminous effects, does not alter very much as the pressure is reduced from 1·6 millimetres to ·8 millimetre, but at about half a millimetre pressure the appearance of the positive column begins to change.

This is illustrated by an experiment with the same tube as before (3 centimetres in diameter, electrodes 22 centimetres apart). At the pressure ·57 millimetre and the comparatively large current $i = 6 \times 10^{-3}$ ampère ($v = 680$ volts) the positive column (15·7 centimetres long) tends to break into striae, or alternately bright and dark sections, at the end near the negative electrode, but the part in contact with the positive electrode remains continuous. With this current the negative glow is visible for a distance of 2·5 centimetres from the negative electrode, and the Crookes dark space is about 4 millimetres wide. With smaller currents the positive column does not tend to become striated at this pressure (·57 millimetre). Thus when the current is 8×10^{-4} ampère the positive column is 16 centimetres in length and quite uniform, the negative glow is visible for a distance of one centimetre from the negative electrode, and the Crookes dark space is about 5 millimetres wide. The tendency of the positive column to become striated increases as the pressure is reduced. Thus when $p = \cdot 37$ millimetre and $i = 10^{-3}$ ampère the positive column is continuous for a distance of 10 centimetres from the positive electrode, and in the remaining 5 centimetres there are three distinct striations. When the current is increased to $6{\cdot}7 \times 10^{-3}$ ampère the whole column is divided into ten striations extending to a distance of 14·7 centimetres from the positive electrode.

At a pressure ·24 millimetre the positive column is completely striated for a large range of currents, the number of striations and the length of the positive column depending on the current. The luminous sections obtained with different currents are as follows :

Discharge in air, $p = \cdot 24$ mm. of mercury, distance between electrodes 22 cm., diameter of tube 3 cm.

Current in ampères.	Number of striations in positive column.	Distance from positive electrode to end of last striation.	Distance from negative electrode to end of negative glow.	v in volts.
$2\cdot 6 \times 10^{-2}$	4	8·5 centimetres	9·5 centimetres	850
$1\cdot 3 \times 10^{-2}$	5	10·5 „	7·5 „	720
$7\cdot 5 \times 10^{-3}$	6	11·2 „	6·7 „	680
6×10^{-3}	7	12·3 „	6·0 „	660

As the pressure is diminished, the positive column contracts still further and eventually disappears when the Faraday dark space reaches the positive electrode.

272. Lengths of the positive column and the negative glow. When longer tubes are used, the position of the negative glow and the two dark spaces remain practically the same for a given current and pressure of the gas, but there is an increase in the length of the positive column equal to the increase in the length of the tube.

A very long column of luminous gas may be thus obtained, which is of uniform brightness when the pressure is of the order of a millimetre, but tends to become striated as the pressure is reduced.

With shorter tubes it is easy to attain a pressure at which the positive column disappears. Thus at ·1 millimetre pressure, and a current 2×10^{-3} ampère, the Crookes dark space is about 1·5 centimetres wide and the negative glow may extend to a distance of 12 or 14 centimetres from the negative electrode, and when the electrodes are less than this distance apart the negative glow reaches the positive electrode.

These latter effects are easily obtained with a tube 3 centimetres wide, in which the electrodes are about 10 or 12 centimetres apart. Also, with a potential of 1,000 volts, discharges may be obtained in a tube of this length at much higher pressures than is possible in the longer tubes.

At pressures of 4 or 5 millimetres, a discharge may take place in which there is no positive column, provided that the current is not very large. With a distance of 11·5 centimetres between the electrodes and a pressure of 4 millimetres, a bright uniform positive column is obtained extending to a distance of 6 centimetres from the positive electrode when the current is 2×10^{-2} ampère. The positive column diminishes in intensity and becomes somewhat shorter as the current is reduced, and it disappears completely when the current is 7×10^{-3} ampère. In the latter case the tube is quite dark, with the exception of the negative glow, which is close to the negative electrode, and the bright layers at the surfaces of the electrodes.

273. Classification of Geissler discharges. The discharges through Geissler tubes may thus be considered to be of four different types, which change gradually from one to another as the pressure is reduced. In air they may all be obtained with a potential less than 1,000 volts in a tube 10 or 12 centimetres in length and 3 centimetres in diameter.

At the pressure of about four millimetres a discharge through air may be obtained which is accompanied by luminous effects at the electrodes, the tube being otherwise quite dark. At a pressure of 1 millimetre, the second type is obtained, of which the characteristic feature is a uniform position column extending from the positive electrode to within 5 or 6 centimetres of the negative electrode. In the third type, at a pressure of a quarter of a millimetre, the positive column becomes shorter and is divided into striations. Finally, the fourth type is obtained; at a pressure of about one-tenth of a millimetre, the positive column disappears and the negative glow fills the greater part of the tube.

274. Distribution of force in Geissler discharges; Graham's experiments. Several investigations have been made to determine the distribution of electric force in the Geissler discharges. The principle usually adopted has been to find the potential difference between one of the electrodes and a wire projecting into the gas at various parts of the discharge. This method obviously cannot give reliable results when the wire is very near one of the electrodes, but it may be assumed that the potentials taken up by an insulated wire at points of the discharge remote from the electrodes are

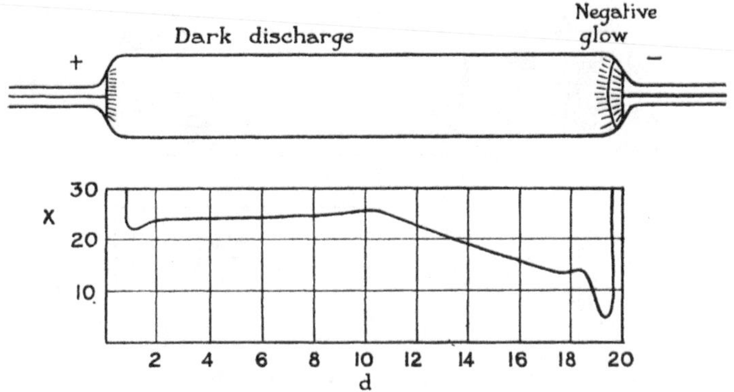

Figure 64 (i). Nitrogen, pressure 2·5 millimetres of mercury. Current $2{\cdot}39 \times 10^{-3}$ ampère. X in volts per centimetre. d the distance from the positive electrode in centimetres.

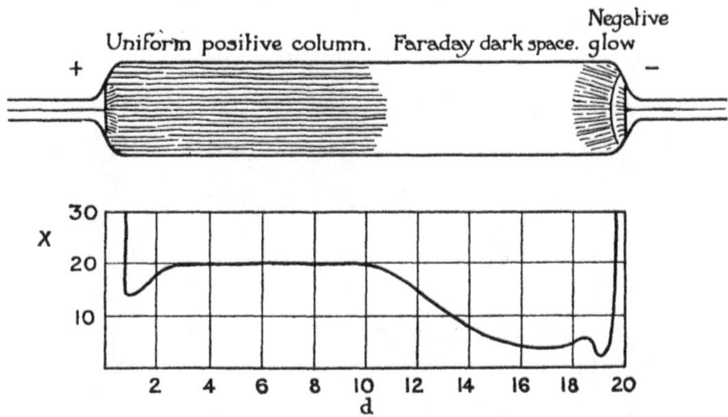

Figure 64 (ii). Nitrogen, pressure ·95 millimetre of mercury. Current $2{\cdot}46 \times 10^{-3}$ ampère. X in volts per centimetre. d the distance from the positive electrode in centimetres.

approximately the same as the potentials at those points when the wire is removed.

In a series of experiments made by Graham this method was improved by using two insulated wires at a short distance a apart, and finding the potential difference E between the wires. The wires were fixed in a movable

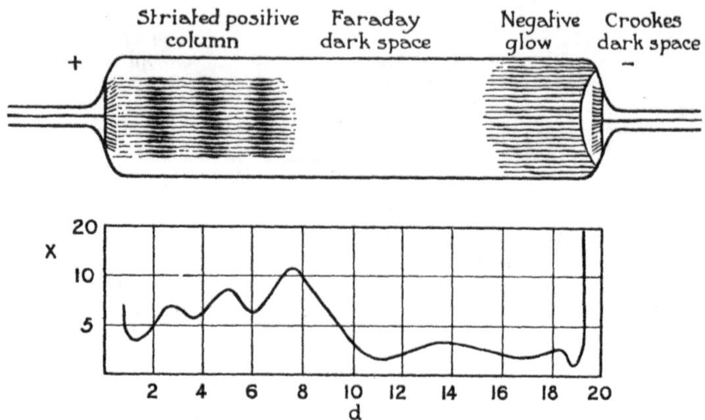

Figure 64 (iii). Nitrogen, pressure ·63 millimetre of mercury. Current $1{\cdot}16 \times 10^{-3}$ ampère. X in volts per centimetre. d the distance from the positive electrode in centimetres.

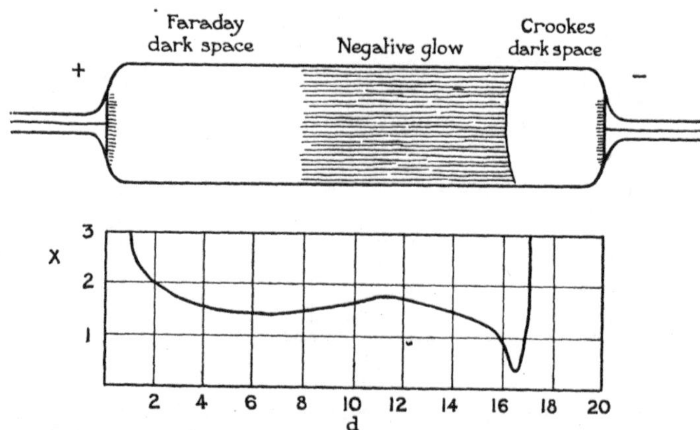

Figure 64 (iv). Nitrogen, pressure very low. Current $3{\cdot}3 \times 10^{-4}$ ampère. X in volts per centimetre. d the distance from the positive electrode in centimetres.

frame placed at different parts of the discharge and the electric force E/a was thus found for the whole distance between the electrodes.

The results obtained by Graham * for discharges in pure nitrogen are given in figures 64 i, ii, iii, and iv. The diagrams show the appearance of the four different types of discharge, the electrodes being plane aluminium discs about 20 centimetres apart. The positive electrode is on the left of the figure and the shading represents the luminous parts of the discharge. The curves give the electric force in volts per centimetre, the abscissae being the distances from the positive electrode in centimetres.

Neglecting the effects quite close to the electrodes, the curves show how the force is distributed in the dark and luminous sections of the gas.

In the first type of discharge, $p = 2{\cdot}5$ millimetres, the force is constant (25 volts per centimetre) for a distance of about 10 centimetres from the positive end. Towards the negative electrode the force diminishes and attains a minimum value of the negative glow.

In the second type the force at the positive end of the tube is constant throughout the length of the positive column, diminishes in the Faraday dark space, and attains a minimum value at the negative glow.

These two types are therefore not very different, and the brightness of the positive column in the second case may be due to the increased velocity of the ions; at the positive end of the tube $X/p = 21$ when the pressure is ·95, but at the pressure 2·5 millimetres the value of X/p is about 10.

In the third type of discharge, where the positive column is striated, the force rises and falls, the maxima occurring in the light striations and the minima in the dark parts between the striations.

In the fourth type the Crookes dark space is about 4 centimetres long and the negative glow extends to within 7 centimetres of the positive electrode, and in this case the positive column disappears.

* W. P. Graham, Wied. Ann. **64**, p. 49, 1898.

Similar investigations have been made by Wilson,* whose experiments with nitrogen and hydrogen are in good general agreement with the results obtained by Graham. Wilson found that the force near the positive electrode was very small and apparently became negative at very low pressures. These results, being obtained from observations of the potentials assumed by insulated wires at different points of the discharge, were not very reliable, and Wilson considered that it is not certain that the force becomes negative.

275. The cathode fall of potential. The numerous researches that have been made to determine the field of force in the neighbourhood of the electrodes have shown that there is a large difference in potential between the negative electrode and points in the negative glow. This potential is known as the 'cathode fall of potential', and is usually measured by placing a wire at the end of the glow near the cathode, but as the potential gradient in the negative glow is so small, the fall of potential is practically the same when the wire is in any position near the end of the glow.

Hittorf's † researches are among the first that were made to determine the field of force in the discharges, and he found that the cathode fall of potential was independent of the current, provided that the negative glow did not cover the cathode. This potential is the normal cathode fall of potential, and Warburg ‡ found that it is independent of the pressure of the gas. In air, the normal cathode fall of potential is about 340 volts. With small currents the glow is not distributed over the whole electrode but is confined to a small area; as the current increases, the glow spreads and covers the whole electrode. When the negative electrode is completely covered with the glow, the cathode fall of potential increases with the current.

* H. A. Wilson, Phil. Mag. (5) **49**, p. 505, 1900.
† W. Hittorf, Wied. Ann. **20**, p. 705, 1883; **21**, p. 133, 1884.
‡ E. Warburg, Wied. Ann. **31**, p. 545, 1887.

The force near the negative electrode was also investigated by Schuster,* who showed that the phenomena indicated that the positive ions were of larger mass than the negative ions. The concentration of the force in the neighbourhood of the negative electrode must arise from a large positive charge in the gas, and this accumulation of positive ions may be explained if the positive ions move more slowly than the negative ions under the electric force.

276. Determinations of the normal cathode fall of potential. Several determinations have been made of the normal cathode fall of potential and the width of the Crookes dark space in different gases when various metals are used as electrodes. The earlier experiments showed that with ordinary metals the cathode fall for any gas is approximately independent of the electrode, but Mey has found that it is much smaller when one of the alkali metals is used as the negative electrode than with other metals. This is also the case with the amalgams of sodium and potassium, or the alloy formed with these metals, when the discharge passes through nitrogen, hydrogen, or helium.

In nitrogen the surfaces of the electrodes become tarnished, owing to a chemical combination of sodium or potassium with the gas, and this causes an increase in the potential fall which disappears when the surface is renewed by reversing the current.

Mey gives a list of the values of the normal cathode fall of potential as obtained in some of the principal investigations,† the numbers being as follows:

* A. Schuster, Proc. Roy. Soc. **47**, p. 541, 1890.

† The table is made up from researches of the following authors, the particular results obtained by each being indicated by the letters given with the numbers:

E. Warburg, Wied. Ann. **31**, p. 545, 1887; **40**, p. 1, 1890.
J. W. Capstick, Proc. Roy. Soc. **63**, p. 356, 1898.
Hon. R. J. Strutt, Phil. Mag. (5) **49**, p. 293, 1900.
W. Heuse, Inaug. Diss., Berlin, 1901.
K. Mey, Ann. der Phys. (4) **11**, p. 127, 1903.

NORMAL CATHODE FALL OF POTENTIAL IN VOLTS.

	Platinum.	Carbon.	Mercury.	Silver.	Copper.	Iron.	Zinc.	Aluminium.	Magnesium.	Sodium.	Sodium and Potassium Alloy.	Potassium.
Oxygen	369 C	—	—	—	—	—	—	—	310 W	—	—	—
Mercury vapour	—	—	340 W	—	—	389 H	—	—	—	—	—	—
Hydrogen, slightly damp	350 W	—	—	—	—	—	—	236 W	207 W	—	—	—
Hydrogen, pure	300 W 289 C	—	—	295 W	280 W	230 W	218 W	190 W	168 W	185 M	169 M	172 M
Nitrogen, slightly damp	260 W	256 W	—	—	252 W	262 W	260 W	224 W	—	—	—	—
Nitrogen, pure	232 W	—	226 W	—	—	—	—	—	207 W	178 M	125 M	170 M
Helium	226 S	—	—	—	—	—	—	—	—	80 M	78·5 M	69 M
Argon	167 S	—	—	—	—	—	—	100 S	—	—	—	—
Water vapour	469 C	—	—	—	—	—	—	—	—	—	—	—
Ammonia	582 C	—	—	—	—	—	—	—	—	—	—	—
Nitric oxide	373 C	—	—	—	—	—	—	—	—	—	—	—

There is thus a distinct difference in the potentials corresponding to the various electrodes, those obtained with the alkali metals being comparatively small.

277. Properties of the alkali metals. The electric properties * of these metals are also remarkable in other respects, since they lose a negative charge very rapidly when exposed to ordinary daylight. The small cathode fall of potential is probably due to the fact that under the action of the radiation from the discharge or of the impacts of the positive ions, the alkali metals emit large numbers of electrons as compared with the numbers that would be emitted under similar circumstances from ordinary metals.

In their recent investigations Elster and Geitel † have found that, with cathodes of the alkali metals, the electric discharge through hydrogen at a low pressure produces a compound on the surface of the electrode which is blue or green in the case of potassium, sodium, rubidium, and cobalt, or yellow-brown with sodium. A much greater photo-electric effect is obtained with these coloured metals than with the pure metals. The colour and sensitiveness to light change slowly in hydrogen.

The surfaces obtained by the discharges through hydrogen retain their sensitiveness to light in an atmosphere of argon or helium, and a large number of electrons are emitted when infra-red radiation falls on the electrode. The currents thus obtained may be measured by a galvanometer.

278. The Crookes dark space, and the cathode fall of potential. Recently a more complete investigation of the discharge in the neighbourhood of the cathode was undertaken by Aston,‡ and the experiments were continued in collaboration with Watson.§ Their researches deal with discharges in which the negative glow covers the whole

* See Section 47.

† J. Elster and H. Geitel, Phys. Zeitschr. **11**, p. 257, 1910; **12**, pp. 609 and 758, 1911.

‡ F. W. Aston, Proc. Roy. Soc. A, **79**, p. 80, 1907.

§ F. W. Aston and H. E. Watson, Proc. Roy. Soc. A, **86**, p. 168, 1912.

surface of the cathode, so that currents below a certain minimum were not used, and the cathode fall of potential exceeded the normal value. Also the gases were reduced to such low pressures that the Crookes dark space was not less than a few millimetres wide, and an accurate measurement of the width was possible.

The apparatus consisted of a wide cylindrical glass vessel containing a pair of large aluminium discs, of the same diameter as the containing vessel, which were used as the electrodes. One of the discs was divided into an inner disc and an outer ring in the same plane, which acted as a guard ring. The air gap between the ring and the disc was very small, and as both were approximately at zero potential they acted as one continuous surface. The inner disc was connected to earth through a galvanometer, in order to measure the current in the central part of the field which was unaffected by the glass surface. The other electrode was movable, and was raised to a high potential by a battery of accumulators. The distance between the electrodes was of the same order as the diameter of the vessel in which they were contained.

When the current was sufficiently great to cover the cathode with a glow and there was no positive light, it was found that the distance between the plates, when it was considerably longer than the dark space, had no measurable effect, either on the dark space, the voltage, or the current. The discharge was therefore of the fourth type described above, in which the Faraday dark space or the negative glow extends to the positive electrode, and the results are in agreement with Graham's experiments, which show that the electric force is very small in the negative glow and in the Faraday dark space.

279. Formulae given by Aston and Watson for D and V. It follows that the fall of potential across the Crookes dark space is practically the same as the potential difference between the electrodes when the negative glow extends to the positive electrode, and the latter potential cannot vary by any great amount with the distance between the electrodes.

Under these conditions it was found that the length of the dark space D and the potential difference between the electrodes V were given in terms of the pressure p and the current density C by the formulae

$$D = \frac{A}{p} + \frac{B}{\sqrt{C}},$$

$$V = E + \frac{F\sqrt{C}}{p},$$

A, B, E, and F being constants.

The results obtained with argon and helium are not accurately represented by these formulae, and the numbers given in the tables for these gases give only approximate values of V and D.

If D be measured in centimetres, p in hundredths of a millimetre of mercury, C in tenths of a milliampère per square centimetre of the cathode, the values of the constants for the different gases are given in the following table. The pressures at which the dark space is one centimetre in length when unit current is flowing ($i = 10^{-4}$ ampère per square cm.) are given in the column P. The potentials required to maintain the discharge under these conditions when the electrodes are 17 centimetres apart are given in the column V, and since the force is very small in the distance of 16 centimetres from the end of the dark space to the positive electrode, these potentials are not much greater than the cathode fall of potential.

	A	B	E	$F \times 10^{-2}$	P	V
Hydrogen	26·5	·43	144	57·3	46·8	266
Helium	36	·49	255	100	70·6	395
Carbon monoxide	10	·42	255	41·5	17·5	489
Nitrogen	6·8	·40	230	23·6	11·3	434
Air	6·5	·42	255	23·0	11·3	457
Oxygen	5·7	·49	290	17·6	11·4	444
Argon	5·4	·34	240	29·4	8·2	594

The influence of the nature of the cathode on the discharges through oxygen and hydrogen has recently been investigated by Aston,* and it was found that the values of V and D were always given by equations of the same form as those obtained with aluminium electrodes. In these experiments the electrodes were circular discs, 10 centimetres in diameter and 11 centimetres apart, an aluminium disc being used as the positive electrode in each case. For simplicity in calculating the current density, the whole surface area was used. This involves an edge error which is probably much the same for each metal.

The following table gives the constants for discharges through oxygen and hydrogen when different metals are used as the negative electrode:

	Oxygen.				Hydrogen.			
	A	B	E	$F \times 10^{-2}$	A	B	E	$F \times 10^{-3}$
Mg.	4·8	·41	310	13·5	23	·39	190	50
Al.	5·7	·43	310	17·5	23	·41	170	66
Fe.	8·5	·38	337	26·0	34	·44	260	91
Cu.	8·9	·40	340	28·5	47	·45	300	130
Zn.	7·3	·43	335	23·5	43	·41	220	118
Ag.	10·8	·36	350	33·0	51	·47	325	140
Sn.	7·9	·40	363	24·0	41	·45	250	120
Pt.	8·8	·40	335	30·0	45	·42	270	120
Pb.	8·7	·41	340	31·5	51	·50	320	166

280. Potential assumed by a wire in a conducting gas. The method of taking the potential at a point in a conducting gas by means of an insulated wire probably gives reliable results when the gas in the neighbourhood of the wire contains a large number of positive and negative ions. But when the wire is placed near one of the electrodes the stream of ions approaching the electrode generally exceeds the stream in the reverse direction, and the wire takes up a charge opposite in sign to the charge on the electrode. Thus when a wire is placed near the positive electrode it becomes negatively charged. In the latter case the final potential assumed by an insulated wire is not the same as the potential in the gas when the wire is

* F. W. Aston, Proc. Roy. Soc. A, **87**, p. 437, 1912.

removed. Thus Skinner * has observed that a wire placed very close to the positive electrode takes up a potential which is about 30 volts less than the potential of the electrode, and it has generally been supposed that the experiments show that there is fall of potential of that order between the anode and the gas. This conclusion is obviously not justified, and there is no reliable experimental evidence to show that there is an abrupt change of potential between the electrode and the gas in its immediate neighbourhood when the current is undisturbed by the wire.†

It is interesting to consider the effects obtained with an insulated wire near the negative electrode. When the wire is placed in the negative glow in air, the normal potential difference between the wire and the electrode is 340 volts. As there are streams of positive and negative ions on each side of the wire, the potential assumed by the wire is probably the same as that of the gas. When the wire is brought into the dark space near the electrode the potential diminishes, so that the potential difference between the wire and the electrode becomes less than the potential required to maintain a current. The gas in the narrow space between the wire and the electrode ceases to conduct, and the wire is exposed on one side to the stream of positive ions approaching the electrode. The wire therefore becomes positively charged, and the potential becomes greater than that of the undisturbed gas. The final potential is attained when the rate at which negative ions are attracted to the wire from the surrounding gas becomes equal to the current of positive ions colliding with the wire.

Thus when the wire is brought very near the negative electrode it attains a potential higher than that of the electrode, but, as in the case of a wire very close to the positive electrode, it is not possible to draw any conclusion as to the potential in the gas near the electrode.

* C. A. Skinner, Phil. Mag. (6) **4**, p. 490, 1902.
† Phil. Mag. (6) **11**, p. 733, 1906.

281. Thomson's experiments on the distribution of force in a striated column. The force in the cathode dark space has been determined by Aston by another method which had been used by Thomson to investigate the force in the positive column. The principle of the method consists in passing a narrow beam of cathode rays from a side-tube across the discharge. The rays are received on a fluorescent screen in another side-tube opposite to that from which they emerged. The side-tubes are at right angles to the axis of the principal discharge tube, and the cathode rays in passing through the conducting gas are deflected by an amount which is proportional to the force.

In order to find the force at different parts of the discharge, the electrodes were mounted at a fixed distance apart on a movable frame, and it was thus possible to place either electrode at any required distance from the line in which the cathode rays traversed the tube.

Thomson * applied this method to test some effects observed by the ordinary method with exploring wires, in an investigation of the distribution of force in the striated positive column in hydrogen. The results obtained by the two methods were in good agreement. At low pressures variations were observed in the intensity of the force in the bright and dark parts of the striations, the electric force being negative on the cathode side, just in front of the bright striation. These negative forces were small and were only observed at low pressures, but when the current was reduced until the discharge was no longer striated, large negative forces were observed in the neighbourhood of the anode. In some cases the region in which the force was negative extended for a long distance from the anode, which may be as much as two-thirds of the distance from the anode to the cathode. These results differ from those obtained by Graham and Wilson, who did not observe reversals in the direction of the force at points in the striated positive column.

* J. J. Thomson, Phil. Mag. (6) 18, p. 441, 1909.

282. Aston's experiments on the force in the Crookes dark space. Aston * examined the distribution of force in a discharge between two large parallel plates of aluminium, 12 centimetres in diameter and 6·5 centimetres apart, using the deflection of a narrow stream of cathode rays as a measure of the intensity of the force. In order that the rays should not be deviated by a large angle in traversing the discharge, it is necessary to use rays travelling with a high velocity. This imposed a limit on the range of pressures at which it was possible to make the experiments, since the pressure in the side-tube in which the cathode rays were generated was the same as the pressure in the main discharge, and it is only at very low pressures that cathode rays with a high velocity can be produced. The pressures, as deduced from the length of the dark space, were of the order of ·1 millimetre in hydrogen, and ·025 millimetre in the other gases—air, oxygen, and argon.

The results obtained are very simple; the force F in the Crookes dark space is proportional to the distance from the edge of the negative glow; in the negative glow the force is inappreciable.

There were no large forces observed in the layers of gas near the electrodes. The force at the positive electrode was inappreciable, and that at the negative electrode was $2V/D$, V being the cathode fall of potential, and D the width of the dark space.

The dark space in these experiments was about three centimetres wide, and the negative glow filled the remainder of the distance between the plates.

Since the force in the negative glow is small, the potential difference between the electrodes must be almost the same for different distances between the electrodes, when the total length of the discharge is a small multiple of the width of the Crookes dark space. This agrees with the previous experiments.† Also, if there are no potential differences between

* F. W. Aston, Proc. Roy. Soc. A, **84**, p. 526, 1911.

† See Section 279.

the gas and the electrodes, the total fall of potential between the electrodes should be approximately the same as the fall of potential obtained by the integration $\int_0^D F dx$. The following table shows that the cathode fall of potential calculated by this method is practically the same as the potential difference between the electrodes given by a voltmeter.

The different potentials given for hydrogen and oxygen correspond to different currents through the gas. In all cases the currents were sufficiently great to cover the cathode with the negative glow.

	Hydrogen.				Air.	Oxygen.			Argon.
Length of dark space in cm.	3·2	3·1	3·1	2·81	2·86	3·05	3·10	2·70	3·02
Calculated potential $\int_0^D F\,dx$	270	369	536	595	590	442	493	602	610
Potential between the electrodes given by a voltmeter	265	375	530	610	583	445	495	610	620

The close agreement between the potential fall in the dark space and the potential difference between the electrodes shows that there cannot be any abrupt potential fall between the gas and either of the electrodes of the order of 30 volts.

283. Force in the uniform positive column; Herz's experiments. The phenomena that occur at the positive end of a long Geissler tube are simpler than those at the negative end when the pressure and current are adjusted so as to produce a uniform positive column. In this case the electric force parallel to the axis, or potential gradient, is constant in the luminous column.

The changes that take place in the force corresponding to changes in the pressure of the gas, the current, and the diameter of the tube have been investigated by Herz,* the gases which were used being hydrogen and nitrogen.

When the pressure is constant, the force diminishes as the current increases, the relation between X and i being given by the formula

$$X = X_0 - b(i - i_0).$$

The force X_0 corresponding to the current i_0 increases with the pressure of the gas, but the factor b is practically independent of the pressure. This factor diminishes as the diameter of the tube increases.

The following table gives the force X in volts per centimetre in nitrogen, in tubes of different diameters $2R$, corresponding to a current of 1·2 milliampères; the values of b are in volts per centimetre when the rise of current $i - i_0$ is measured in milliampères:

ELECTRIC FORCE IN POSITIVE COLUMN IN NITROGEN.

p in mm. of mercury.	$2R$.			
	10 mm.	15 mm.	20 mm.	25 mm.
8	—	156·8	—	—
7·5	—	148·4	—	—
7	144·4	140·1	—	—
6·5	139·2	131·9	—	—
6	132·6	123·8	—	—
5·5	116·1	115·8	—	—
5	118·2	107·8	—	—
4·5	109·4	99·9	97·7	—
4	99·7	92·2	89·3	—
3·5	89·2	84·5	80·5	—
3	77·7	76·1	71·2	—
2·55	—	66·2	61·5	60·2
2	—	55·4	51·4	48·7
1·5	—	43·6	40·8	37·5
1	—	—	29·8	26·9
b	10·0	8·5	3·5	3·4

The following are the forces in the positive column in a tube 15 millimetres in diameter containing hydrogen at different pressures, the current being one milliampère:

* A. Herz, Wied. Ann. **54**, p. 244, 1895.

ELECTRIC FORCE IN POSITIVE COLUMN IN HYDROGEN.

p	X
8	117·0
7	105·0
6	92·6
5	79·5
4	64·5

$b = 8{\cdot}0$

The determinations of the temperature of the gas in the positive column show that the decrease in the potential gradient which accompanies an increase of the current is to some extent due to the diminution of the density of the gas.*

284. Wilson's determinations of the force in the positive column. The values of the force in the positive column in air, oxygen, and hydrogen at lower pressures have been determined by Wilson.†

In air at pressures less than one millimetre, the force is nearly constant for the small currents that may be used, but with higher pressures the force diminishes as the current increases. The experiments were made in a tube 2·1 centimetres in diameter, and the following forces (in volts per centimetre) were obtained in air at 2·81 millimetres pressure:

$$X = 57 \qquad i = 4{\cdot}05 \times 10^{-3}$$
$$X = 54 \qquad i = 7{\cdot}85 \times 10^{-3}$$
$$X = 50{\cdot}6 \qquad i = 11{\cdot}5 \times 10^{-3}.$$

With small currents of the order 10^{-3} to 10^{-4} ampère the force is independent of the current, and increases as the square root of the pressure.

The ranges of pressure, for the different gases, within which the ratio $X/\sqrt{p}$ was found to be constant, are

Air $X/\sqrt{p} = 34{\cdot}9$, from $p = {\cdot}2$ to $p = 2{\cdot}82$ mm.
Oxygen $X/\sqrt{p} = 26{\cdot}9$, from $p = {\cdot}139$ to $p = {\cdot}85$ mm.
Hydrogen $X/\sqrt{p} = 28{\cdot}0$, from $p = {\cdot}25$ to $p = 1{\cdot}36$ mm.

* See Section 287.
† H. A. Wilson, Proc. Camb. Phil. Soc. 11, pp. 249 and 391, 1902.

285. Wilson's investigations of the Hall effect. Wilson also investigated the Hall effect in the uniform positive column with the same pressures and currents as those for which the values of X were determined. When a transverse magnetic force H acts on the streams of positive and negative ions moving in opposite directions with velocities U_+ and U_- under an electric force X, the two streams are deflected in the same direction, but the positive ions are deflected less than the negative ions when U_+ is less than U_-. The ions thus become separated and an electrostatic force Z is produced (in a direction at right angles to the directions of X and H), which tends to diminish the deflection of the negative ions and to increase the deflection of the positive ions. The two streams, in fact, tend to keep together owing to the mutual attraction of the positive and negative charges. The force Z may be observed by placing two insulated wires in the same section of the discharge so that they both assume the same potential when the magnetic force is zero. When a transverse magnetic force is applied to the conducting gas, the two wires assume different potentials, the sign indicating that the negative ions move faster than the positive ions.

The potential difference between the wires was found to be independent of the current and inversely proportional to the pressure. The following are the values of the force Z in volts per centimetre, obtained from observations of the Hall effect in the positive columns in air, oxygen, and hydrogen:

Air . . . . $Z = \cdot 0248 \times \frac{H}{p}$.

Oxygen . . . $Z = \cdot 0379 \times \frac{H}{p}$.

Hydrogen . . . $Z = \cdot 0205 \times \frac{H}{p}$.

It is difficult to arrive at an accurate estimate of the velocity of the electrons from these observations, as no satisfactory theory has been given to account for the properties of the current in the uniform positive column.

286. Variation of the temperature of the gas in a discharge tube: Wood's experiments. When a current passes through a gas, a large proportion of the kinetic energy acquired by the ions under the electric force is transferred to the molecules of the gas and appears as heat, which is shown by the rise in temperature along the path of the discharge.

Under ordinary conditions the principal effect of the increase of temperature is due to the diminution of the density of the gas, and the temperature is not sufficiently high to cause the glows that appear in discharge tubes or to account for the chemically active state of the gas. If the discharge tube is connected with other vessels, or if there are spaces in an exhausted vessel which are not traversed by the current, some of the gas is expelled from the path of the discharge, and the general effect is similar to that obtained by a reduction of pressure; that is, at the higher pressures the resistance of the gas is diminished, and at the lower pressures it is increased by an increase of the current. If the current passed between two plane electrodes at the ends of a sealed tube, an increase of temperature has a smaller effect, which corresponds to a diminution of the density at the hotter parts of the tube and an increase of density where the temperature is not so high.

Several determinations of the distribution of temperature in discharge tubes have been made by inserting thermometers at various distances from the electrodes, but probably the most accurate observations are those made by Wood,* who used a bolometer of fine platinum wire. In these experiments the spiral of wire was fixed in a movable frame, and the distribution of temperature in a Geissler discharge was observed by placing the wire at various distances from one of the electrodes.

The lowest temperature observed was in the Faraday dark space, and the highest was in the negative glow near the negative electrode. The following numbers, obtained with a tube 18 centimetres long and 1·5 centimetres in diameter,

* R. W. Wood, Wied. Ann. **59**, p. 238, 1896.

give the distribution of the temperature in a discharge through nitrogen at 1·5 mm. pressure, the current being ·001 ampère and the temperature of the room 26° C.

In the positive column, at one centimetre from the electrode the temperature was 37°, at 6 centimetres from the electrode 38·5°, and at 12·5 centimetres 37°; in the Faraday dark space the temperature fell and attained a minimum 34° at a distance 14·5 centimetres from the positive electrode.

In the uniform positive column the temperature is nearly constant, but when the pressure is reduced and the column becomes striated, the general rise of temperature is accompanied by local rises coinciding with the striations, the temperature being from ·5° to 1·5° higher in the luminous sections than in the intermediate dark spaces. The variation of temperature observed in passing from the positive column to the Faraday dark space is similar to the change in the potential gradient, and shows that the amount of heat generated in those sections of the gas is approximately proportional to the potential gradient. In the negative glow, however, the potential gradient is small and the temperature very high, which is due either to conductivity of heat from the Crookes dark space, or to electrons that acquire a high velocity in the Crookes dark space and lose their energy by colliding with molecules in the negative glow. Also, the glow in the gas shows that the molecules are set in vibration by particles travelling with a high velocity.

287. Rise of temperature in the positive column. Wood also determined the rise of temperature in the unstriated positive column, for different pressures and currents. When the pressure is constant, the rise of temperature θ is approximately proportional to the current i, the ratio θ/i tending to diminish as the current increases. According to the theory suggested by Warburg, the rise of temperature in the positive column is proportional to the product iv where v is the potential gradient, so that when v is constant θ is proportional to i. The force in the positive column diminishes as the density is reduced, as is seen from the experiments on the variation of v with the pressure. If the discharge tube is connected with a pump or pressure gauge while the current is flowing, some of the gas is expelled from the tube and the potential gradient diminishes, so that the ratio θ/i diminishes as the current increases. When the discharge tube was disconnected from other vessels so as to maintain the gas at a constant volume, Wood found that the ratio θ/i was very nearly constant. The following table gives the values of θ, obtained under these latter conditions, in a discharge through nitrogen at ·3 millimetre pressure, the diameter of tube being 1·5 centimetres:

$i \times 10^3$	1·5	1·7	2·0	2·6	2·9	3·4
θ	13	15·2	17·4	21·9	24·2	28·5
$\theta/i \times 10^{-2}$	87	89	87	84	83	84

The ratio θ/i increases with the pressure, the values of θ obtained with a current ·0015 ampère in nitrogen being as follows:

p	·3	1·8	2	3
θ	13	21·7	23·3	31

In hydrogen the increase of temperature was about one-tenth that observed under similar conditions in nitrogen.

288. Warburg's theoretical investigation of the rise of temperature. According to the theory given by Warburg,* the rise of temperature may be calculated on the hypothesis that the energy acquired by the ions in the uniform positive column appears as heat in the gas, and the steady state is reached when the energy iv is equal to the loss of heat conducted through unit length of the glass tube per second. The rise of temperature of the gas in a tube 1·5 centimetres internal diameter and 1·7 centimetres external diameter was calculated, using the numbers given by Herz for the potential gradient v in the positive column, the outer surface of the tube being supposed to be at the temperature of the room. The highest rise of temperature θ_0 at the axis of the tube

* E. Warburg, Wied. Ann. **54**, p. 265, 1895.

is about double the mean rise $\bar{\theta}$. The following numbers, given by Warburg, represent the results thus obtained for a current ·0012 ampère in dry nitrogen, p being the pressure of the gas in millimetres of mercury.

NITROGEN, $i = \cdot 0012$ ampère.

p	v	θ_0	$\bar{\theta}$
8	156	59·2	30·4
5	107	41·3	21·1
3	76	29·4	14·9
2	55	21·6	11·0

In hydrogen the rise of temperature is smaller, owing to the higher conductivity of that gas and the small potential gradient v.

HYDROGEN, $i = \cdot 001$ ampère.

p	v	θ_0	$\bar{\theta}$
8	117	6·0	3·9
5	79·5	4·1	2·1

The temperatures found by Wood are in good agreement with these calculations.

289. Cathode rays. The phenomena that accompany the discharges in high vacua are quite different from those obtained in Geissler tubes. As the pressure is reduced, the cathode glow recedes from the cathode, and the luminous effect produced by rays traversing the gas becomes more distinct. Eventually the cathode glow disappears, and the cathode rays impinge on the surface of the tube with a high velocity, a green fluorescence appearing on the surface of the glass; also, the potential required to start the discharge increases as the pressure is reduced, as may be seen by the increase of the length of the equivalent spark in air.

Experiments have been made by Villard* with a view to finding the action on which the emission of the cathode rays depends, and he concluded that the rays are emitted from the points of the cathode on which the positive ions impinge. At other points of the surface a very large electric force may be established, but no electrons are emitted. A number

* P. Villard, Journal de Physique (3) 8, p. 1, 1899.

of different experiments may be explained on these hypotheses. In an ordinary cylindrical tube, with a plane circular cathode extending to within a short distance of the surface of the glass, the cathode rays, which at first emanate from the whole electrode, become confined to an area at the centre of the electrode as the pressure is reduced, and in a high vacuum a narrow beam of rays, coinciding with the axis of the tube, is obtained. This effect has been attributed to a positive charge on the glass, which repels the positive ions towards the centre of the cathode. This tendency of insulated bodies to become positively charged is shown by placing an auxiliary electrode in front of the cathode and at a few millimetres from it. When a high vacuum is obtained, corresponding to a spark length of 7 or 8 centimetres, the potential difference between the anode and the auxiliary electrode is very small, corresponding to a spark length of a few millimetres, while the potential difference between the cathode and the auxiliary electrode is nearly as great as the total fall of potential between the cathode and anode.

290. Villard's theory of the action of positive ions on the cathode. The action of the positive ions is shown more definitely by means of a tube constructed as shown in figure 65. The cathode C is a large plane disc 4·8 centimetres in diameter, at the end of a tube 5 centimetres in diameter, and a metal diaphragm D of the same diameter, with two small apertures a and b, is set up at a distance of 1·5 centimetres from the cathode. The diaphragm is provided with a metallic connection d sealed in the glass. A movable electrode E is placed near one of the apertures, and may be set at any required distance from the aperture a by tilting the tube. Two additional electrodes A and A' are sealed into short side-tubes, and either of these may be used as anode.

If the tube be gradually exhausted (A being the anode and D insulated), then so long as the dark space does not extend to the diaphragm D the current flows from the whole surface of the cathode. As the dark space expands, the negative glow disappears from the space between C and D, and the emission from the cathode becomes concentrated at a' and b' opposite the apertures a and b. Eventually, with a high vacuum, two narrow luminous beams are obtained passing through the apertures, and the rest of the space between C and D is nearly dark. The two beams mark the path of the cathode rays through the gas, and a bright green fluorescence is produced at the points f and g where the rays impinge on the glass. The beams are deflected by a transverse magnetic force in a direction which shows that they consist of negatively charged particles travelling with a high velocity.

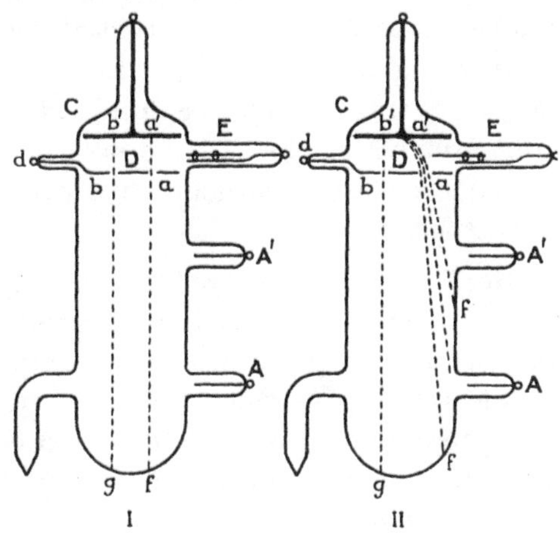

Figure 65.

The following explanation of these results is given by Villard. In the early stages, before the cathode glow is cut off from the diaphragm, the positive ions generated throughout the space between C and D impinge on the cathode, and electrons are emitted uniformly from the whole surface. As the pressure is reduced the number of ions generated between C and D diminishes; but some of the positive ions generated in the gas beyond the diaphragm pass through the apertures, and a large number of electrons are emitted from the points opposite the apertures on which the positive ions impinge.

The experiments show that a large electric force at the

surface is not the only condition required in order that electrons shall be given off from the surface, since there is a large force at all points of C when the cathode rays are developed at a' and b'. In fact, if the diaphragm be connected to the anode, the form of the discharge remains unaltered, but is less marked. In this case, very few positive ions are directed towards the apertures a and b by the small forces in the field below the diaphragm.

If the electrode E be moved towards the aperture a, the discharge takes the form shown in figure 64, ii, where E is positively charged. It is convenient in these experiments to regulate the potential of the electrode E by means of a connector to the tube itself: the potential may be raised in steps by connecting E to D, to A', or to A. The positive charge repels the positive ions approaching the cathode from a, and the point from which the cathode rays start is repelled to a'. At the same time the cathode rays are attracted by E and become dispersed.

291. Theoretical investigation of the currents between parallel plates. The experimental investigations show that the discharges in rarefied gases are very complicated, and even if all the electrical properties of ions corresponding to every force and pressure were known, it would be extremely difficult to give an accurate theory to account for many of the observed phenomena.

The simplest method of arriving at an explanation of some of the properties of discharges is to consider the current obtained between parallel plates. The investigations may at first be confined to the smaller currents obtained with pressures above the critical pressure,* as in this case the ions may be considered to be generated in the gas by the collisions of positive and negative ions with the molecules. When the current is large the phenomena become more complicated as the gas becomes heated, and the density varies along the path of the current. At the lower pressures, other processes

* The critical pressure, corresponding to the minimum sparking potential, is inversely proportional to the spark length S. In air, the critical pressure in millimetres of mercury is $\cdot 65/S$ approximately.

of ionization increase in importance, and in addition to the ions generated by collisions with molecules of the gas, cathode rays are emitted from the negative electrode, and radiations which ionize the gas are produced inside the tube.

292. Comparison of the sparking potential with the potential required to maintain a discharge. A remarkable property of the discharge is observed when a battery of a large number of accumulators is connected through a high resistance to a pair of parallel-plate electrodes. It is found that no current takes place until the potential of the battery is raised to a certain value V, which is the sparking potential, but when the current is flowing the potential difference v between the electrodes may be much less than the potential V required to start the current. The following experiment* affords an illustration of this phenomenon, the gas being air at 4·31 millimetres pressure between parallel plates at 8 millimetres apart.

At first no current was observed when a potential difference between the electrodes was 601 volts, but a spark took place when the potential was raised to 603 volts. A current of ·0052 ampère then continued to flow, and the potential difference between the plates fell to 350 volts, the remainder of the battery potential being taken up in sending the current through the large external resistance. In this case the potential required to maintain the current was 253 volts less than the potential required to start the current.

These effects are obtained when the pressure exceeds the critical pressure † and the current is not reduced to a very small value by a high external resistance.

In the discharges from points and wires, a small current is obtained when the sparking potential is reached, and in order to increase the current it is necessary to increase the potential. A theory to account for this increase of the

* Phil. Mag. (6) **8**, p. 750, 1904.

† In discharges at pressures below the critical pressures, the potential v required to maintain a current exceeds the spark potential and increases as the current increases. J. A. Brown, Phil. Mag. (6) **12**, p. 210, 1906.

potential with the current has been given in the preceding chapter,* but it does not apply to discharges in which the ionization takes place in the whole space between the electrodes. In the former cases a large charge, which may be either positive or negative, accumulates in those parts of the gas where the electric force is small, and tends to reduce the electric force in the region in which ionization takes place. In the latter cases the negative ions are in the electronic state along the whole length of the discharge, and the principal disturbance of the field is due to the accumulation of the slowly moving positive ions.

293. Conditions for the maintenance of a current by ions generated by collisions. The difference between the potential required to start a current in a uniform field and the potential required to maintain it may be explained by considering the relation connecting the value of α and β along the line of a discharge, which expresses the condition that a sufficient number of ions are generated by collisions to maintain the current.†

Let u be the velocity of the electrons, n_1 the number per cubic centimetre of the gas, v and n_2 the corresponding quantities for the positive ions. When a current is passing through the gas, the positive ions and the electrons are distributed unequally in the space between the electrodes, and this gives rise to a charge in the gas which disturbs the uniformity of the electric field between the plates. The force X at any point varies with the distance x from the negative electrode, so that u and v are not constant. Similarly, the quantities α and β vary from point to point in the gas. Since the velocities u and v are large, the rate of recombination $\theta n_1 n_2$ may be neglected in comparison with the rate at which ions are generated $\alpha n_1 u + \beta n_2 v$, when the currents are not very large. In the steady state n_1 and n_2 are constant with respect to the time and the equations of continuity become

* See Section 257.

† Phil. Mag. (6) **9**, p. 289, 1905; (6) **11**, p. 729, 1906.

$$\frac{dn_1}{dt} = -\frac{d}{dx}(n_1 u) + \alpha n_1 u + \beta n_2 v = 0,$$

and
$$\frac{dn_2}{dt} = \frac{d}{dx}(n_2 v) + \alpha n_1 u + \beta n_2 v = 0.$$

But $n_1 u + n_2 v = c$, ce being the current per unit area of the electrodes, which is a constant, and on substituting for $n_1 u$ the value $c - n_2 v$, the following equation for $n_2 v$ is obtained:

$$n_2 v = BZ(x) - cZ(x)\int_0^x Z(x)^{-1}\alpha\, dx,$$

where $Z(x)$ represents the integral $e^{\int_0^x (\alpha - \beta)\, dx}$.

Since all the ions are generated in the gas, the following conditions are obtained at the electrodes.

At the negative electrode $x = 0$, $n_1 = 0$, since there are no ions coming from the electrode. Similarly, at the positive electrode $x = S$, $n_2 = 0$. These conditions give $B = c$ and

$$1 = \int_0^S Z(x)^{-1}\alpha\, dx.$$

Hence the equation

$$\int_0^S \alpha e^{\int_0^x (\beta - \alpha)\, dx} dx = 1$$

represents the condition which must be satisfied by the values of α and β along the path of the discharge in order that a continuous current may be maintained. This may also be shown to apply to a discharge between conductors of any shape.

When the current is very small, so that the charge in the gas may be neglected, the field of force between the plates is uniform, and α and β are constant. In this case the above relation reduces to

$$1 = \frac{\alpha}{\beta - \alpha}\left(\epsilon^{(\beta - \alpha)S} - 1\right),$$

or
$$\alpha = \beta\epsilon^{(\alpha - \beta)S}.$$

This is the equation that determines the spark length when α and β are known, and the investigation shows that the sparking potential may be defined as the potential which is required to maintain a very small current.

294. Distribution of ions in a current. In order to determine the effect of the increase of current on the potential, it is necessary to consider how the force between the plates is affected by the separation of the positive ions from the electrons. The velocity of the positive ions is small compared with that of the electrons, so that there is a greater number of positive than negative ions in the gas at any time. This excess of positive electricity causes an appreciable disturbance in the uniformity of the field as the current rises, and since all the positive ions must pass through the gas at the negative electrode the positive charge is greatest in the neighbourhood of that electrode. The effect of this charge is to increase the force near the negative electrode and to diminish it in other parts of the field.

When the current is small and the force is approximately uniform, the distribution n_1 of electrons is given by the equation

$$n_1 u = \frac{c\beta}{\alpha - \beta}\left(e^{(\alpha-\beta)x} - 1\right).$$

A similar expression is obtained for the product $n_2 v$, and the ratio $n_2 v / n_1 u$ becomes

$$\frac{n_2 v}{n_1 u} = \frac{\alpha}{\beta}\,\frac{1 - \epsilon^{(\beta-\alpha)(S-x)}}{\epsilon^{(\alpha-\beta)x} - 1}.$$

The values of this ratio for a given force and pressure may be calculated from the values found for α and β, and it may be seen that $n_2 v$ exceeds $n_1 u$ for the greater part of the discharge. Hence, since v is less than u, the number of positive ions exceeds the number of electrons except at points in a layer of a gas near the positive electrode where the negative charge is not very large.

As an example, the currents in hydrogen at 9·27 millimetres pressure between plates 8 millimetres apart may be considered.* In this case the sparking potential is 487 volts

* See Section 225.

and the values of α and β are 6.03 and ·05 respectively. The values of the ratio n_2v/n_1u for different distances x from the negative electrode are as follows:

x	0	·1	·2	·3	·4	·5	·6	·7	·8
$\frac{n_2v}{n_1u}$	∞	143	50	22·7	11·1	5·3	2·0	·83	0
$\frac{en_2v}{i}$	1	1	·98	·96	·92	·84	·66	·45	0

Since v is less than u, the second line of figures shows that the positive charge n_2e exceeds the negative charge n_1e for most of the distance between the plates. Near the negative electrode, where n_1 is small compared with n_2, the positive charge is proportional to the numbers in the last line, which represents the ratio

$$n_2v/(n_1u+n_2v) = en_2v/i,$$

i being the current per square centimetre of the electrodes.

295. Effect of concentration of the force near the negative electrode. Hence there is a positive charge in the gas which tends to concentrate the force near the negative electrode, and as the current rises the uniformity of the field is disturbed. The variation of the force, given by the equation

$$dX/dx = 4\pi e\,(n_1 - n_2),$$

depends on the velocities of the ions and the electrons, so that a complete investigation becomes difficult and the field of force can only be represented by a complicated formula.

The decrease in the potential difference between the electrodes which accompanies the rise in the current may, however, be shown by an approximate method. It is due to the concentration of the force near the negative electrode; and this effect may be investigated by considering a field made up of two parts, in each of which the force is constant, the value of the constant being greater in the part near the negative electrode.

Let α and β have the values α_1 and β_1 corresponding to the force X_1 in the layer of gas of thickness a near the cathode, and the values α_2 and β_2 corresponding to a force X_2 in the remainder of the field of thickness b.

Negative electrode $x = 0$	$\leftarrow a \rightarrow$ α_1, β_1 X_1	$\leftarrow b \rightarrow$ α_2, β_2 X_2	Positive electrode $x = S = a + b$

In this case the condition for the maintenance of the current between the plates becomes

$$\frac{\alpha_1 - \beta_1 \epsilon^{(\alpha_1 - \beta_1) a}}{\alpha_1 - \beta_1} + \frac{\beta_2 - \alpha_2 \epsilon^{(\beta_2 - \alpha_2) b}}{\beta_2 - \alpha_2} = 1.$$

In order to apply this formula to a definite case, let the gas between the plates be hydrogen at 10 millimetres pressure, and let the force be 80 volts per millimetre in a layer extending to a distance of 2 millimetres from the negative electrode and 50 volts per millimetre in the rest of the field. The values of $\alpha_1, \beta_1, \alpha_2, \beta_2$ as obtained from the curves giving α/p and β/p in terms of X/p are as follows:

$$\alpha_1 = 10{\cdot}0, \quad \beta_1 = {\cdot}081, \quad \alpha_2 = 3{\cdot}6, \quad \beta_2 = {\cdot}021,$$

so that the distance b may be found from the above equation, a being given equal to ·2. The value of b thus obtained is ·8 centimetre, and hence the total fall of potential between the electrodes is $80 \times 2 + 50 \times 8 = 560$ volts, which is about 20 volts less than the sparking potential in hydrogen at 10 millimetres pressure between plates one centimetre apart.

Larger differences between the potentials are obtained when further increases are made in the force near the cathode. Thus if $X = 200$ volts per millimetre in a layer one millimetre thick at the negative electrode ($a = {\cdot}1$), and $X_2 = 40$ volts per millimetre in the rest of the field, a similar calculation shows that a continuous current will be maintained between the electrodes if $b = 3{\cdot}25$ millimetres, so that the fall of potential between the electrodes is 330 volts. This potential is 55 volts less than the sparking potential corresponding to the distance 4·25 millimetres between the plates.

Hence the potential required to maintain a current in the gas diminishes as the current increases, owing to the increase in the force near the negative electrode.

296. Simple experiments on the concentration of the force. The following experiments with rarefied air confirm the theory on which the explanation is based. Instead of depending on the charge on the positive ions to intensify the field near the negative electrode, a gauze of fine wires may be used to increase the force.

The rise of temperature due to the current, which also improves the conductivity at these pressures by reducing the density of the gas, is thus eliminated.

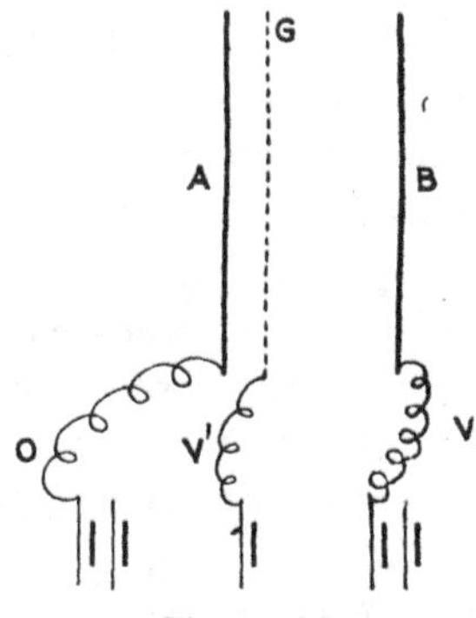

Figure 66.

Two electrodes *A* and *B*, figure 66, were set up in a glass tube and a grating of fine wire *G* was placed near the negative electrode, the connection to the grating being made by a wire sealed in a side-tube. A potential difference V' was established between the negative electrode and the grating, and the sparking potential was found by increasing the potential difference V between *A* and *B* until a discharge took place. The sparking potential was found to depend on the potential of the grating. The following table gives the values of V required to produce a discharge between the plates when the gauze was raised to the potential V':

V'	100	150	170	210	240
V	700	680	640	600	550

Thus the sparking potential between A and B was lowered by 150 volts by concentrating the force near the cathode.

This accounts for the principal characteristic of a spark passing between two large conductors. The conductivity improves as the current rises and the potential difference between the conductors after a spark has passed is reduced to a small fraction of the sparking potential.

297. Effect of the cathode fall of potential in a Geissler discharge. The theory which has been explained in the preceding sections may also be applied to the phenomena that occur near the negative electrode in discharge tubes.

It may be seen that as the current between two parallel plates continues to increase, a point is reached when no further reduction can be obtained in the potential v. Let a be the distance between the plates corresponding to the minimum sparking potential V' when the pressure is p. In the discharge between parallel plates at a distance apart S greater than a, the potential at some point at a distance y from the negative electrode will exceed that of the electrode by the potential V'. As before, in order to simplify the investigation, let the force have the constant value V'/y in the layer near the cathode and $(v-V')/(S-y)$ in the remainder of the gas. As the current increases, the force becomes concentrated near the cathode, so that y diminishes and approaches the value a. When y becomes equal to a the quantity $\alpha_1-\beta_1 e^{(\alpha_1-\beta_1)a}$ vanishes and a sufficient number of ions are generated in this layer to maintain the current. This stage corresponds to the case in which the normal fall of potential is established. A further increase of the current tends to narrow the layer y in which the ions are generated, but the fall of potential required to maintain a current in a layer narrower than a is greater than V', so that the potential v required to maintain the current tends to increase as the current exceeds a certain value.

298. Currents obtained between parallel plates. These

results may be compared with the properties of discharges between parallel plates when the current is maintained by a battery connected to the electrodes through a resistance. In general there are two distinct forms which the discharge may assume when the resistance R is large and the potential of the battery exceeds the sparking potential.

In the first type of discharge the current fills the space between the plates, and the potential difference v_1 between the electrodes is not much less than the sparking potential V. Also the current i per unit area of the plates has a small value, $i = (E - v_1)/RA$ (E being the potential of the battery and A the area of the plates), and the field of force between the plates is only slightly disturbed.

When R is diminished v_1 diminishes, and as the current increases the force becomes concentrated near the negative electrode, and eventually the normal cathode fall of potential is obtained.

The other type of discharge consists of a larger current, and the discharge as shown by a narrow luminous column is confined to a small area B of the plates. The potential difference between the electrodes v_2 is much less than v_1, and the current density $(E - v_2)/RB$ in the discharge is comparatively large. In this case the normal fall of potential is developed in the narrow luminous column, and as R diminishes v_2 remains approximately constant, but the section of the current increases and eventually the discharge fills the whole space between the plates.

The stability of these discharges depends on the value of R; with the smaller resistances, the discharge is usually of the second type and represents what takes place in a discharge tube under ordinary conditions before the cathode glow covers the whole electrode.

As the resistance increases, the section of the current contracts and a point is reached when the width of the section becomes comparable with the length of the cathode dark space. The charge in the gas is then insufficient to maintain the cathode fall at its normal value, and the potential required to maintain the discharge increases. Thus the current may be stopped completely by increasing the resistance unless the potential of the battery is as great as the sparking potential.

299. Effect of the nature of the electrode on the cathode fall of potential. The above theory also applies to the phenomena that take place in the neighbourhood of a plane cathode in a discharge tube, when the distance of the negative glow from the electrode is small compared with the linear dimensions of the electrode and the number of ions that come into contact with the surface of the tube may be neglected.

It is thus possible to account for the development of the normal cathode fall of potential, which in ordinary cases is nearly the same as the minimum sparking potential,* but it would obviously be impossible to obtain a complete explanation of these phenomena without taking into consideration the rise of temperature of the gas and the distribution of the force between the Faraday dark space and the negative electrode. Also, in a complete theory it would be necessary to estimate the effect of electrons set free from the cathode as well as those generated from molecules of the gas, since the cathode fall of potential depends on the nature of the cathode, being much less with alkali metals than with the metals of which electrodes are usually made.

With the alkali metals, large numbers of electrons are set free from the surface by the action of light when under similar conditions no effect is obtained with zinc or aluminium electrodes. It is possible that, with alkali cathodes, electrons are set free by the impacts of positive ions travelling with velocities much smaller than those required to set free electrons from other metals or from molecules of the gas. Under these conditions a sufficient number of ions to maintain a current would be obtained with a smaller cathode fall of potential than is required in those cases in which a small proportion of the electrons is supplied by the surface of the cathode. Hittorf observed a similar effect

* See Section 297.

with cathodes raised to a high temperature, the cathode fall being reduced to a very small value. This is no doubt due to the fact that electrons emitted by the cathode contribute largely to the current.

Since the force is greatest at the surface of the negative electrode when the normal cathode fall of potential is established, the velocity with which the positive ions collide with the negative electrode is greater than the velocity with which they collide with molecules of the gas, and also greater than the velocity of the positive ions when a discharge in a uniform field is produced with the minimum sparking potential.

The effect of electrons set free from the electrode would be greater in the former case, and for this reason the cathode fall of potential may depend on the nature of the electrode, while the minimum sparking potential is practically independent of the electrodes when ordinary metals such as copper and zinc are used. These considerations show that at very low pressures, when sparking is obtained with high potentials, and the electrons liberated from the surface of the cathode by the impacts of the positive ions contribute largely to the current, the potentials would depend on the nature of the electrode.

The various determinations of the cathode fall of potential are not in very good agreement; but the order of the forces involved may be obtained from the numbers given by Aston for a current of 10^{-4} ampère per square centimetre. With aluminium plates the cathode fall of potential in hydrogen is 266 volts, and the dark space one centimetre wide at a pressure of ·46 millimetre of mercury. The force in the dark space is proportional to the distance from the cathode glow, so that at the negative electrode the force is 532 volts per centimetre. With the same pressure, ·46 millimetre, the minimum sparking potential for parallel plates (which is about 275 volts) is obtained when the distance between the plates is between 2·5 and 3 centimetres, so that the force in the space between the plates would then be of the order of 100 volts per centimetre.

300. Current in the positive column. The supply of ions necessary for a continuous current in a discharge tube is maintained by the ionization that takes place at the cathode end of the tube, and the phenomena that take place in the rest of the tube may be treated as if a stream of electrons is supplied to the gas at a certain point P which may be considered to be at the end of the positive column or at a short distance from it in the Faraday dark space.

The effect produced by inserting an electrode at P so as to divide the tube into two sections shows that the stream of positive ions entering the Faraday dark space is small compared with the total current, and the conductivity at the cathode end of the tube is not appreciably affected by cutting off the positive stream.

In the original discharge between the two electrodes A and B, let V_1 be the potential at the point P, the potential of the negative electrode A being zero, and that of the positive electrode B being V. If the electrode inserted at P is maintained at the potential V_1 by a battery connection, the current will flow from A to P as before, but in general no current will flow from P to B. (In air at one millimetre pressure the force in the positive column is about 35 volts per centimetre, so that if the positive column is less than 9 centimetres long the potential difference $V - V_1$ would be less than that required to maintain a current under any conditions.) Since positive ions are not given off from metals at ordinary temperatures, there are no positive ions entering the gas at P in the discharge between the electrodes at A and P; it follows that the number of positive ions crossing the section at P in the original discharge is very small.

The conductivity of the gas in the positive column cannot be investigated by the same method as in the space near the cathode, by comparing the phenomena to those obtained between parallel plates, where the sides of the vessel have no effect on the currents. With a plane cathode the widths of the dark space and of the negative glow are small compared with the diameter of the tube or of the cathode when the pressure is sufficiently large, and the dark space and the glow are not affected by the length of the discharge tube. But with the positive column, the length of the column may be much greater than the diameter of the tube, and the force is the same at all sections when the column is uniform. This effect must be due to the charges on the sides of the tube and in the volume of the gas, which become distributed in such a manner as to give rise to an electric force whose component parallel to the axis is constant.

If X be the force parallel to the axis of the tube, and R be the force along the radius arising from a charge in the gas, the difference between the numbers of positive ions and electrons at any point of the gas is given by the equation

$$4\pi e\,(n_1 - n_2) = \frac{1}{r}\frac{d}{dr}\left(r\frac{dR}{dr}\right) + \frac{dX}{dx}.$$

301. Number of ions in the positive column. The simplest case is that of the uniform positive column in which the force X is constant. If it be assumed that R is small, the numbers n_1 and n_2 become approximately equal.

For a first approximation n_1 and n_2 may be taken as being equal, in order to find the order of the quantity n_1. Since the electrons move with a velocity which is large compared with that of the positive ions, $n_2 v$ may be neglected in comparison with $n_1 u$ and the current i per square centimetre of the tube becomes $i = e n_1 u$. Taking as an example the case of a current 10^{-3} ampère in air in a tube 2·1 centimetres in diameter, the force X is 35 volts per centimetre when the pressure is one millimetre of mercury. The velocity of the electrons in dry air, when $X/p = 35$, is $1{\cdot}3 \times 10^7$ centimetres per second, which gives $n_1 = 1{\cdot}5 \times 10^8$ (the value taken for e being $4{\cdot}5 \times 10^{-10}$ electrostatic unit).

The number of positive ions is approximately equal to the number of electrons throughout the volume of the gas, except at the ends of the column. In the Faraday dark space there is a negative charge in the gas, as is shown by the variation of electric force, and at the positive electrode, since there are no positive ions moving from the electrode into the gas, the number of electrons must exceed the number of positive ions. Owing to the negative charge in the gas, the force at points near the positive electrode is greater than the average force in the positive column, and a comparatively large number of molecules are ionized near the positive electrode. The positive ions thus produced move into the gas and the number per cubic centimetre becomes approximately equal to n_1 at a short distance from the electrode.

302. General theory of the uniform positive column. Thus up to a certain point some of the properties of the discharge may be explained, but no satisfactory method has been given of calculating the value of the force X from theoretical considerations.

The chief difficulty arises from a want of any precise knowledge as to what takes place when the electrons and positive ions come into contact with the surface of the glass tube.

If n_1 represents the mean value of the number of electrons per cubic centimetre of the gas over a section of the tube at a distance x from the positive electrode, the equation of continuity when the steady state is reached becomes

$$-\frac{d}{dx}(n_1 u) = \alpha n_1 u + n_1 Y - \theta n_1 n_2 - n_1 S_1,$$

α being the number of molecules ionized by an electron per centimetre of its path, $\theta n_1 n_2$ the rate at which electrons disappear by recombination with positive ions, $n_1 S_1$ the rate at which they disappear by contact with the sides. Since the force is small, the effect of positive ions in generating others by collisions may be neglected; but if ions are generated by other processes, such as radiation from the gas or from the sides of the tube, it is necessary to include them, so that the term $n_1 Y$ is added to represent the rate at which electrons are generated by such processes.

A similar equation is obtained from the equation of continuity for positive ions

$$\frac{d}{dx}(n_2 v) = \alpha n_1 u + n_1 Y - \theta n_1 n_2 - n_2 S_2,$$

the first three terms on the right of the equation being the same as before, since positive ions and electrons are generated simultaneously and also disappear together when they recombine. The last term, $n_2 S_2$, represents the rate at which positive ions are lost by colliding with the surface. It is different in form from the term $n_1 S_1$ occurring in the first equation, but the two terms must be equal. Since the current $i = e\,(n_1 u + n_2 v)$ must be the same at all sections of the tube when the steady state is reached, the differential coefficient di/dx must vanish. But the equations of continuity give

$$\frac{d}{dx}(n_1 u + n_2 v) = n_1 S_1 - n_2 S_2.$$

Hence $n_1 S_1 = n_2 S_2$.

The rates at which positive ions and electrons are lost by colliding with the surface of the tube must therefore be the same at each section of the discharge. But the rate of diffusion of the electrons is much greater than that of the positive ions, and if n_1 were equal to n_2 the electrons would probably be lost more rapidly than the positive ions, unless they rebound from the surface. In order that the current may be the same at all sections, it would be necessary to suppose that there is a positive charge in the gas which repels positive ions towards the surface. Thus the number n_2 is probably slightly greater than n_1.

The appearance of the glow and the conductivity will be the same at each section of the tube if the differential coefficients $d(n_1 u)/dx$ and $d(n_2 v)/dx$ vanish, and the equations of continuity then reduce to

$$\alpha u + Y - \theta n_1 - S_1 = 0.$$

The terms of this equation involve the force X, the pressure of the gas and the dimensions of the tube, so that it is possible to obtain a mathematical equation to determine X. The term Y may also involve the current, and the other terms, since they depend on the temperature of the gas, are affected indirectly by the current.

The term θn_1 is usually very small, and the effect of recombination in discharges at low pressures may as a rule be neglected.

Thus, taking as an example the positive column in air at one millimetre pressure when the current is 10^{-3} ampère in a tube 2·1 centimetres in diameter, the number of electrons or positive ions per cubic centimetre as found above is about $1{\cdot}5 \times 10^8$. According to Langevin's investigations the rate of recombination of positive and negative ions is proportional to the pressure, the coefficient θ at atmospheric pressure being 3200 e. If the coefficient of recombination of positive ions and electrons were of the same order as the coefficient of recombination of positive and negative ions, then θn_1 in the above equation would be of the order ·3, which is very small compared with αu, since $u = 1{\cdot}3 \times 10^7$ and α is of the order ·01 when $X = 35$ and $p = 1$. (The value of α corresponding to $X/p = 35$ has not yet been determined accurately, and in this case it would depend on the temperature, since the density diminishes as the gas becomes heated.)

303. H. A. Wilson's investigation of the uniform positive column. An investigation of the positive column has been made by H. A. Wilson,* who has adopted the hypothesis that the gas is ionized by a radiation arising from the recombination of positive and negative ions. That such radiations may exist is shown by the researches of E. Wiedemann † and J. J. Thomson, ‡ but their investigations were not carried so far as to indicate to what extent the ionization in the positive column may be due to this effect.

The cases considered by Wilson were those in which the currents were very small (of the order 10^{-8} ampère). The loss of ions by diffusion to the sides was not taken into account, and the condition for uniformity along the axis was

* H. A. Wilson, Phil. Mag. (6) **6**, p. 180, 1903.
† E. Wiedemann, Zeitschr. f. Electrochemie, p. 159, 1895.
‡ J. J. Thomson, Proc. Camb. Phil. Soc. **10**, p. 74, 1899.

obtained by equating the loss due to recombination to the ionization produced by radiation and the collisions of electrons with molecules. An equation of the same type as that given above was thus obtained, S_1 being taken as zero. Wilson found it necessary to assume that the remaining three terms were of the same order of magnitude, so that it is possible to estimate the order of the coefficient of recombination of positive ions and electrons which is thus obtained. Thus in air at ·66 millimetre pressure when the force is 54 volts per centimetre in the positive column and the current is 3×10^{-9} ampère, the quantity θn_1 is of the same order as αu. In this case $\alpha = \cdot 28$, $u = 2\cdot 4 \times 10^7$, and the current $n_1 eu = 9$, e being expressed in electrostatic units. Hence θ is of the order $\frac{\alpha u^2 e}{nue} = 1\cdot 8 \times 10^{13} \times e$. Since this number is so large compared with $4 \times e$, the coefficient of recombination of positive and negative ions in air at one millimetre pressure, it is difficult to consider that the investigations are conclusive while the results remain unconfirmed by further experimental evidence.

If the rate of recombination of positive ions and electrons is as rapid as the above number indicates, the quantity θn_1 would be very large compared with αu when the currents are of the order 10^{-3} ampère, and in order to supply the loss due to recombination it would be necessary to assume that the ionization in the positive column is of an order very much higher than that obtained by collisions.

304. Thomson's theory of the uniform positive column. Another theory of the positive column has been given by Thomson* which is founded on a principle that he has introduced in his description of the theory of ionization by collision and applied to various cases of conductivity in rarefied gases. The feature of his hypothesis, which is different from those adopted by other physicists, is that the electrons are supposed to disappear by combining with

* J. J. Thomson, 'Conduction of Electricity through Gases', 1906 edition, pp. 271, 274, 497, 602.

neutral molecules of the gas. The loss of electrons by this process is considered to be much greater than the loss by recombination with positive ions, the latter effect being neglected in the investigations of currents at low pressures. An electron which has combined with a neutral molecule is treated as if it ceases to act as an ionizing agent, and also it ceases to be available for carrying the current.

The effects taking place in the positive column are obtained by an adaptation of the method applied to the currents between parallel plates when ultra-violet light falls on the negative electrode. In the positive column of a discharge tube, electrons are supplied from the Faraday dark space and a certain proportion are lost by diffusion to the sides of the tube, but with parallel plates there is no loss by diffusion, so that the conditions are simpler. In the latter case, when the force is small, ionization by the collisions of positive ions may be neglected, and the equation of continuity for the electrons when the steady state is established becomes

$$\frac{d}{dx}(nu) = nu\,(A - B);$$

n being the number of electrons per cubic centimetre, u the velocity in the direction of the electric force, nuA the number of electrons generated per cubic centimetre per second in the gas, and nuB the number of electrons that disappear per cubic centimetre per second by combining with neutral molecules.

The above equation gives

$$nu = n_0\,u\epsilon^{(A-B)\,x},$$

nue being the quantity of negative electricity passing in unit time through unit area of a plane at right angles to the force, at a distance x from the origin 'and can be measured by a metal plate at this distance connecting it with an electrometer and measuring by means of this instrument the rate at which negative electricity is reaching the plate'.

One objection to this formula is that when A is less

than B (A varies with the force and B is apparently independent of the force, so that with some forces A is less than B) the charge acquired per second by the electrometer diminishes with the distance x when the supply $n_0 u$ at the origin $x = 0$ is constant. There is no experimental evidence in support of this result, as it has been found that in the conductivity produced by ultra-violet light the charge acquired by the positive electrode increases with the distance between the plates, except when very small forces are used, and the charge received per second is then constant and equal to $n_0 u$.

It may also be seen that the currents obtained by this theory are not the same at all distances from the electrodes when the steady state is reached. For if m be the number of positive ions per cubic centimetre and v their velocity under the electric force X, the equation of continuity for the positive ions when the steady state is reached is

$$-\frac{d}{dx}(mv) = nuA,$$

since positive ions are generated simultaneously with electrons and are not lost by combining with neutral molecules. The variation of the current i with the distance x is given by the equation

$$\frac{d}{dx}\left(\frac{i}{e}\right) = \frac{d}{dx}(nu + mv) = -nuB,$$

which shows that i is not constant, but diminishes with the distance x from the negative electrode. As there are no electrical discharges in which the current at the positive electrode differs from the current at the negative electrode when the steady state is attained, the general principle of the theory must be considered unsatisfactory.

In the investigation of the positive column in a discharge tube, the above equation of continuity for negative electricity is adopted and a term is added to represent the loss of electrons by diffusion to the sides of the tube. When the variation of the negative stream, $d(nu)/dx$, along the axis of the tube vanishes, the force X is obtained by equating to zero the remaining terms in the equation. Assuming that the quantity A which represents the rate at which electrons are produced is proportional to X, and that B, the rate at which they disappear by combining with neutral molecules, is independent of X, Thomson finds that the force X in the positive column is given by the equation

$$\frac{X}{p} = C + \frac{D}{d^2 pX},$$

C and D being constants, and d the diameter of the tube. Thus when d is large, the force X is proportional to p.

It is very improbable that the solution of the problem is represented by this equation. For when d is large the problem is reduced to the investigation of conductivity between parallel plates, in which the diffusion to the sides is neglected, the quantities A and B are equal, and the current through a section at a distance x from the negative electrode becomes $i = i_0 - nueBx$ which diminishes with x. The general objections which have been raised to the theory may be avoided by equating C (which is proportional to B) to zero, but the equation then gives a value of the force $X = \sqrt{D}/d$ which is independent of the pressure.

305. Application of the collision theory to positive and negative discharges from points and cylinders. The theory which has been given to account for the difference between the sparking potential and the potential required to maintain a current between parallel plates also explains the difference between the potentials required to produce positive and negative discharges from points or from the surface of a cylinder.

Thus in the currents through air at atmospheric pressure from a cylindrical surface or from a rounded end of a wire, the starting potential is in some cases less for positive than for negative discharges, but when the pressure is reduced a negative discharge is obtained with a much smaller potential than a positive discharge.

The equation which expresses the condition that a discharge should be maintained by ions generated in the field of force between two conductors has been given in Section 293. If the integration be taken along a line of force S from one electrode A at which $s = 0$ to another electrode B at which $s = S$, the equation becomes

$$1 = \int_0^S \alpha \epsilon^{\int_0^s (\beta - \alpha)\, ds} ds,$$

if A is the negative electrode.

When the forces are reversed, and A becomes the positive electrode, the corresponding condition is obtained by interchanging the quantities α and β. The two conditions are not the same, since the integral is not symmetrical with respect to α and β. It follows that in general the potential required to start a negative discharge is not the same as the potential required to start a positive discharge.

Taking the values of α and β that have been determined experimentally, it may be seen by the method employed in Section 295 that if the condition for sparking is satisfied when the larger values of α and β are in the field near the negative electrode, the forces will not be sufficient to maintain a discharge when they are reversed in direction.

This result is not due to the fact that β is small compared with α, but it depends rather on how rapidly the quantities α and β increase with the force. This point may be examined by considering whether it is more advantageous to increase the force near the negative electrode or near the positive electrode when the potential difference between the two electrodes is not quite sufficient to produce a discharge. If α and β be increased to $\alpha + \delta\alpha_1$ and $\beta + \delta\beta_1$ for a distance δS near the positive electrode, the condition for sparking then becomes

$$1 = \int_0^S \alpha \epsilon^{\int_0^s (\beta - \alpha)\, ds} ds + \epsilon^{\int_0^S (\beta - \alpha)\, ds} \delta\alpha_1 \delta S.$$

It will be observed that in this case no advantage is gained by the increase $\delta\beta_1$.

If α and β are increased to $\alpha + \delta\alpha_2$ and $\beta + \delta\beta_2$ for a distance δS near the negative electrode, the condition for sparking becomes

$$1 = \delta\alpha_2 \delta S + \int_0 \alpha \epsilon^{(\delta\beta_2 - \delta\alpha_2)\,\delta S + \int_0^s (\beta-\alpha)\,ds} ds.$$

Multiplying by the factor $1 + (\delta\alpha_2 - \delta\beta_2)\,\delta S$, this equation becomes

$$1 = \int_0^S \alpha \epsilon^{\int_0^s (\beta-\alpha) ds} ds + \delta\beta_2 \delta S.$$

Hence it will be more advantageous to increase the force near the negative electrode if $\delta\beta_2$ exceeds

$$\delta\alpha_1 \epsilon^{\int_0^S (\beta-\alpha)\,ds}.$$

For simplicity, the case of a uniform field may be considered. When the potential approaches the value required to produce a spark, the integral $\epsilon^{\int_0^S (\beta-\alpha)\,ds}$ is approximately equal to β/α, so that a small increase in the force near the negative electrode will have more effect than an equal increase near the positive electrode if $\frac{\delta\beta}{\delta X}$ exceeds $\frac{\beta}{\alpha}\frac{\delta\alpha}{\delta X}$, or if

$$\frac{1}{\beta}\frac{d\beta}{dX} > \frac{1}{\alpha}\frac{d\alpha}{dX}.$$

It will be found that with the values of α and β that have been given in Chapters VIII and IX this condition is satisfied; thus when

$$X/p = 325, \; \frac{1}{\alpha}\frac{d\alpha}{dX} = \frac{\cdot 0036}{p}, \text{ and } \frac{1}{\beta}\frac{d\beta}{dX} = \frac{\cdot 0140}{p}.$$

Hence when the forces in the field in which the ions are generated are of the same order as those for which these determinations have been made, that is when X/p exceeds 150, the negative discharges from points or cylinders will probably be obtained with smaller potentials than the positive discharges.

306. Positive and negative discharges between coaxial cylinders. The forces in the field in which the ions are generated may be determined when the discharge takes place between coaxial cylinders. At the higher pressures the ionization takes place within a short distance of the inner cylinder. If a be the radius and E the charge per unit length of the inner cylinder, the ionization takes place at points where the force exceeds a certain value $2E/c$, and the ratio of the maximum to the minimum force in the field is c/a. The ratios c/a for cylinders of different diameters may be obtained from the values of a and $c-a$ given in Section 253 for discharges through air at 760 mm. pressure. Thus for a cylinder of 5 millimetres radius the ratio c/a is 1·33, and for a cylinder ·5 millimetre radius the ratio increases to 2.5. The largest force is 75 kilovolts per centimetre at the surface of the wire of ·5 millimetre radius, and the corresponding value of X/p is 100 approximately.

In these cases there was very little difference between the starting potentials for positive and negative discharges.

At low pressures of the order of one millimetre, the ionization takes place throughout the whole space between the inner and outer cylinder, and the ratio of the maximum to the minimum force in the field is A/a, A being the radius of the outer cylinder. In the experiments on the discharges through gases at low pressures described in Section 259, it was found that with an outer cylinder of fixed diameter, 4 centimetres, there was a considerable difference between the potentials required to produce positive and negative discharges which was very marked when the inner cylinder was of small diameter. The curves, figure 62, show that even where the inner cylinder is comparatively large, 15·95 millimetres in diameter, the potentials differ by 44 volts when the air is at one millimetre pressure. In this case the ratio A/a was 2·5, and the value of X/p at the surface of the inner cylinder was 480, corresponding to the potential 354 volts between the cylinders.

The values of X/p in these cases come within the range of values for which it may be shown that $\frac{1}{\beta}\frac{d\beta}{dX}$ exceeds $\frac{1}{\alpha}\frac{d\alpha}{dX}$.

307. Positive and negative discharges from points. In point discharges, as in the discharges between coaxial cylinders, the difference between the potentials required to produce positive and negative discharges depends on the values of the quantity X/p in the field of force.

In air at atmospheric pressure, the potential required to start a discharge is less for sharp points than blunt points, but the force at the sharp point is greater than that at the blunt point. This may be shown by considering the potentials given in Section 263 for points of definite shape obtained by rounding the ends of wires of different diameters. Thus the potential required to start a positive discharge from the end of a wire ·174 millimetre in diameter to a plate at 1 centimetre distance is 2,850 volts, and for a wire ·244 millimetre in diameter at a distance 1·5 centimetres from the plate the starting potential is 3,375 volts. The ends of the wires were hemispherical, so that the surfaces were exactly similar, and since the distances from the plate are nearly in the same ratio as the radii of the wires, the fields of force near the points in the two experiments were almost exactly similar. Under these conditions the forces at corresponding points are directly proportional to the potentials and inversely proportional to the linear dimensions. Hence the ratio of the force at the surface of the small point to that at the surface of the large point is 1·19 : 1. Thus the ratio of the maximum to the minimum force in the field in which the ions are generated increases as the radius of the point diminishes, the minimum in these cases being taken as 30 kilovolts per centimetre in air at atmospheric pressure. Also, the average value of the force in the field in which the ions are generated is larger with sharp points than with blunt points.

At atmospheric pressure, the potential required to produce a positive discharge from a sharp point or from the end of a fine wire is greater than the potential required to produce a negative discharge, but with a blunt point, such as the rounded end of a wire ·5 millimetre in diameter, the two potentials are the same. This difference between the sharp points and blunt points is due most probably to the fact that the force at the end of the sharp point is greater than the force at the end of the blunt point.

The following experiment shows that the nature of the phenomena is determined by the values of the ratio X/p. If the rounded end of a wire ·5 millimetre in diameter be set up at a distance of 1·5 centimetres from a plate, the following are the values of the potentials required to produce positive and negative discharges at different pressures:

Pressure of air in millimetres	763	548	352	106	31	11
Starting potential in kilovolts for positive discharges	4·78	4·20	2·96	1·64	1·05	·850
Starting potential in kilovolts for negative discharges	4·80	4·32	3·01	1·49	·76	·610

Thus the potentials diminish, and a considerable difference between the positive and negative potentials is obtained when the pressure is reduced, the changes being similar to those obtained when the pressure is constant and the diameter of the wire is reduced.*

This result is in agreement with the general theorem given in Section 251, from which it follows that the force required to produce a discharge from the end of a wire of definite shape is connected with the diameter a and the pressure p by a simple relation. If F_+ and F_- be the forces required to produce positive and negative discharges, then

$$aF_+ = \phi_1 (ap), \text{ and } aF_- = \phi_2 (ap).$$

Hence the ratio $\frac{F_+}{F_-} = \phi_3 (ap)$, which shows that if a difference in the two forces is obtained by reducing the radius, a similar difference must be obtained by reducing the pressure.

Also, it is easy to see by the method used in Section 260 that as the pressure is reduced the potential required to produce a discharge between any two surfaces cannot fall off as rapidly as the pressure, hence in all cases of discharge the value of X/p at the surface of a conductor increases as

* See tables of starting potentials given in Section 263.

the pressure is reduced. When the starting potential for negative discharges from points is less than that for positive discharges, the average value of the ratio X/p is comparatively large. But as X/p diminishes, the potentials become equal, and when ap exceeds a certain value the positive potential is slightly less than the negative. It will be seen from Section 305 that these effects obtained by reversing the forces depend on the relative values of the quantities $\frac{1}{\beta}\frac{d\beta}{dX}$ and $\frac{1}{\alpha}\frac{d\alpha}{dX}$, and the conclusions to be drawn from the experiments with cylinders and with points are that for the larger values of X/p the ratio $\frac{1}{\beta}\frac{d\beta}{dX}$ exceeds $\frac{1}{\alpha}\frac{d\alpha}{dX}$, but with the smaller values of X/p (from 40 to 100) the two ratios become nearly equal. The latter result shows that for a certain range of forces and pressures the ratio α/β is nearly constant.

CHAPTER XII

CATHODE RAYS AND POSITIVE RAYS

308. Early experiments with cathode rays. The discovery of the properties of the cathode rays has been an important factor in establishing the modern theory of electrical phenomena. According to Kelvin * the cathode rays were discovered by Varley in 1871 and rediscovered eight years later by Crookes. In Varley's experiments a discharge was passed through hydrogen between two electrodes in the form of rings, and from the effects obtained in a magnetic field on the distribution of the glow in the gas he concluded that 'attenuated particles of matter' were projected from the negative pole in all directions. The experimental evidence in favour of this hypothesis, which was given by Varley,† is not very convincing, and his experiments on the effects of a magnetic field on the discharge are not so extensive as those previously made by Plücker ‡ and Hittorf.§

309. Crookes' investigation of the velocity of the rays. A marked advance was made by Crookes,‖ who studied the process of the development of the cathode rays as the pressure is reduced, and made the first investigations of some of their well-known properties.

In one series of experiments, the relative velocities of rays obtained in gases at various small pressures were

* Lord Kelvin, Phil. Mag. (6) **8**, p. 534, 1904.
† C. F. Varley, Proc. Roy. Soc. **19**, p. 238, 1871.
‡ J. Plücker, Pogg. Ann. **103**, p. 88, 1858.
§ W. Hittorf, Wied. Ann. **134**, pp. 1 and 197, 1869.
Full accounts of Plücker's and Hittorf's researches are given in 'Ions, Électrons, Corpuscules'.
‖ W. Crookes, Phil. Trans. **170**, p. 135, 1879.

investigated. The apparatus used consisted of a vacuum tube with a plane aluminium electrode, in front of which was placed a mica screen with a small hole in the centre. The plane of the screen was parallel to the surface of the electrode and distant 2·5 centimetres from it. The rays which were projected through the aperture in the mica were received on a glass plate, and the small area on which they impinged was marked by a green fluorescence.

The narrow beam of rays passing through the aperture in the mica was deflected by a transverse magnetic force, and the effect was observed by the displacement of the spot of light on the glass plate. Measurements were made of the displacements produced by a magnet in a fixed position when the discharge passed through gases at different pressures.

With air in the tube, no rays were observed to pass through the aperture until the pressure was reduced to ·077 millimetre. At this pressure the dark space between the cathode and the negative glow was 12 millimetres wide, and the displacement of the luminosity on the plate by the fixed magnet was 13·3 millimetres.

When the pressure was reduced to ·023 millimetre (dark space 16 millimetres), the spot of light became more brilliant and more sharply defined, the magnetic displacement being 11·5 millimetres.

With a pressure ·02 millimetre (dark space 25 millimetres), the magnetic displacement was 9·8 millimetres, and with a pressure ·007 millimetre the displacement was ·3 millimetre. In the latter case the negative glow had disappeared.

Hence the magnetic deflection diminishes as the pressure is reduced, which shows that the velocity of the particles increases.

The high velocity acquired by the particles was shown by the rise of temperature which is produced when the rays impinge on a solid body. When a hollow cathode is used to focus the rays on a small area, a strip of platinum may be raised to a white heat.

310. Perrin's experiment to determine the charge. Thus, according to Varley and Crookes, the cathode rays are formed of material particles charged with negative electricity, which are repelled from the cathode and acquire a very high velocity in the electric field; but many physicists did not accept this explanation of the phenomena, and maintained that the rays were a form of undulatory motion of the ether. The corpuscular theory depends on the hypothesis that the rays are negatively charged, and the first experiments which were made to detect the charge were inconclusive, and in some cases gave quite unexpected results.

This problem was successfully investigated by Perrin,* and his experiments were generally accepted as proving that the cathode rays consisted of negatively charged particles travelling with a high velocity. In the apparatus

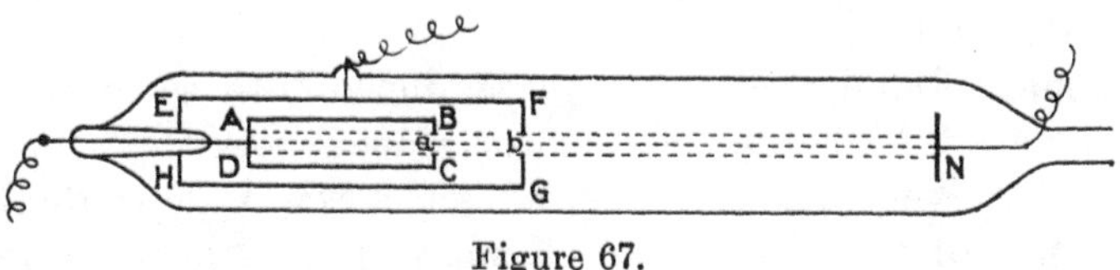

Figure 67.

used by Perrin (figure 67), the cathode rays entered an insulated metallic cylinder *ABCD* through a small aperture *a*, the cylinder being screened from external charges by an outer cylinder *EFGH* maintained at zero potential.

Two small apertures were bored in the larger cylinder, one *b* facing the cathode *N*, and the other at the opposite end of the cylinder, through which the inner cylinder was connected with an electroscope. The outer cylinder served as the positive electrode, and when a discharge was passed through the tube some of the cathode rays from *N* passed through the apertures *a* and *b*, and the inner cylinder acquired a negative charge.

When the discharge was produced by an induction coil,

* J. Perrin, Comptes rendus, **121**, p. 1130, Dec. 1895; Ann. de Chim. et de Phys. (7) **11**, p. 496, Aug. 1897.

the charge acquired by the inner cylinder was in some cases as great as 600 electrostatic units for each interruption of the primary circuit of the coil.

The cathode rays may be deflected from the axis of the tube by a transverse magnetic field, so that they shall not fall near the aperture *b*. The deflection of the rays was observed by a fluorescent powder on the surface *FG* of the outer cylinder, and when the luminous area was displaced from the centre the electroscope connected to the inner cylinder remained uncharged when the discharge passed through the tube.

In the absence of the magnetic field, the inner cylinder became charged when the aperture at *b* was a small pin-hole, or when the aperture was covered by thin metal foil.

(The penetration of the cathode particles through thin sheets of metal was discovered by Hertz, and this property of the rays was used by Lenard * to obtain rays in the air outside the tube.

Lenard found that the rays produced intense ionization in the air. A charged conductor placed near the tube lost its charge very quickly, and the effect was independent of the sign of the charge, which shows that the numbers of positive and negative ions in the gas were practically the same. The number of cathode particles which traverse the gas is therefore small compared with the number of molecules that are ionized, otherwise a conductor would have lost its charge more rapidly when it was positively charged than when it was negatively charged.)

In Perrin's experiments the gas was at a very low pressure, and comparatively few ions were generated in the space between the two cylinders. In this case, therefore, the negative charge of the cathode rays is the principal effect; but as the inner cylinder becomes charged the positive ions are attracted to it, and the direct effect of the cathode rays is reduced.

Many experiments have been made to show that the rays are deflected by an electric force. The method adopted

* See Section 46.

by Perrin may be seen by the arrangement of the discharge tube shown in figure 68 *a*. The cathode rays, after passing through a perforated positive electrode, traversed the gas in the large sphere and cast a shadow of the wire *W* on the surface of the sphere. When the wire was charged negatively, the rays were deflected as shown in figure 68 *b*, and when the wire was charged positively the deflection was as shown in 68 *c*.

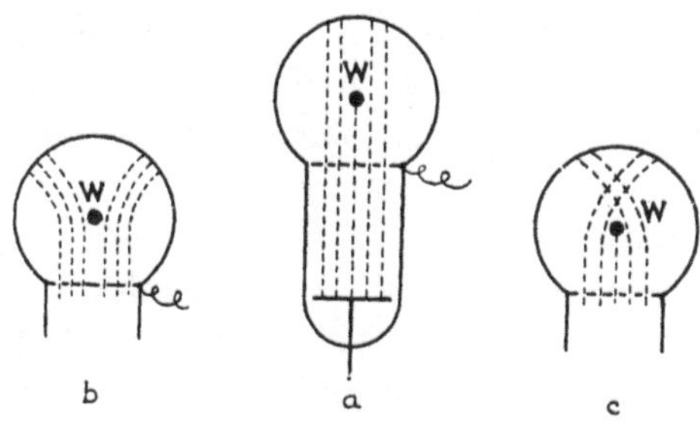

Figure 68.

311. Secondary cathode rays. When the cathode rays fall on a metal surface Röntgen rays are generated, also secondary rays consisting of negatively charged particles are emitted in all directions from the metal. The secondary rays may be observed in a side tube connected to the main discharge tube, by placing a metal plate in the path of the cathode rays opposite the opening to the side tube. These rays were observed by Thompson,* who found that they were of the same nature as the cathode rays, since they cast a shadow of a wire on the surface of the glass, and were easily deflected by a magnetic force.

The properties of the secondary cathode rays were also investigated by Campbell Swinton,† and the negative charge carried by the rays was demonstrated by Perrin's method, in which the rays enter an insulated cylinder contained inside another cylinder maintained at zero potential. The reflection of a beam of rays was thus found

* S. P. Thompson, Phil. Trans. **190**, p. 471, 1897.
† A. A. Campbell Swinton, Proc. Roy. Soc. **64**, p. 377, 1899.

to be largely diffuse, some particles being emitted in all directions; but there was a considerable concentration along the direction in which an ordinary beam of light would have been reflected.

The charge received by the reflector depends on the inclination of the primary beam to the normal. When the incidence is normal the metal receives a negative charge which diminishes as the angle of incidence increases, and becomes zero when the angle is about 45°. When the rays are inclined at large angles to the normal the reflector becomes positively charged, which shows that in this case the number of electrons in the secondary rays is greater than the number of the primary absorbed by the metal.

Lenard * found that the diffuse secondary rays from a metal surface are emitted with comparatively small velocities, of the same order as that with which the electrons are given off from a surface under the action of ultra-violet light.

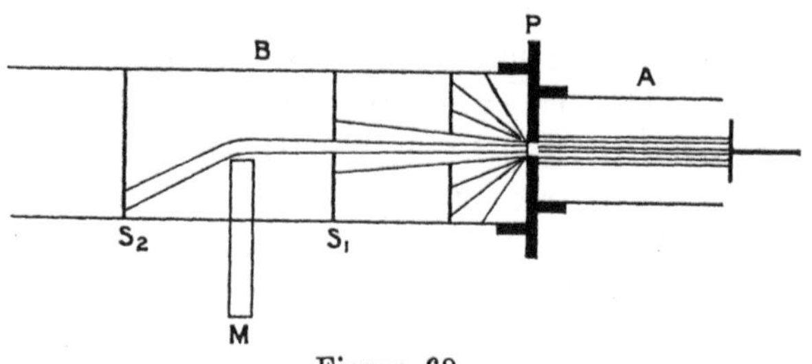

Figure 69.

312. Passage of the rays through gases. The extent to which the cathode rays penetrate gases and thin sheets of metal has been investigated by Lenard,† who found that the distance to which the rays penetrate is inversely proportional to the density of the medium. The principle of the method adopted by Lenard ‡ to investigate the velocity is shown in figure 69. The two tubes A and B were fixed on opposite sides of a metal plate P, with

* P. Lenard, Ann. der Phys. (4) **15**, p. 485, 1904.
† P. Lenard, Wied. Ann. **51**, p. 225, 1894.
‡ P. Lenard, Wied. Ann. **52**, p. 23, 1894.

a small hole in the centre covered by aluminium foil, so that it was possible to have the gases at different pressures in the two tubes. The rays generated in the tube A fell on the aluminium window and penetrated into the gas in the tube B. This tube contained two movable screens, S_1 and S_2, and a narrow beam of the rays diverging from the aluminium foil passed through a hole in the screen S_1 and produced a small fluorescent spot on the screen S_2. The magnetic force was produced by a horseshoe magnet M placed midway between the screens, with the poles on opposite sides of the tube, so that the spot of light is deflected towards the edge of the screen S_2.

The magnetic deflection of the rays depended on the pressure of the gas in A. When the pressure was reduced, the potential required to produce the discharge increased, and the magnetic deflection of the rays passing between S_1 and S_2 was diminished. This shows that the velocity of the rays increases with the potential difference between the cathode and anode in the tube in which they are generated.

But when the pressure in A is constant, and rays of a definite velocity pass through the aluminium window into the tube B, they retain their velocity to a remarkable extent, no appreciable change being observed in the magnetic deflection when considerable changes were made in the pressure of the gas in B. Thus when B contained air and the screens were at distances of 10 and 20 centimetres respectively from the aluminium window, the deflection of the rays was the same for all pressures between ·02 millimetre and 33 millimetres.

These experiments show that when the cathode-ray particles have acquired a high velocity, they may make a large number of collisions with molecules of a gas before the speed becomes appreciably reduced.

313. The ratio of the charge to the mass of a particle. The problem which is of most importance is to determine the ratio of the charge e to the mass m of the particles in a stream of cathode rays. The ratio E/M, where E is the charge on an atom of hydrogen in a liquid electrolyte and M the mass of an atom of hydrogen, may be easily found. Since a current of one electromagnetic unit flowing through an electrolyte gives off ·00010357 gramme of hydrogen per second, the ratio of the charge on an atom of hydrogen to the mass of the atom is approximately 10^4, or if the charge be expressed in electrostatic units the ratio E/M is 3×10^{14}. Hence, when the ratio e/m is known for any particle its mass may be found in terms of the mass of an atom of hydrogen if the charges E and e are the same.

Different methods have been used to determine the ratio e/m for small particles, but most of the calculations depend on some experimental investigation of the effect of a magnetic force on the motion of the particle. The simplest case is that of a particle moving in a vacuum in which the electric force is zero and the magnetic force H is perpendicular to the direction of motion. If V be the velocity of the particle, the force HeV acting on it is in a direction at right angles to the direction of motion and to the magnetic force, so that when H is constant the particle moves in a circle with a constant velocity. The radius r of the circle is obtained by equating the centrifugal force to the force HeV along the normal to the trajectory. Hence

$$\frac{mV^2}{r} = HeV,$$

or

$$\frac{e}{m} = \frac{V}{Hr}.$$

In a discharge tube containing a gas at a very low pressure, the electric force in the neighbourhood of the cathode is large, so that the particles set free from the cathode acquire a high velocity and may penetrate considerable distances without much loss of energy by collisions with molecules. If the electrodes are fixed at one end of the tube, the rays move with a velocity which is approximately constant for the remainder of their path, and the curvature $1/r$ of their trajectory produced by a magnetic force H may easily be found, so that one relation between the two quantities e/m and V is thus obtained. If, in addition, the velocity V is known, or some other relation connecting e/m and V, the values of both these quantities may be obtained.

In the early investigations made by Schuster,* the value attributed to V was much too small, and the numbers obtained for e/m were not much greater than the values for charged atoms.

314. Wiechert's determination of the ratio e/m for the cathode rays. Wiechert,† in January 1897, first showed that the velocity of the cathode rays is in some cases about one-tenth of the velocity of light, and that the ratio e/m for a cathode-ray particle is between 4,000 and 2,000 times as great as the value of E/M corresponding to an atom of hydrogen. He attributed the large value of e/m to the smallness of the mass m, and considered the charges e and E to be the same.

Wiechert, working with rays in hydrogen at a very low pressure, measured the curvature $1/r$ of their trajectory produced by a known magnetic force H, and obtained the value of Hr for substitution in the formula

$$\frac{e}{m} = \frac{V}{Hr}.$$

The velocity V was determined by a direct method, in which the period of oscillation of a condenser, discharging through a circuit of known self-induction, was used to estimate the short interval of time required by the rays to traverse a given distance in the discharge tube. This principle had previously been used by Des Coudres, who found that the cathode rays had a velocity exceeding 2×10^8 centimetres per second. Guided by this result, Wiechert designed an apparatus by means of which it was possible to compare the time in which the rays traversed a distance

* See A. Schuster, Proc. Roy. Soc. **47**, p. 526, 1890.

† E. Wiechert, Sitzungsber. der physikal.-ökon. Ges. zu Königsberg, **38**, p. 1, 1897; Wied. Ann., Beiblätter, **21**, p. 443, May 1897.

of about 20 centimetres with the period of discharge T of a condenser, T being between 10^{-8} and 10^{-7} second. The arrangement of the apparatus is shown by the illustration of the discharge tube (figure 70). In front of the cathode K and at a distance 25 centimetres from it was placed a plate of glass G that fluoresced brightly under the action of the rays. Two metal screens, B_1 and B_2, were placed between the cathode and the glass plate. The screen B_2 was 5 centimetres from the glass plate and had a slit in the centre a few millimetres wide. The other screen, B_1, 7½ centimetres from the cathode, extended across the lower part of the tube, and its edge was parallel to the slit in B_2. The positive electrode, which is not shown in the figure, was in the form of a ring and was placed between the

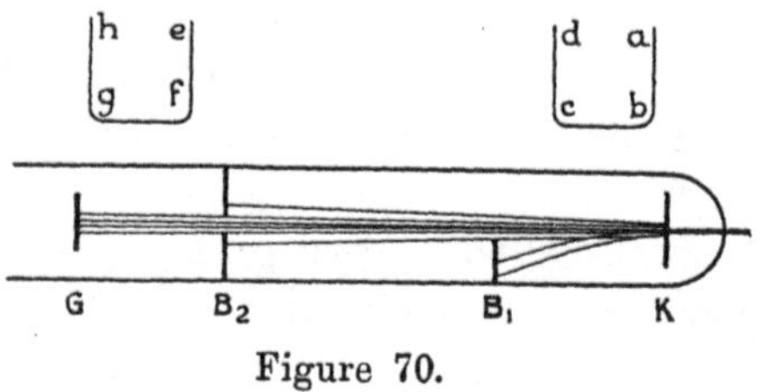

Figure 70.

cathode and screen B_1. The discharge was produced by the secondary circuit of a Tesla transformer, and the rays coming from the centre of the cathode passed over the edge of the first screen and through the slit in the second. A narrow fluorescent strip, a few millimetres wide, marked the points on the surface of the glass G on which the rays impinged.

When a magnet M is placed in a suitable position near the cathode most of the rays are bent down and fall on the screen B_1, and only a slight fluorescence is seen on the glass plate.

The two wires *abcd* and *efgh* formed part of the circuit of the oscillatory discharge of a condenser, which was charged inductively by the Tesla apparatus used to produce the cathode rays. Thus a current flowed through the wire *abcd*, and at the same instant the high potential was established between the electrodes. When the wire is brought close to the tube, as shown in the figure, the magnetic force due to the current in *bc* counteracts the effect of the magnet *M*, when the cathode rays are emitted, so that some of the rays pass over the edge of B_1. An increase is thus produced in the fluorescence at *G*, due to rays which pass between *K* and B_1 when the alternating current in the condenser circuit is in a certain phase.

The effect of the current on the rays as they pass from the slit in B_2 to the glass plate is then observed by bringing the wire *efgh* near the tube. Let *t* be the time in which the rays travel from B_1 to *G*, *T* the period of oscillation of the condenser discharge; then if *T* is large compared with *t* the deflection produced by the current in *fg* is in the same direction as the deflection produced by *bc*. If the period of the condenser discharge is reduced until $T/4 = t$, the deflection produced between B_2 and *G* becomes very small. Thus by observing the displacements of the fluorescence of the glass plate, obtained by reducing the period *T*, the time *t* may be estimated.

It was thus found that with rays for which the value of Hr was 150, the velocity *V* was about 3×10^9 centimetres per second, and the value e/m was about 2×10^7; the true values being possibly greater than these numbers.

Wiechert also considered the possibility of determining e/m from measurements of the potential difference *W* between the electrodes. An upper limit of the velocity of the electrons in the tube may be obtained on the hypothesis that the rays start from the negative electrode and move freely under the electric force. The maximum kinetic energy acquired by the charged particles is then

$$\frac{mV^2}{2} = eW,$$

and $\frac{e}{m}$ is given by the equation $\frac{e}{m} = \frac{2W}{H^2r^2}$.

The upper limit of the value of e/m thus obtained was 4×10^7.

Subsequently* the arrangement of the apparatus for measuring the velocity of the rays was improved, and the most probable values of e/m were found to be between

$4{\cdot}64 \times 10^{17}$ and $3{\cdot}04 \times 10^{17}$ in electrostatic units,

or $1{\cdot}55 \times 10^{7}$ and $1{\cdot}01 \times 10^{7}$ in electromagnetic units.

These investigations are important, as the velocity of the electrons is found directly by measuring the time of passing over a certain distance. But it is improbable that very exact results can be obtained by this method when the charged particles travel with such high velocities, as it is difficult to obtain an oscillatory discharge of very short period which is not highly damped.

315. Kaufmann's determination of the ratio e/m for the cathode rays. Kaufmann† in 1897 made a series of measurements of the ratio e/m for cathode rays and obtained a result which is in good agreement with the most recent determinations. His method depends on the simple principle that in a gas at sufficiently low pressures the kinetic energy acquired by the electrons in passing from the cathode to the anode is eW, W being the potential difference between the electrodes. Under these conditions the value of e/m is given by the formula $\frac{e}{m} = \frac{2W}{H^2 r^2}$, r being the radius of curvature of the trajectory in a transverse magnetic field H. This implies that the loss of energy of the electrons due to collisions is very small, and the investigations show that any such effect must have been negligible. A large number of experiments were made in which the gas was at different small pressures, and the potential difference between the electrodes W required to produce the discharges varied from 3,000 to 14,000 volts. The smaller values of W correspond to the larger pressures, where the collisions between the rays and the molecules are most numerous, so that if the velocity were appreciably diminished by collisions, the error in the formula would

* E. Wiechert, Wied. Ann. **69**, p. 739, 1899; also 'Ions, Électrons, Corpuscules', vol. ii, pp. 1029-53.

† W. Kaufmann, Wied. Ann. **61**, p. 544, July 1897.

increase as W diminished. Since there was no appreciable difference in the values of the quantity W/H^2r^2 for the different pressures at which the experiments were made, it may be concluded that no serious error was introduced by taking eW as the kinetic energy of the rays.

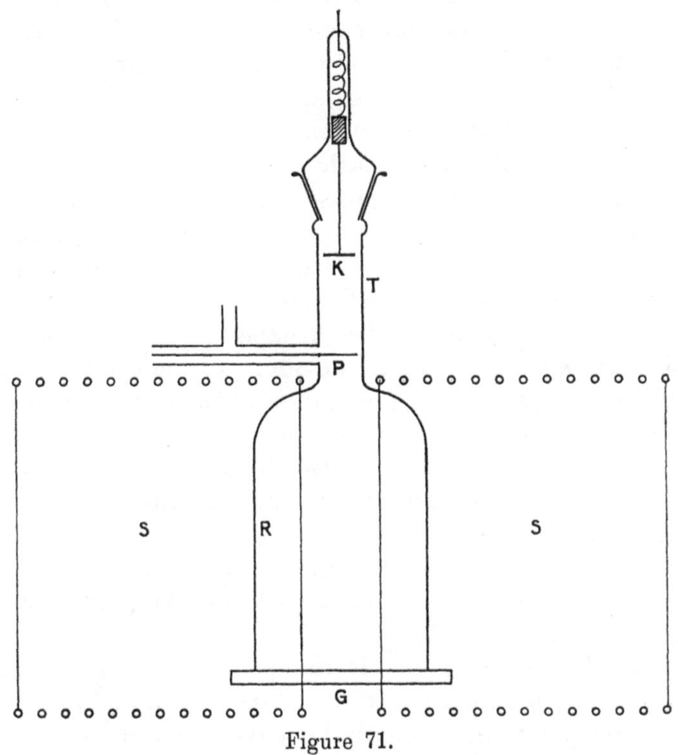

Figure 71.

The apparatus used by Kaufmann is shown in figure 71: the glass tube R, 11 centimetres long and 6·5 centimetres wide, was closed with a glass plate G, the electrodes K and P being contained in the tube T. The cathode K was raised to a high potential by a Wimshurst machine, and the potential difference W between the electrodes was measured by an electrostatic voltmeter. The case of the voltmeter and the positive electrode P were connected to earth. A thin layer of chalk, which fluoresced under the action of the rays, was spread over the plate G, and the electrode P, which was a platinum wire half a millimetre in diameter, cast a shadow on the fluorescent plate. The magnetic force H was established in the space between P and G by the current in the solenoids S. The deflection of the rays was measured by the displacement δ of the shadow of the wire, and since δ was small compared with the distance $P\,G$, the radius of curvature r of the trajectory was inversely proportional to δ ($2r\delta = d^2$ approximately, d being the distance PG). Experiments made with a copper electrode at K gave the following results.

With air at different pressures, ·03 to ·07 millimetre, the potentials W required to produce the discharge varied from 10,630 volts to 3,260 volts, but the quantity $\sqrt{W}/Hr$ remained constant, the mean values being proportional to 393, 398, and 406, in a series of experiments in which the cathode was placed at various distances from the wire P.

With coal-gas the mean value 401·5 was obtained in experiments in which W varied from 6,410 to 11,850 volts.

In hydrogen and carbonic acid the quantity $\sqrt{W}/Hr$ was found to be proportional to 404 and 398, the potentials W ranging from 4,000 to 14,000 volts.

An aluminium cathode was also used with air in the tube, and the results were the same as those obtained with the copper electrode.

Thus the value of e/m is independent of the pressure, the distance between the electrodes, and the nature of the gas. An approximate calculation gave $e/m = 10^7$, e being in electromagnetic units.

In order to obtain an exact value it was necessary to take into account the fact that the field is not absolutely uniform and to make accurate measurements of the force H along the line from P to G due to a given current in the coils S.

When the rays traverse a distance x in the transverse magnetic field, the velocity v at right angles to the original direction of motion is

$$v = \int_0^x \frac{He}{m}\,dx,$$

so that the (small) deflection δ on a screen at a distance d from the origin is

$$\delta = \int_0^d v dt = \frac{e}{mV}\int_0^d dx \int_0^x H dx$$

$$= \sqrt{\frac{e}{2mW}}\int_0^d dx \int_0^x H dx.$$

A complete investigation of the magnetic field was made later, and in a subsequent memoir Kaufmann* gives the value of $1{\cdot}77 \times 10^7$ for e/m for cathode rays.

Kaufmann's investigations were subsequently repeated by Simon † with some improvements in the apparatus, and the number $1{\cdot}865 \times 10^7$ was obtained.

316. Thomson's determinations of the ratio e/m for cathode rays. Thomson,‡ in 1897, also investigated the cathode rays and found, by two different methods, values of e/m which are in general agreement with the results obtained by Wiechert and Kaufmann.

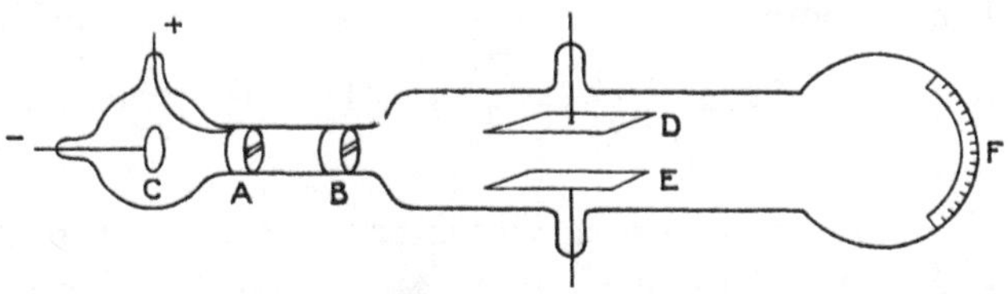

Figure 72.

In one of these methods the velocity of the rays is found by comparing the deflections produced by transverse electric and magnetic fields. The apparatus used is shown in figure 72. The rays starting from the cathode C passed through the slits in the metal diaphragms A and B, and fell on the surface of the glass tube at F. Both diaphragms were connected to earth, and A acted as the positive electrode. A potential difference was established between the electrodes D and E by connecting them to the terminals

* W. Kaufmann, Wied. Ann. **62**, p. 598, Dec. 1897.
† S. Simon, Wied. Ann. **69**, p. 589, 1899.
‡ J. J. Thomson, Phil. Mag. (5) **44**, p. 293, Oct. 1897.

of a battery of accumulators, so that an electric force X acted on the rays as they passed between the plates.

Each particle moves in the direction of the force with an acceleration eX/m for the time l/V, V being the velocity of the rays parallel to the axis of the tube, and l the length of the plates. The velocity thus acquired in the direction of the force is eXl/mV, and on emergence from the space between the plates the direction of motion is inclined at a small angle θ to the original direction. The angle θ, which is equal to eXl/mV^2, was measured by observing the displacement of the luminosity due to the rays at F.

The effect of a magnetic force H acting for the same distance l was found by placing the tube between two coils of diameter l. The force H due to a current C in the coils was determined by the induced electromotive force in a secondary coil obtained by reversing the current C. In this case the particle moves with an acceleration eHV/m for the time e/V, and on emergence from the field the direction of its motion is inclined at the small angle ϕ to the original direction, ϕ being given by the equation $\phi = Hel/mV$.

Thus $\dfrac{e}{m} = \dfrac{V\phi}{Hl}$; and $\dfrac{e}{m} = \dfrac{V^2\theta}{Xl}$, so that the values of e/m and V may be found.

In the particular case in which the electric and magnetic forces Xe and HeV are equal and opposite, there is no deflection of the rays when both forces are acting and $V = X/H$.

The mean value of the ratio e/m for air, hydrogen, and carbonic acid obtained by this method was $7{\cdot}7 \times 10^6$.

In the second method used by Thomson the velocity of the rays was estimated by finding the heat T generated when the rays fell on a thermo-electric couple formed by the junction of two metals. The thermo-electric couple was set up inside a small insulated cylinder, and the temperature to which it was raised was measured by the deflection of a galvanometer which was connected to the couple by a pair of insulated wires. The small cylinder was surrounded by a larger metal cylinder connected to earth and the rays entered through small apertures in the ends of the cylinders facing the cathode. The charge Q received by the inner cylinder in a given time is Ne, where N is the number of electrons that enter, and the kinetic energy $T = \frac{1}{2}mNV^2$ was measured by the rise of temperature of the thermo-junction, assuming that the greater part of the energy is converted into heat.

Hence $$\frac{e}{m} = \frac{V^2Q}{2T}.$$

Combining this equation with the relation between e/m and V given by the magnetic deflection, Thomson found the number $1{\cdot}17 \times 10^7$ as the value of e/m for cathode rays.

317. Determinations of e/m for particles set free by ultra-violet light, and for the negatively charged particles emitted by incandescent solids. The earlier investigations of the negatively charged particles obtained by different methods in gases at very low pressures show that the ratio e/m was probably exactly the same in all cases. Lenard,* who used a method similar in principle to that used by Kaufmann, determined the ratio of the charge to the mass of a particle set free by the action of ultra-violet light from a metal surface and obtained the number $1{\cdot}15 \times 10^7$.

Thomson † also examined these particles by a special method which is applicable to cases in which the particles start with a small velocity from a negatively charged surface.

The particles move under an electric force, Z, between two parallel-plate electrodes at a distance S apart, and a magnetic force F parallel to the axis of x is also established in the space between the electrodes. When negatively electrified particles are set free by the action of light from the surface of the negative electrode, they begin to move normally to the surface in the z direction. As their velocity increases they become deflected by the magnetic force and

* P. Lenard, Sitzungsberichte Akad. Wien, **108**, Abt. 11[a], p. 1649, Oct. 1899.

† J. J. Thomson, Phil. Mag. (5) **48**, p. 547, Dec. 1899.

the trajectories of the particles are cycloids in planes normal to the axis of x.

Under these conditions the equations of motion of a particle are

$$m\frac{d^2z}{dt^2} = Ze - Fe\frac{dy}{dt},$$

$$m\frac{d^2y}{dt^2} = Fe\frac{dz}{dt}.$$

The solution of these equations may be expressed in the form

$$z = \frac{Z}{F\omega}(1 - \cos\omega t),$$

$$y = \frac{Z}{F\omega}(\omega t - \sin\omega t),$$

where $$\omega = \frac{e}{m}\cdot F,$$

the constants being chosen to satisfy the condition that the particle starts from rest from the origin at the time $t = 0$.

The particle reaches its maximum distance

$$\frac{2Z}{F\omega} = \frac{2Z}{F^2}\cdot\frac{m}{e},$$

from the negative electrode when $\cos\omega t = -1$, and then moves back towards the electrode.

When the distance S between the plates exceeds the value $2Z/F\omega$ the conductivity ceases, since the negatively charged particles do not reach the positive electrode. If the distance S or the magnetic force F be reduced so that the product SF becomes less than $2Z/\omega$, the positive electrode begins to receive a charge, and the value of e/m is given by the equation

$$\frac{e}{m} = \frac{2Z}{F^2S}.$$

The mean value of e/m found in these experiments is $7{\cdot}3\times10^6$.

This method also gave rather a low value $8{\cdot}7\times10^6$ of e/m for the ions set free from an incandescent carbon filament in hydrogen at a very small pressure.

318. Accurate determinations of the value of e/m for electrons obtained under various conditions. Many careful investigations have recently been made of the ratio e/m, and the results are in good agreement with the value $1\cdot77 \times 10^7$ originally found by Kaufmann for cathode rays. The following are some of the recent determinations, e being expressed in electromagnetic units:

Slowly moving Becquerel rays, by magnetic and electrostatic deflection.

[a] Kaufmann $1\cdot884 \times 10^7$; [b] Bucherer $1\cdot763 \times 10^7$; [c] Neumann $1\cdot765 \times 10^7$.

Cathode rays.

[d] Bestelmeyer $1\cdot72 \times 10^7$, by magnetic and electrostatic deflection.

[e] Malassez $1\cdot769 \times 10^7$, by magnetic deflection and potential difference between electrodes.

Cathode rays from glowing oxides, by magnetic deflection and potential difference between electrodes.

[f] Classen $1\cdot776 \times 10^7$; [g] Bestelmeyer $1\cdot766 \times 10^7$.

Photo-electric effect by magnetic deflection and potential difference between the electrodes.

[h] Alberti $1\cdot756 \times 10^7$ and $1\cdot766 \times 10^7$.

Zeeman effect.

[i] Weiss & Cotton $1\cdot767 \times 10^7$; [k] Stettenheimer $1\cdot791 \times 10^7$; [l] Gmelin $1\cdot771 \times 10^7$.

[a] W. Kaufmann, Ann. der Phys. (4) **19**, p. 487, 1906.
[b] A. H. Bucherer, Ann. der Phys. (4) **28**, p. 513, 1909.
[c] See Cl. Schaefer, Verh. d. D. Phys. Ges. **15**, p. 935, 1913.
[d] A. Bestelmeyer, Ann. der Phys. (4) **22**, p. 429, 1907.
[e] J. Malassez, Ann. de Chim. et de Phys. (8) **23**, pp. 231 and 397, 1911.
[f] J. Classen, Phys. Zeitschr. **9**, p. 762, 1908.
[g] A. Bestelmeyer, Ann. der Phys. (4) **35**, p. 909, 1911.
[h] E. Alberti, Ann. der Phys. (4) **39**, p. 1133, 1912.
[i] P. Weiss and A. Cotton, Journal de Physique (4) **6**, p. 429, 1907.
[k] A. Stettenheimer, Dissertation Tübingen, 1907.
[l] P. Gmelin, Ann. der Phys. (4) **28**, p. 1079, 1909.

The charge of the negative ions obtained in high vacua has usually been assumed to be the same as the charge on a univalent ion in a liquid electrolyte.

This result has been found experimentally, by the methods described in Chapter V, for ions set free by ultra-violet light and also for negative ions generated by various other processes. The methods are only applicable to cases in which the ions diffuse in gases at comparatively high pressures, but since the charge is not affected by the pressure, there is good reason to believe that the charge on an ion is the same at all pressures, down to the very lowest.

The large value of e/m obtained for negative ions in high vacua must therefore be attributed to the smallness of m, which is less than the mass of an atom of hydrogen in the proportion 1 : 1830.

Thus the atoms of negative electricity, or the electrons obtained by ionizing molecules, have the same charge and the same small mass whatever be the nature of the molecule from which the electron is derived, or the process employed to ionize the molecule.

319. Values of e/m when the velocity of the electron approaches the velocity of light. When the velocity of the electrons is small compared with the velocity of light the value of e/m as found experimentally is constant. Some of the β particles emitted by radio-active substances move with velocities much greater than the cathode rays, and Kaufmann found by accurate experiments that the effective mass increases with the velocity. Thus the number $1{\cdot}77 \times 10^7$ applies to cases in which v is small compared with the velocity of light (3×10^{10} cm. per second), but the ratio e/m diminishes from the small-velocity value to $1{\cdot}31 \times 10^7$ when the velocity is $2{\cdot}36 \times 10^{10}$, and to $\cdot 63 \times 10^7$ when the velocity is $2{\cdot}83 \times 10^{10}$.

These investigations have an important bearing on electromagnetic theory. When an electric charge is in motion there is a certain amount of electromagnetic energy resident in the surrounding field, and the charge when accelerated exhibits the phenomena of inertia even when supposed devoid of ordinary or Newtonian mass.

For slow speeds the electromagnetic mass m_0 of an electron is constant, being of the order e^2/a where e is the charge in electromagnetic units, and a the radius of the electron; but when the velocity approaches that of light the mass increases.

For accelerations in the direction of motion the charge behaves as though it had a mass m_l (longitudinal electromagnetic mass), while for accelerations at right angles to the direction of motion it appears to have a different mass m_t (transverse electromagnetic mass).

According to the theoretical investigations given by Abraham and Lorentz the longitudinal mass is greater than the transverse mass. Of these Abraham's theory considers the electron as rigid, and Lorentz's theory, for a special reason, considers it as contracting in the direction of motion.*

Lorentz's theory leads to the following formulae for m_l and m_t in terms of the velocity v of the particle:

$$m_l = \frac{m_0}{(1-v^2/c^2)^{\frac{3}{2}}}; \quad m_t = \frac{m_0}{(1-v^2/c^2)^{\frac{1}{2}}};$$

both masses being equal to m_0 when v is small compared with c, the velocity of light.

Kaufmann's original determinations of the transverse electromagnetic mass and the recent experiments of Bucherer † with β rays and those of Hupka ‡ on fast cathode rays are in agreement with this theory.

The experiments therefore support the view that the mass of an electron is entirely electromagnetic.

320. The Kanalstrahlen or positive rays. The Kanalstrahlen or positive rays were discovered by Goldstein,§

* See 'Ions, Électrons, Corpuscules', vol. i; and M. Abraham, Theorie der Elektrizität, vol. ii.

† A. H. Bucherer, Ann. der Phys. (4) **28**, p. 513, 1909. (Bucherer's experiments have been repeated and extended by G. Neumann. See Cl. Schaefer, Verh. d. D. Phys. Ges. **15**, p. 935, 1913.)

‡ E. Hupka, Ann. der Phys. (4) **31**, p. 169, 1910.

§ E. Goldstein, 'Ions, Électrons, Corpuscules,' i. 243.

who studied the luminous appearance presented by different layers of the gas in the neighbourhood of the cathode in a Geissler discharge.

The space between the bright negative glow and the cathode, which is usually known as the Crookes dark space, has a slight bluish appearance, and immediately in contact with the cathode there is a narrow layer of gas that glows brightly.* The colour of the glow depends on the gas: in nitrogen or air it is yellow; in hydrogen it is pink, or rose coloured.

With a plane disc as cathode, slightly smaller in diameter than the glass tube, the yellow glow is very narrow and covers the whole surface when the air is at a certain pressure, but as the pressure is reduced the glow recedes from the edges and becomes thicker over an area near the centre of the cathode. At low pressures the glow may extend to a distance of 2 centimetres from the surface of the electrode.

Goldstein found that if holes are bored in the cathode a similar yellow glow extends from each aperture into the space at the back of the cathode, the light being distributed in straight narrow columns which are slightly divergent. At low pressures, when the yellow glow in front of the cathode is concentrated towards the centre, the rays at the back of the cathode proceed only from those apertures which are near the centre and are covered by the yellow glow in front. Also, the rays are not obtained through a thick plate when the holes are not perpendicular to the surface, since particles moving normally to the surface cannot pass through without colliding with the metal.

The luminous effects produced by the positive rays differ from those obtained in the negative glow and the positive column; and both of these, as Pellat has shown, are due to collisions between the electrons and the molecules of the gas. The colour depends on the velocity with which the electrons collide with the molecules. In the negative glow the velocity is very high, and the intensity of the violet rays in the spectrum of the gas is relatively high. In the

* See Section 270.

positive column the electrons move comparatively slowly, and the relative intensity of the orange and yellow rays increases, so that the positive column and the negative glow appear to be of different colours.*

321. Charge carried by the rays. Many researches have been made to discover the nature and properties of the *Kanalstrahlen*, and it has been found that the phenomena may be explained on the hypothesis that the positive ions in a discharge tube approach the cathode with a high velocity, and when they arrive at an aperture in the cathode they pass through into the space at the back of the cathode.

Perrin † showed that the rays which traverse an aperture in the cathode are positively charged. With the same apparatus (figure 67) which was used to investigate the charge on the cathode rays, it was found that when the cylinder $EFGH$ acted as the negative electrode, the inner cylinder $ABCD$ became positively charged when a discharge passes through the tube. The positive charge is small, but may be detected with an electrometer when the tube is not highly exhausted. As the pressure is reduced, the charge increases and may be observed with an electroscope connected to the inner cylinder. It was found possible to deflect the rays by a magnet, so that the inner cylinder did not receive a positive charge.

The first determinations of e/m for the positive rays were made by Wien ‡ from observations of the deflections produced by magnetic and electrostatic forces. In order to obtain a measurable deflection, a very large magnetic force is required, and it is necessary to limit the magnetic field so as not to disturb the discharge in the tube in which the rays are generated.

The results of the first experiments gave the velocity $3{\cdot}6 \times 10^7$ centimetres per second and the ratio of the charge to the mass of the particle $3{\cdot}2 \times 10^{-2}$. If the charge is the atomic charge, this number shows that the mass is

* H. Pellat, 'Ions, Électrons, Corpuscules,' vol. ii, p. 447.
† J. Perrin, Comptes rendus, **121**, p. 1130, Dec. 1895.
‡ W. Wien, Wied. Ann. **65**, p. 440, 1898.

approximately the same as that of a molecule of oxygen or nitrogen.

The phenomena, however, are not so simple as this result appears to indicate. A beam of positive rays contains particles moving with a high velocity, some of which are charged and some uncharged. It is therefore a much more complicated system than a beam of cathode rays in which each particle has the same charge for the whole length of its trajectory. With the positive rays the deflections of different particles produced either by electric or magnetic fields are not the same, but vary from zero to a maximum value. This is shown by the fact that a narrow beam producing a small spot of light on a fluorescent screen is in general spread out under the action of a transverse force, and produces a bright line which originates from the point of the screen on which the undeflected rays impinged.

322. Wien's determinations of e/m for different gases. In a subsequent investigation, made by Wien,* the deflections produced in an electric and magnetic field were measured by an apparatus which is shown in figure 73. The tube A in which the rays were generated was cemented to an iron plate K, which acted as the cathode, and the rays passed through a hole 2 millimetres wide into the vessel B, where the effects of the transverse electric and magnetic forces were observed. The vessel B was placed between the poles N and S of an electro-magnet, and the discharge tube A was screened from magnetic forces by the plate K and an iron cylinder C. The rays passed between two aluminium electrodes $a\,a$ at about 5 millimetres apart and 6 centimetres long, the ends of the electrodes being 5 centimetres from the glass plate P that closed the wide end of the vessel.

The electric force X obtained by connecting the two electrodes $a\,a$ to the terminals of a battery of cells was in the same direction as the magnetic force H. The deflections due to these forces were at right angles to each other and

* W. Wien, Ann. der Phys. (4) **8**, p. 244, 1902.

were measured by the displacements of the fluorescence produced on the plate P, in the two directions parallel and perpendicular to the force X.*

In traversing the distance l_1 between the electrodes $a\,a$, the particles are displaced through the distance $\frac{1}{2}\frac{Xe}{m}\cdot\frac{l_1^2}{V^2}$, and acquire the velocity $\frac{Xe}{m}\cdot\frac{l_1}{V}$ in the direction of the electric force. The displacement x of the fluorescence on the plate P is

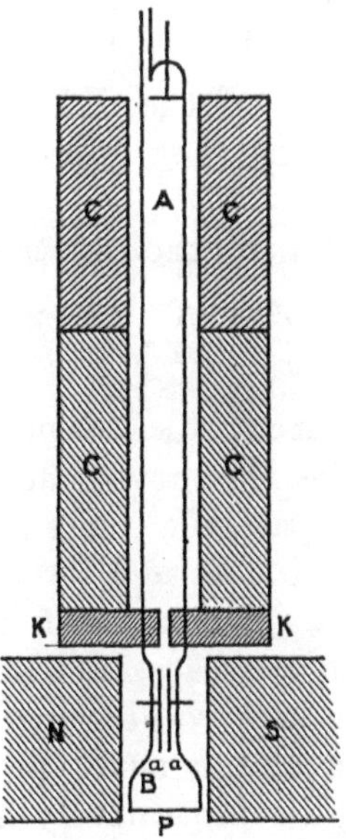

Figure 73.

$$x = \frac{1}{2}\frac{Xe}{m}\cdot\frac{l_1^2 + 2\,l_1 l_2}{V^2},$$

l_2 being the distance of the ends of the electrodes $a\,a$ from the plate.

* This arrangement of the magnetic and electric forces had previously been used by Kaufmann to investigate the values of e/m for the β particles from radium, which travel with high velocities approaching that of light. See Section 319.

The displacement y due to the magnetic force H is

$$y = \frac{1}{2}\frac{He}{m} \cdot \frac{l^2}{V},$$

l being the distance the particles travel under the action of the force, which in this case is $(l_1 + l_2)$.

The velocity V of the particles is therefore given by the ratio of the two deflections

$$\frac{y}{x} = \frac{H}{X} \cdot \frac{l^2 V}{l_1^2 + 2 l_1 l_2}.$$

Hence if the velocities of the different particles are the same, the points at which they impinge on the plate are in a straight line.

The ratio e/m is given by the equation

$$\frac{e}{m} = \frac{y^2}{x} \cdot \frac{2X}{H^2} \cdot \frac{l_1^2 + 2 l_1 l_2}{l^4}.$$

Hence if the rays consist of a system of particles travelling with different velocities, the curves traced out by the particles on the plate will be a parabola if the ratio e/m is constant.

Different values were obtained for the ratio e/m as given by the above formula. The highest was of the order 10^4, which corresponds to the case in which m is an atom of hydrogen. When the tube was exhausted after having been filled with air or oxygen, a small proportion of the rays were deflected through comparatively large distances, corresponding to the value of $e/m = 10^4$. This effect becomes very feeble when special precautions are taken to remove hydrogen or water vapour from the gas. The presence of these rays is most probably due to small residues of hydrogen or water vapour given out from the electrodes or the surface of the glass.

323. Wien's investigations of the general properties of positive rays. Villard,* from his early experiments, concluded that the positive ions become discharged as they move through the gas, as he found that the rays are deflected

* Villard, Journal de Physique (3) 8, pp. 5 and 140, 1899.

by an electric force as they approach the cathode, but after they have traversed the cathode and moved for some distance through the gas they are not affected by an electric force. These phenomena have been investigated more completely by Wien, who has shown that a large proportion of the particles in a beam of rays may be uncharged, but there is always a certain proportion of charged particles which are deflected either by an electric or a magnetic force.

Hence the numbers obtained for e/m correspond to the average values of the charge of the particle while it is under the action of the transverse forces. The maximum deflections are obtained when a particle remains charged for the whole length of the field in which the transverse forces are acting. The undeflected particles are those which lose their charge before they enter the field and do not acquire a charge in their passage through the gas. Thus any deflection may be obtained varying from zero to the maximum.

A small proportion of the particles are deflected as if they were negatively charged atoms or molecules, the ratio of the charge to the mass being small, as in the case of the positive rays.

The transformations that occur as the stream of particles moves through the gas were determined from experiments * on the deflections produced in a long wide tube, so that it was possible to establish strong magnetic fields in separate sections of the tube at different distances from the cathode.

One form of apparatus that was used for this purpose is shown in figure 74. The rays were generated by a discharge in the tube A and passed through the aperture in the iron cathode K. The discharge tube was connected, by a capillary tube C, to the long wide tube B in which the properties of the rays were investigated. This tube was provided with two iron cathodes, K_1 and K_2, which were connected to K and maintained at zero potential. The poles N_1 and S_1 of an electro-magnet were placed on opposite sides of the tube

* W. Wien, Ann. der Phys. (4) **27**, p. 1025, 1908.

so as to produce a strong magnetic field in the section between K_1 and K_2; a second field of force was similarly produced at a further distance from the cathode by the poles N_2 and S_2. In some of the experiments the lines of force in the first field were perpendicular to those in the second,* but for these investigations the forces may be inclined at any angle to each other, provided they are normal to the axis of the tube. Part of the apparatus was surrounded by an iron screen so that there should be no magnetic force in the space between K and K_1.

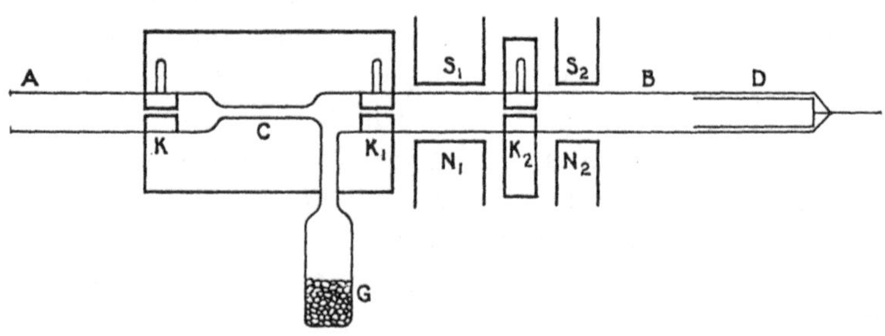

Figure 74.

The rays emerging from the cathode K passed through the capillary tube C and the apertures in K_1 and K_2, and were received in a cylinder D connected to a sensitive galvanometer by which the positive charge on the rays was measured.

324. Experiments on the passage of rays through two magnetic fields. When the first magnetic field is established, the particles which are charged while the beam of rays passes between the poles N_1 and S_1 are deflected, and the uncharged particles continue their motion parallel to the axis of the tube and are received in the cylinder D. Two streams of particles may thus be obtained, the original stream s and the stream s_1 obtained by removing charged particles from s by the first magnetic field.

The galvanometer gives a deflection δ when the stream s

* Experiments on this principle were also made by J. J. Thomson; see Phil. Mag. **18**, p. 821, Dec. 1909.

enters the cylinder, and this is reduced to δ_1 by the removal of charged particles in the first magnetic field. The deflection δ_1 is a large proportion of δ, which shows that either the original stream s must contain positively charged particles that are not deflected by a magnet or some of the uncharged particles in s_1 acquire positive charges as they move through the gas.

The effect produced by the second magnetic field on the streams s and s_1 affords evidence in favour of the latter explanation of the phenomena.

When the second magnet is excited, ions are removed from the stream s and the deflection δ is reduced to δ_2. An exactly similar effect takes place with the stream s_1; the deflection δ_1 is reduced by the second field to δ_{12} where $\delta_{12}/\delta_1 = \delta_2/\delta$. Thus the streams are reduced in the same proportion by the second field, so that on entering the fields of force the proportion of charged particles to uncharged particles is the same in the two cases.

The following table gives the results of a series of experiments made on this principle. The numbers in the column δ are the deflections of the galvanometer due to a beam of rays entering the cylinder *D* when no magnetic force is acting, δ_1 when the beam is reduced by the first field of force, δ_2 when the beam is reduced by the second field of force, δ_{12} when the beam is reduced by both fields.

	δ	δ_1	δ_2	δ_{12}	δ_2/δ	δ_{12}/δ_1
Air	37	29	16	12	·43	·41
,,	46	32	18	13	·39	·41
,,	16	10	5	3	·31	·30
Hydrogen	120	80	42	28	·35	·35
,,	150	100	54	32	·36	·32
,,	155	110	48	32	·31	·39

In the experiments with hydrogen the intensity of the first field was greater than with air.

The number of ions removed by a magnetic field from a beam of rays depends on the pressure of the gas, and also on the length of the section of the tube in which the magnetic force is acting. Some of the uncharged particles entering the field become charged before they leave it, so that when the field extends over a long section of the tube a large number of particles are deflected from the beam that enters the cylinder. In order to investigate this effect it is necessary to make variations in the pressure of the gas in the tube *B*, but it is undesirable to make large reductions in the pressure of *A* as it is difficult to obtain discharges at very low pressures. Since the rays will not traverse thin sheets of metal the communication between *A* and *B* must be open, and in order to obtain different pressures in *A* and *B* a special arrangement of the apparatus is required. For this purpose the tubes *A* and *B* were connected through the capillary tube *C* as shown in the figure. The vessel *G* containing charcoal was immersed in liquid air, and the air in *B* was rapidly absorbed as it entered through the capillary *C*. The pressure in *B* was thus reduced below that of *A*, and, in order to maintain a constant pressure in *A*, air was supplied through a fine capillary tube at the same rate at which it escaped through *C*. It was thus found that the number of particles removed by a given field of force from a beam of rays was diminished when the pressure was reduced. The following table gives the results of some experiments with air in which the pressure in *B* was very low. The numbers in the first column are the potentials required to produce the discharge, and the other numbers are the galvanometer deflections obtained under the conditions described above.

Potential in volts.	δ	δ_1	δ_2	δ_{12}	δ_2/δ	δ_{12}/δ_1
30,000	104	70	73	50	·70	·70
24,000	122	93	88	66	·76	·75
28,000	100	65	70	46	·70	·70
3,000	10	6	5	3	·60	·60

325. Arrangement of the apparatus for experiments at low pressures. The apparatus with which these determinations were made was not suitable for the investigation of the phenomena in hydrogen at very low pressures, since hydrogen is not rapidly absorbed by the charcoal in the vessel *G*. Wien* found that by using a Gaede pump, instead of the charcoal tube, it was possible to obtain a very high vacuum in the vessel *B*, while the discharge tube *A* contained sufficient gas to enable the discharge to pass, when potentials of the order of 30,000 volts were used. The arrangement of the apparatus for adjusting the pressures was somewhat more complicated than that used in the first investigation but the principle remained the same.

In these experiments different methods were used to measure the intensity of the beam of positive rays. A thermo-couple was fixed at the end of the tube *B*, and the rise of temperature is proportional to the number of particles which are undeflected from the axis of the tube by a transverse force. The brightness of the light produced by the rays was also measured. It was found that the relative intensities of any two beams, as measured by either of these methods, was the same as the ratio of the positive charges carried into a cylinder, as measured by a sensitive galvanometer.

Since the rise of temperature of the thermo-couple is proportional to the total number of charged and uncharged particles in the beam, it follows from these experiments that the numbers of positively charged particles in the beam is proportional to the total number of particles. If the charged particles are removed by a magnetic field, the equilibrium state between the charged and uncharged particles is re-established after the beam traverses a certain distance in the gas, and a positive charge is conveyed by the rays moving with a high velocity along the axis of the tube.

The results obtained with hydrogen at very low pressures were similar to those obtained in air. The number of particles removed from a beam by a magnetic field diminishes as the pressure is reduced. Also, a beam which has been

* W. Wien, Ann. der Phys. (4) 30, p. 349, 1909.

reduced by a magnetic field is further reduced by a second field approximately in the same proportion as the original beam. In these experiments the two sections of the tube in which the magnetic fields were established were a long distance apart, about 70 centimetres, so that even at very low pressures a large number of collisions occur with molecules of the residual gas and the equilibrium distribution between the charged and uncharged particles becomes re-established, after the former have been removed by the first field of force.

326. Rates at which particles become charged and discharged. Wien* has recently made a more complete investigation of the changes that take place between the neutral and electrified states of the particles in a stream of rays. When all the charged particles are removed from a beam by a transverse force the uncharged particles that remain tend to become charged as they move along the axis of the tube, and the final stage is reached when the ratio of the numbers of charged and neutral particles attains a definite value. In general let n and n' be the numbers of charged and uncharged particles in the beam at any position x, measured along the axis of the tube. The changes that tend to occur are represented by the equations

$$\frac{dn}{dx} = \gamma' n' - \gamma n = -\frac{dn'}{dx},$$

where γn represents the rate at which the charged particles lose their charge, and $\gamma' n'$ the rate at which the uncharged particles acquire charges. The mean distance l through which a charged particle moves before it loses its charge is $1/\gamma$. Similarly the mean distance l' that a neutral particle travels before it acquires a charge is $1/\gamma'$. In the equilibrium state

$$\frac{n}{n'} = \frac{\gamma'}{\gamma} = \frac{l}{l'}.$$

* W. Wien, Ann. der Phys. (4) **39**, p. 519, 1912.

The solution of the above equations which satisfies the condition $n = 0$ when $x = 0$ is

$$n = n_1 [1 - \epsilon^{-(\gamma+\gamma')x}],$$

$$n' = n_1' + n_1 \epsilon^{-(\gamma+\gamma')x},$$

the constants n_1 and n_1' being the values of n and n' when x is large.

In order to test this theory experimentally, and to find the values of l and l', ten pairs of parallel plates one centimetre long and one millimetre apart were arranged in series along the axis of a tube so that the rays passed through each pair in succession. The distance between the ends of each consecutive pair of plates was one millimetre. The rays were received on a thermopile at the end of the tube, and the reduction of the stream, due to the removal of the charged particles by an electric force in the space between a pair of plates, was determined by the effect on the thermopile, which was proportional to the number of particles falling on it. It was found that the reduction of the stream, due to charging one pair of plates, at first increased with the force between the plates, and a maximum effect was obtained when the potential difference between the opposite plates was 200 volts. All the particles which were charged when passing through a section of the tube were thus removed from the beam that fell on the thermopile.

It was thus found that in hydrogen at ·005 millimetre pressure, 15 per cent. of the total number of particles are charged when the equilibrium state is established.

When all the charged particles are removed by the first pair of parallel plates the ratio of the number of charged particles in the beam to the total number, at distance x from the first pair of plates, is given by the formula

$$\frac{n}{n_1 + n_1'} = \frac{n_1}{n_1 + n_1'} [1 - \epsilon^{-(\gamma+\gamma')x}].$$

The ratio $n/(n_1 + n_1')$ is measured by the reduction of the effect on the thermopile obtained by charging a second pair of plates at a distance x from the centre of the first pair.

The constant $\frac{\gamma}{\gamma+\gamma'} = \frac{n_1'}{n_1+n_1'}$ is the fraction of the original beam that falls on the thermopile, when the charged particles are removed by the field between one pair of plates. Thus γ/γ', and $\gamma+\gamma'$ may be found from the experiments, and the mean distances l and l', which are the reciprocals of γ and γ', obtained.

The following examples of the experiments in hydrogen at ·0058 millimetres pressure illustrate the method of investigation. The numbers in the first line indicate the pair of plates which were used to produce the second field of force.

Condenser producing second field of force.	2nd	4th	6th	8th	10th
$n/(n_1+n_1')$	·018	·056	·09	·114	·120

The effect on the thermopile is only reduced by a small amount so that a high degree of accuracy cannot be obtained in measuring the number of charged particles at different distances from the first pair of plates, but the experiments show that this number increases as the rays move along the tube and is finally equal to 15 per cent. of the total.

The following values of l and l' were obtained from the experiments:

HYDROGEN.

Pressure ·0051 mm.; $n_1/(n_1+n_1') = ·15$.

$l = 9·2$ cm.; $l' = 52$ cm.

Pressure ·039 mm.; $n_1/(n_1+n_1') = ·274$.

$l = 2·75$ cm.; $l' = 7·3$ cm.

NITROGEN.

Pressure ·0013 mm.; $n_1/(n_1+n_1') = ·134$.

$l = 14$ cm.; $l' = 96$ cm.

327. Königsberger and Kutschewski's experiments on the effects obtained over a large range of pressures. The mean distances l and l' have also been determined by Königsberger and Kutschewski,* for different gases. The following table gives the values they have obtained for rays in hydrogen at various pressures p, the velocity of the rays being $2{\cdot}4 \times 10^8$ centimetres per second.

$p \times 10^3$	·2	·3	·6	1·0	1·8	2·4	3·2	5·5	6·5	7·5	8·7	11
l	238	140	86	66	55	47	42	31	29	24	21	16
l'	330	195	111	91	76	66	56	44	40·5	33·5	29·0	22·5

The ratio l/l' is greater than that obtained by Wien.

If the charging and discharging of the particles were a direct effect of the collisions with the molecules the ratio $n_1/(n_1+n_1') = l/(l+l')$ would be independent of the pressure and l and l' would be inversely proportional to the pressure.

The numbers in the above table show that $l/(l+l')$ is constant, but the second condition indicated by the theory is not satisfied.

328. Wien's investigation of the variation of the velocity with the discharge potential. It has generally been supposed that the particles in a stream of positive rays are the positive ions generated in the discharge tube, which acquire a high velocity when moving under the electric force towards the aperture in the cathode. Under these conditions the velocity would be less than that corresponding to the discharge potential since the ions are not necessarily generated at the anode, and for some parts of their paths before they reach the cathode they may be uncharged. Wien's† investigation of the velocity of rays produced under different conditions are in agreement with this theory. The velocity of the rays was found to depend on the voltage V of the discharge tube; it was smaller than the velocity corresponding to the potential V and with the larger potentials the velocity was proportional to $\sqrt{V}$.

* J. Königsberger and J. Kutschewski, Ann. der Phys. (4) **37**, p. 161, 1912.

† W. Wien, Ann. der Phys. (4) **33**, p. 871, 1910.

This result was obtained by finding the deflection of the rays produced by a transverse electric field; a determination of e/m being unnecessary.

The velocity v' corresponding to the voltage V is given by the formula $mv'^2 = 2\,eV$.

If an electric force X acts on the rays for a distance S in the space behind the cathode, and the deflection is measured by the displacement δ of the fluorescence produced on a screen at a distance S' from the electric field, the kinetic energy of the particle while it moves in the electric field is

$$\frac{mv^2}{2} = \frac{Xe(S^2 + 2\,SS')}{4\,\delta}.$$

Thus the ratio of the velocities v and v' is obtained.

Wien found that mv^2/e was approximately equal to V, or that the kinetic energy of the rays is equal to that produced by the action of a potential $V/2$ on a particle having a constant charge e. In these experiments the discharge potential V varied from 30,000 to 40,000 volts.

329. Thomson's investigations of the properties of the rays. The velocity of the rays has also been investigated by Thomson, who arrived at different conclusions as to the nature of the rays and the action by which the rays acquire a high velocity. These investigations* are remarkable from the fact that in many cases the greater portion of the rays appeared to consist of particles for which the value of e/m was approximately 10^4, being the same from whatever gas the rays were derived. Thus in a discharge tube filled with oxygen there may be no appreciable number of particles in the positive rays with masses equal to that of an atom of oxygen. It was suggested that the process of ionization, for instance ionization by Röntgen rays, takes place by the emission of neutral doublets with high velocities from the molecules of the gas. The doublet is supposed to be the same for all gases and consists of a positive unit and an

* J. J. Thomson, Phil. Mag. **13**, p. 561, 1907; **16**, p. 657, 1908; **18**, p. 821, 1909.

electron. Some of the doublets break up into their components, which may subsequently attach themselves to molecules of the gas.

Thus at the start from the cathode the canal rays may include a large number of neutral bodies, if indeed they do not wholly consist of them.

The velocity* with which these neutral particles are emitted from molecules is approximately 2×10^8 cm. per second, which is the velocity corresponding to a potential fall of 20,000 volts when the value of e/m is 10^4. Thus it appeared that when the discharge potential was 3,000 volts some of the particles moved as fast as the fastest obtained when the pressure was very low and the discharge potential was 40,000 volts.

From the results of these and subsequent investigations Thomson † concluded that the canal rays may be divided into three classes.

1. Rays that are undeflected by electric or magnetic forces.
2. Secondary positive rays produced by rays of the first type as they pass through the gas and collide with the molecules. The rays of this type have a constant velocity 2×10^8 cm. per second, whatever may be the potential difference between the electrodes, and a constant value of $e/m = 10^4$. At the higher pressures and when the discharge tube is small these rays predominate, but get fainter as the pressure is reduced below a certain amount.
3. Rays characteristic of the gases in the discharge tube, the velocity depending on the discharge potential and the value of e/m being inversely proportional to the atomic weight of the gas.

The first two types of rays independent of the nature of the gas have not been observed by other physicists. Wien found that the rays for which the value of e/m is equal to 10^4 become fainter and fainter as the gas is purified and

* J. J. Thomson, Phil. Mag. **19**, p. 424, March 1910.
† J. J. Thomson, Phil. Mag. **20**, p. 752, Oct. 1910.

traces of hydrogen are removed. He was unable to confirm Thomson's observations of rays travelling with a velocity greater than that corresponding to the discharge potential. These conclusions must therefore be accepted with some reserve.

330. Thomson's determinations of the ratio e/m. It has been shown (section 322) that when the rays are deflected by an electric force in the direction parallel to the axis of x and by a magnetic force in the direction parallel to the axis of y the value of e/m is given by the formula

$$\frac{e}{m} = \frac{y^2}{x} \cdot \frac{2X}{H^2} \cdot \frac{l_1^2 + 2 l_1 l_2}{l^4},$$

or

$$\frac{e}{m} = \frac{y^2}{x} \cdot \frac{X}{H^2 L^2},$$

x and y being the displacements on the screen on which the rays impinge, and L a length depending on the dimensions of the apparatus.

Thus when e/m is constant particles travelling with different velocities impinge on points of the screen lying on a parabola. Thomson * has used this principle in order to determine the nature of the particles of which the rays are composed.

In some experiments the rays were received on a photographic plate and the distribution of the deflected particles was obtained from measurements of the bright patches on the photograph. In another series of experiments the rays were received on a metal plate having a narrow slit in the form of a parabolic curve given by the equation $y^2 = ax$, so that the rays fall on the slit when the forces X and H satisfy the relation

$$\frac{e}{m} \cdot \frac{L^2 H^2}{X} = a.$$

The rays passing through the slit were received in an insulated vessel connected to an electroscope, and the

* Sir J. J. Thomson, Proc. Roy. Soc. A **89**, p. 1, 1913.

intensity of the beam was measured by the charge acquired by the vessel in a given time. When the beam of rays contained a number of particles for which the values of e/m were different, the various particles were brought on to the slit by changing the magnetic force, and the ratio e/m was obtained from the above formula.

The values thus obtained correspond to positively charged atoms of the different elementary gases, the charges being in some cases single and in other cases double, except in hydrogen, where only single atomic charges were observed. With mercury vapour the charge on the atom may be as great as eight times the atomic charge. Also singly charged molecules of the gases were observed.

In some cases rays were observed for which the ratio e/m was one-third of that corresponding to an atom of hydrogen, which seems to show that there is an allotropic form of hydrogen in which the molecule consists of three atoms. These rays were observed in the hydrogen evolved from metals or metallic salts when exposed to the action of cathode rays.

331. Hammer's determination of the ratio e/m for positive rays in hydrogen. The positive rays have also been investigated by Hammer, on the principles already used by Wiechert to determine the velocity and the ratio e/m for cathode rays. In this method the velocity of the rays is found directly by comparing the time in which a group of particles traverse a given distance with the period of the oscillatory discharge of a condenser.

A high degree of accuracy may be obtained in measuring the velocity of the positive rays by this method, and the determination is not affected by the charging and discharging of the particles as they move through the gas. The value of e/m for positive rays in hydrogen obtained from the first experiments * was 10,040 and a more accurate investigation † gave the number 9,775, which is in close

* W. Hammer, Phys. Zeitschr. **12**, p. 1077, 1911.
† W. Hammer, Ann. der Phys. (4) **43**, p. 653, 1914.

agreement with the value corresponding to a hydrogen atom with a single atomic charge.

332. Discovery of the Döppler effect. A series of investigations of much interest, particularly in connection with the theory of the emission of light from conducting gases, has been made on the spectrum of the positive rays. Stark * discovered that part of the spectrum of the rays exhibits the Döppler effect, which shows that some of the particles emitting the light move with a high velocity. The lines in the spectrum are displaced towards the violet when the positive rays are observed from a position in which the rays move towards the spectroscope, and the displacement is towards the red when the rays recede from the spectroscope. The change in the wave length of the light emitted by the moving particles is given by the equation

$$\delta\lambda = \lambda v/c,$$

λ being the wave length which is reduced to $\lambda - \delta\lambda$ when the particle moves with a velocity v towards the spectroscope, and c the velocity of light.

The values of v obtained by this method are less than those corresponding to the discharge potential, but of the same order. Some lines in the spectrum are not displaced, which shows that part of the spectrum is given out by molecules or atoms in the path of the rays, the vibrations being excited by the impacts of the moving particles.

Recently Wilsar † and Fulcher ‡ have studied the effects obtained when the positive rays generated by a discharge through one gas pass through another gas. The two gases cannot be kept apart by having them in closed tubes separated by a diaphragm since the rays would not pass through even the thinnest sheet. The method devised by Wien was adopted, in which the gases are supplied in slow streams to the discharge tube and the observation tube, the two tubes being connected by a co-axial capillary tube

* J. Stark, Phys. Zeitschr. **6**, p. 892, 1905.
† H. Wilsar, Phys. Zeitschr. **12**, p. 1091, 1911.
‡ G. S. Fulcher, Phys. Zeitschr. **13**, p. 224, 1912.

through which the rays pass. The gases are pumped out from the apparatus through a side tube in the centre of the capillary by a Gaede pump.

The spectrum of the rays in the observation tube contains lines characteristic of both gases. The lines corresponding to the gas in which the rays are generated show the Döppler effect, but the lines corresponding to the gas in the observation tube are not displaced from their normal positions in the spectrum.

INDEX

Oxford: Printed at the Clarendon Press by Horace Hart M.A.

www.ingramcontent.com/pod-product-compliance
Lightning Source LLC
LaVergne TN
LVHW091633100826
845152LV00001B/20

* 9 7 8 1 4 2 7 6 1 4 5 9 9 *